Forest Lost

```
F    O    R    E    S    T

F    O    R    E    S

F    O    R    E

F    O    R

F    O

F


                    L    O    S    T

                    L    O    S

                    L    O

                    L
```

Producing
Green
Capitalism
in the
Brazilian
Amazon

MARON E. GREENLEAF

DUKE UNIVERSITY PRESS
Durham and London
2024

© 2024 DUKE UNIVERSITY PRESS
All rights reserved
Printed in the United States of America on acid-free paper ∞
Designed by A. Mattson Gallagher
Typeset in Minion Pro and Chaparral Pro
by Westchester Publishing Services

Library of Congress Cataloging-in-Publication Data
Names: Greenleaf, Maron, [date] author.
Title: Forest lost : producing green capitalism in the Brazilian Amazon / Maron E. Greenleaf.
Description: Durham : Duke University Press, 2024. | Includes bibliographical references and index.
Identifiers: LCCN 2024003703 (print)
LCCN 2024003704 (ebook)
ISBN 9781478031086 (paperback)
ISBN 9781478026853 (hardcover)
ISBN 9781478060079 (ebook)
Subjects: LCSH: Capitalism—Environmental aspects—Brazil—Amazonas. | Carbon offsetting—Brazil—Amazonas. | Carbon sequestration—Brazil—Amazonas. | Economic development—Environmental aspects—Brazil—Amazonas. | Greenhouse gas mitigation—Brazil—Amazonas. | Sustainable development—Brazil—Amazonas. | Emissions trading—Brazil—Amazonas. | BISAC: SOCIAL SCIENCE / Ethnic Studies / Caribbean & Latin American Studies | SOCIAL SCIENCE / Anthropology / Cultural & Social
Classification: LCC HC188.A485 G74 2024 (print) | LCC HC188.A485 (ebook) | DDC 333.750981/1—dc23/eng/20240716
LC record available at https://lccn.loc.gov/2024003703
LC ebook record available at https://lccn.loc.gov/2024003704

Cover art: Feijó, Acre, Brazil, 2014. Photograph by the author.

*To Hal and Etta, who remade my world.
I hope this book helps make yours a bit better.*

CONTENTS

Abbreviations ix
   Preface: Green Capitalism xi
      Acknowledgments xv

         Introduction 1
            1. Carbon Boom 33
               Interlude I. Highway Landscapes 57
            2. Producing the Forest 64
               Interlude II. The Flood 83
            3. Robin Hood in the Untenured Forest 86
               Interlude III. The Rural Road, Part 1 111
            4. Beneficiaries and Forest Citizenship 114
               Interlude IV. The Rural Road, Part 2 128
            5. The Urban Forest 131
               Afterword. Carbon Bust 153

                  Notes 165
                     Bibliography 231
                        Index 271

ABBREVIATIONS

BAU	Business as usual
CARB	California Air Resources Board
CBNRM	Community-based natural resource management
CCBA	Climate, Community and Biodiversity Alliance
CIFOR	Center for International Forestry Research
CIMI	Conselho Indigenista Missionário (Indigenous Missionary Council)
CO_2e	Carbon dioxide equivalent
GCF Task Force	Governors' Climate and Forest Task Force
GDP	Gross domestic product
GHG	Greenhouse gas
IMC	Instituto de Mudanças Climáticas e Serviços Ambientais (Institute for Climate Change and Environmental Services [Acre])
INPE	Instituto Nacional de Pesquisas Espaciais (National Institute for Space Research)
IPAM	Instituto de Pesquisa Ambiental da Amazônia (Amazonian Environmental Research Institute)

MOU	Memorandum of understanding
MST	Movimento dos Trabalhadores Rurais Sem Terra (Landless Workers' Movement)
NGO	Nongovernmental organization
NTFP	Non-timber forest product
PES	Payment for ecosystem services
PPCD	Plano Estadual de Prevenção e Controle de Desmatamento no Acre (State Plan for the Prevention and Control of Deforestation of Acre)
PT	Partido dos Trabalhadores (Workers' Party)
REDD+	Reductions in emissions from deforestation and forest degradation
REM	REDD Early Movers
Rio+20	United Nations Conference on Sustainable Development (2012)
SEAPROF	Secretaria de Extensão Agroflorestal e Produção Familiar (Secretariat of Agroforestry Extension and Family Production [Acre])
SISA	O Sistema de Incentivos a Serviços Ambientais (the System of Incentives for Environmental Services [Acre])
tCO_2e	Tons of carbon dioxide equivalent
TFS	Tropical Forest Standard (California)
UFAC	*Universidade Federal do Acre*, Federal University of Acre
UNFCCC	United Nations Framework Convention on Climate Change
ZEE	Zoneamento ecológico-econômico (ecological-economic zoning)

PREFACE

Green Capitalism

The climate is changing in ways that demand significant and immediate response. Efforts to address climate change, however, cannot target the climate directly. Instead, they must work through the myriad environmental, economic, infrastructural, political, cultural, and multispecies dynamics that have caused climate change and the associated climate crisis.[1] In this, such climate solutions necessarily impact much more than the climate itself. Yet it can be easy to ignore these impacts, particularly in places that seem distant from us—places, perhaps, like the Brazilian Amazon, which is the focus of this book. But we *should* pay attention. Efforts to address climate change are increasingly important in everyday life, reshaping people's relationships with each other, other species, the state and other institutions, and the landscapes in which they live. And these relationships are also critical to the efficacy of climate solutions themselves.

Until recently, at least, climate change itself has also been relatively hard for most of us to pay attention to, and therefore to think about, write about, care about, study, and address.[2] This is in part because some climate impacts can be slow-moving, and because of the steady drip, drip, drip of greenhouse gas (GHG) emissions, year after year.[3] Understandably, the significant and mundane joys and struggles of daily life can make it hard to pay attention to more subtle climate impacts, especially for those usually insulated from their front lines.[4] Among these impacts is the way that climate change is making it harder for the rainforest in the Brazilian Amazon to survive and regenerate day-to-day, even where no one is cutting or burning it. The rainforest may soon reach a "tipping point," scientists warn, in which significant portions of it give way to grassy savanna. The consequences for the climate—not to mention for local ecosystems, humans, and other species—are dire.[5]

Climate change, though, is becoming increasingly obvious in many people's day-to-day lives, including my own. As I write these words in spring 2023, particulates from burning Western Canadian forests move through the altitudes high above me, thousands of miles away on the East Coast of the United States. I sense them in both vibrant sunsets and dulling midday haze. Just a few weeks later, smoke originating from nearby Eastern Canadian fires pollutes not just the air aloft but also the air near the ground that my children and I breathe. A few weeks after that, the neighboring state of Vermont floods, owing to exceptional rainfall. These fires and floods are notable not because they are exceptional, but because, increasingly, they are not. On almost any given day, I could write about a fire or flood (or heat wave or storm or drought) happening, if not close by, then somewhere not too far away—somewhere connected to me by atmosphere or affect. For an increasing number of people, addressing the climate crisis feels urgent and personal.

But what do efforts to address the climate crisis actually entail, and what do they do in everyday life? In this book, I explore this question by examining a preeminent contemporary climate solution: making living forests and the carbon they sequester monetarily valuable. This so-called nature-based solution was widespread during the time in which I researched and wrote this book. I examine it as a form of "green capitalism": the use of capitalist logics, forms, and practices—such as those of monetization and markets—to address the climate crisis and other forms of environmental degradation.[6]

Green capitalism has been the dominant climate mitigation strategy of what I have come to think of as "early climate change"—a term that David McDermott Hughes used in a 2023 talk I attended.[7] Like the terms *late industrialism* and *late capitalism*, early climate change helps me to think beyond the present. It lets me see that this current moment does not encapsulate all that climate change will be, that what it means and how we relate with it will also change, that we are at a beginning. I have come to understand early climate change as a period of uncertain duration when many people with some degree of power or privilege (derived, in part, from the burning of fossil fuels) approached climate change as a market failure that could be fixed with some elegant, internal, and even profitable tinkering. I first encountered versions of this alluring idea in the 2000s while in law school (largely a schooling in systems that maintain that power and privilege): if we could just tax carbon emissions or make corporations trade those emissions or pay for the prevention of those emissions in landscapes like the Brazilian Amazon, then we could solve climate change in ways that would simultaneously shore up the economy, but without fundamentally changing it.

That such iterations of green capitalism have not worked as planned—as I explore in this book—does not mean they can be easily dismissed as just the latest accumulation strategy or form of corporate greenwashing. Instead, ethnographically examining them as often genuine and culturally and materially meaningful efforts to address the climate crisis may be a way to understand this period of early climate change, in all its patchy unevenness and inequity.[8] Perhaps some years from now, during what might be known as middle or late climate change, someone may thumb through a dusty, surviving copy of this book, or scroll through its text on their screen, or download their AI assistant's summary of it as an account of a seemingly distant and perhaps naïve past when some people thought they could fix climate change simply by using the logics and practices that have been its primary cause.

ACKNOWLEDGMENTS

Like a tree, a book can seem singular. It appears to exist on its own, bound within its covers. As I explore, though, neither trees (nor forests nor the carbon that they store) actually do exist on their own. Rather, they exist through complex social and environmental relations. So too do books, including this one. *Forest Lost* would not exist without many people who have been part of my life through the researching and writing of this book, in ways large and small. This is my opportunity to name some of them. All faults, of course, are mine alone.

First, I want to thank those I met in Acre for their generosity, patience, and openness. The friendships of Hamilton Araújo, Mara Braga, Fronika de Wit, George Dobré, Ronizia Goncalves, Ana Paula Kanoppa, Souza Oliveira, Leo Ribeiro, Kaline Rossi, and Gabriela Severino were particularly important in helping me to understand Acre and its forests, and in offering support and companionship when I was there. My deep thanks to the staff of the Institute for Climate Change (particularly Monica de los Rios, Pavel Jezek, and Magaly Medeiros) and of the Secretariat for Agroforestry Extension and Family Production (particularly Ronei Santana) for enabling me to spend time at their institutions. In Feijó, I especially want to thank the people I refer to here as Luis, Flávia, and Manuel, who let me stay in their homes and get to know their families. I am grateful for their generosity and hospitality. Other institutions and people in them were also important for my research. Working with the Center for International Forestry Research (CIFOR) on the Global Comparative Study on REDD+ was a great way to begin fieldwork, particularly because of Amy Duchelle, Denyse Mello, and the amazing Acrean research team I joined. Thanks to Naiara Bezerra, Karina Costa, and Thayna Souza for your work on interview transcriptions, and Fernanda and Souza Oliveira for your help with research. *Saudades*!

While at Stanford University, I benefited tremendously from studying capitalism with Sylvia Yanagisako. Her incisive feedback and questions helped me to cut through my own confusion to the heart of my analysis. Lisa Curran's understanding of the ecological and social dynamics of tropical forests and unconquerable good humor also buoyed me through my early research. Jim Ferguson's scholarship inspired me to become an anthropologist in the first place, and his penetrating yet gentle feedback and mentorship helped me to think differently about concepts like neoliberalism and distribution that are essential to *Forest Lost*. My thanks too to Andrew Mathews and Susanna Hecht for their feedback and excellent scholarship on forests, which inspires my own. Many other people at Stanford were also crucial for my development as an anthropologist and as a person. They include Jess Auerbach, Emily Beggs, Firat Bozcali, Thomas Blom Hansen, Samil Can, Shelly Coughlan, Jacob Doherty, Bill Durham, Paulla Ebron, Patrick Gallagher, David Gilbert, Miyako Inoue, Yasemin Ipek, Lochlann Jain, Jamie Jones, Liisa Malki, Jenna Rice, Johanna Richlin, Nethra Samarawickrema, and Mandy Wetsel. Thank you all.

I also want to thank those who introduced me to anthropology—Eric Worby, Patricia Pessar, and Thomas de Zengotita—and those with whom I studied development and environmental law—Kevin Davis, Ricky Revesz, Jacob Werksman, and Katrina Wyman. In more recent years, mentorship and feedback from, and good conversations with, many people have also been incredibly valuable. These people include Damiano Benvegnù, Jay Blair, William Boyd, Sienna Craig, Amelia Fiske, Susanne Freidberg, Matt Garcia, Jenny Goldstein, Kregg Hetherington, Colin Hoag, Jeff Hoelle, Hiroko Kumaki, David McDermott Hughes, Sarah Kelly, Donald Kingsbury, David Kneas, Marianne Lien, Pam McElwee, Melanie Hughes McDermott, Carole McGranahan, Marcos Mendoza, Amelia Moore, Laura Ogden, Montserrat Perez Castro, Sayd Randle, Geneva Smith, Colleen Scanlon-Lyons, Kathleen Stewart, Eric Thomas, Daniel Tubb, and Wendy Wolford. Thank you particularly to Ben Orlove for being such a supportive mentor during my postdoctoral fellowship. *Forest Lost* also benefited from opportunities to present and receive feedback in various forums, including the Post/extractivism Working Group, Columbia University, Johns Hopkins University, and Rutgers University.

Dartmouth College, where I now work, has offered a wonderful place at which to complete *Forest Lost*. I want to thank everyone in the Department of Anthropology for welcoming me into the fold and for all of the support and friendship since: Sabrina Billings, Elizabeth Carpenter-Song, Jesse Cassana, Sienna Craig, Jerry DeSilva, Nate Dominy, Raquel Fleskes, Julie Gilman, Sergei Kan, Chelsey Kivland, Deb Nichols, Laura Ogden, Zane Thayer, John Watanabe, and Jiajing Wang. At Dartmouth, *Forest Lost* also benefited from research assistance from a few excellent students: Nadine Lorini Formiga, Ruan Magalhaes Rodrigues, Alida Angafor, and Ana Furtado. Dartmouth's Dickey Center for International Understanding supported a generative manuscript workshop, at which comments from Laura Ogden, Susanne Friedberg, Matt Garcia, Geneva Smith, Melody Berkins, Kregg Hetherington, Jeff Hoelle, and Pamela McElwee helped me to improve the book tremendously. My thanks too to the many members of Dartmouth's Environment and Ethnography Lab, who provided insightful feedback on versions of some chapters. Finally, thanks to Sienna Craig, Susanne Freidberg, and especially Laura Ogden for their inspiring scholarship and generous mentorship. Laura, you have expanded my understanding of environmental anthropology and helped me to take myself seriously as a writer.

Many institutions gave me the time and money to complete *Forest Lost*. The Social Science Research Council, the Wenner-Gren Foundation, and grants from Stanford University supported my fieldwork. A Mellon Foundation Dissertation Fellowship from the Stanford Humanities Center and fellowships from Stanford's Institute for Innovation in Developing Countries and Institute for Research in the Social Sciences enabled me to finish my dissertation. I began writing the book manuscript during a postdoctoral fellowship at Columbia University's Earth Institute, and I completed a draft during a Wenner-Gren Hunt Fellowship. Dartmouth College support was also essential for completing the manuscript, including from the Department of Anthropology's Goodman Fund. Along the way, I have published previous versions of parts of the book in articles elsewhere: chapter 2 in *Development and Change* 51, no. 1 (2020); chapter 3 in *Journal of Peasant Studies* 47, no. 2 (February 2020); chapter 4 in *American Anthropologist* 123, no. 2 (2021); and chapter 5 in a 2021 Society for Cultural Anthropology Hot Spots essay. Those who reviewed these articles helped to make *Forest Lost* better—thank you!

Forest Lost has also benefited from a number of editors, reviewers, and others in the publishing world. Earlier in the process, Megan Pugh's editing helped to improve the text. More recently, working with Duke University Press has been a wonderful experience. Thank you to my editor, Gisela Fosado, for the encouragement and good advice throughout the process, to Alejandra Mejía for addressing all my many questions along the way, and to two anonymous reviewers, whose comments and suggestions helped make this book so much better. The excellent copyeditors and production team were also key players in transforming *Forest Lost* from a manuscript to a book.

Finally, I want to thank my family. My mother, Ditty Greenleaf, offered me unquestioning support in this and all my other endeavors, though she did not live to see this book's completion. Newcomb Greenleaf, my father, was my earliest editor and interlocutor, someone who has always encouraged my curiosity and love of learning. Thanks too to others in my family—Michael, Jeanine, Tony, Catherine, Lyn, Tom, and the rest of the Greenleafs, Lords, and Moranos. And to my godmothers, Sunu and Denny, who provided unrivaled inspiration, stability, and adventure in my youth. My two children—Halilee and Estera—arrived somewhere in the middle of this project. Your curiosity, companionship, and endless kisses and cuddles kept me going through to the end. And finally, to Justin, my partner in all things, thank you for your unwavering love and support from the start.

Introduction

At the United Nations' (UN) climate conference in 2007,[1] Norway's Prime Minister Jens Stoltenberg articulated a simple idea that had been circulating for a few years in international policy circles: "Through effective measures against deforestation we can achieve large cuts in greenhouse gas emissions—quickly and at low cost. The technology is well known and has been available for thousands of years. Everybody knows how not to cut down a tree."[2]

This idea has proved to be attractive and influential, and it has shaped climate mitigation and conservation around the globe.[3] It is often known by the acronym REDD+, which stands for "reductions in emissions from deforestation and forest degradation." Since the early to mid-2000s, there have been extensive REDD+ and related efforts to commodify or otherwise monetize reductions in GHG emissions from forests, particularly tropical ones. In carbon markets and other market-based mitigation schemes and practices, polluters may seek to "offset" their own emissions or otherwise monetarily compensate tropical forest regions for forgoing deforestation.[4] Through this monetization, the carbon sequestered in many forests—or forest carbon—can come to have monetary value itself, or at least the potential of it. This effort to make forest carbon monetarily valuable is a key part of contemporary green capitalism—efforts to use capitalist logics, forms, and practices to mitigate environmental damage.

In the years after the 2007 UN conference where Stoltenberg spoke, green capitalism seemed to pervade discussions about the climate and other environmental issues. It was central, for example, to the 2012 Earth Summit in Rio de Janeiro (Rio+20). The first Earth Summit, hosted in the city twenty years earlier, had not only articulated growing alarm about climate change. It had also urgently spotlighted the extensive Amazonian deforestation occurring at the time, deforestation rendered visible by Brazil's innovative new remote sensing–based deforestation monitoring system.[5] But at Rio+20, things seemed different. There was some optimism in the air, in part because Brazil seemed to have "decoupled" Amazonian deforestation and agribusiness-linked economic growth, which had been long understood as inextricably linked.[6] Deforestation in the Amazon had dropped dramatically since the mid-2000s even as the region's gross domestic product (GDP) grew and poverty declined. Preeminent scientists envisioned the end of illegal deforestation in the region, and there were claims that Brazil had done more than any other country in the world to address climate change by reducing Amazonian deforestation.[7] Building on this momentum, many people in the public and private sectors were working to give monetary value to the approximately 100 billion metric tons of carbon estimated to be sequestered in the Amazon forest.[8] Rio+20 was abuzz.

At the conference, officials from Acre—a small and relatively poor Amazonian state located in Brazil's far west on the border with Peru and Bolivia—offered the state as a prime example of this Amazonian environmental and economic success. Deforestation in Acre was down some 80 percent since its 2005 peak, social development indicators were up, the economy was growing, and the state had an ensconced left-of-center government known for prioritizing forest conservation. Speaking alongside other Amazonian governors at an International Forest Day celebration at Rio+20 that I attended, Acrean governor Tião Viana spoke to a crowded hall about how the state's environmental leadership had put Acre "ahead of all the states" in the Brazilian Amazon. Acre had achieved higher economic growth and lower deforestation by breaking the dominant deforestation-linked cattle "paradigm" and making the state's timber industry "sustainable." He then positioned the state's new forest carbon program as the latest element of this successful effort. That program was the leading component of SISA— the state's System of Incentives for Environmental Services, adopted into state law in 2010.

Viana was not alone. In the early 2010s, SISA received international praise for creating one of the world's most advanced REDD+ programs.[9] The new law advanced the state's international reputation as "a global pioneer in forest protection," as one outside funding institution put it.[10] Through SISA, the Acrean government could sell forest carbon credits (representing reductions or removals in GHG emissions), receive payments for emissions reductions, and otherwise pursue monetary compensation for protecting the rainforest.[11] At the time, Acre seemed poised to sell forest carbon credits as offsets in the state of California's prominent carbon market—which would have been an important first for REDD+—as well as to other buyers. In 2012 and 2013–2014, I studied SISA and related efforts in Acre to understand how green capitalism was being enacted there—how reductions in Acrean forest carbon emissions were being commodified and otherwise monetarily valued, and the social and political effects of doing so.

This book offers an ethnographic account of what I found.[12] In conversations, meetings, and documents, my interlocutors in the Acrean government and allied NGOs often described their work to make the forest valuable as "valorizing" (*valorizar*) the forest—not the dead forest that is so often valued in the form of timber or cleared space, but the living one (*a floresta em pé*) that traditionally has been accorded little monetary worth. This valorization was both monetary and cultural—forms of value that they saw not as in competition, but rather more like David Graeber's description of them as "refractions of the same thing."[13] Additionally, this valorization sought to link forest protection and rural development—to improve the well-being of rural people and create a legibly productive rural economy in ways that relied on forest conservation rather than forest destruction.[14] My study of SISA involved spending time in government and NGO offices and with the people who worked in them in Acrean cities, as well as with the smallholders they were trying to entice to forgo deforestation in rural areas. In this book, I trace many of the material, governmental, and multispecies relations I encountered—relations that undergirded this effort to make forest carbon valuable.

When I started researching the valorization of the Acrean forest and its carbon, I anticipated studying things like forest carbon's standardization, privatization, and sale in new carbon markets. This orientation was based both on training I received as a student in law school and on some critical scholarship about neoliberal environmentalism that I had encountered in my PhD program in anthropology. I imagined tracing how these processes simplified landscapes and excluded smallholders, nonforest species and land uses, and tree species that did not maximize carbon sequestration. Yet what I found was, in some ways, different than what I expected.

In this book, I show how, contrary to my expectations, forest carbon's valorization engendered an environmentally premised welfare state and environmentally negotiated citizenship, instead of the type of market relations I was expecting. Rather than making forest carbon into private property, that state made it into a form of public property and wealth, much of which it redistributed as benefits to forest beneficiaries. Rather than another uniform monocrop, forest carbon appeared more like a multicrop—one made indirectly through the production of other forest and field-based products. Rather than gaining value as a form of generic, standardized carbon, Acrean forest carbon appeared valuable because of its singularity. Rather than exclusively centering the forest's monetary value, the Acrean state paired it with efforts to make the forest culturally valuable. These dynamics and the relations entangled in them were different than the paradigmatic forms of neoliberal dispossession and simplification I expected to encounter. Yet they were no less uneven, contingent, and contested. I trace these relations—both their negotiations and tensions—in the pages that follow. In so doing, I also elucidate broader efforts to create capitalist processes suited to this era often called the Anthropocene, and those efforts' alluring promises and vexing failures.[15]

Green Capitalism in Tropical Forests

By the time I arrived in Acre in 2012, green capitalism had been touted as a solution to the climate crisis and deforestation for some time. Market-linked environmental programs had become increasingly central to what Hannah Appel calls the "project" of capitalism, because of both the threat to continued accumulation that environmental degradation poses and the potential for profit that working to address it promises.[16] Green capitalism seeks to "internalize externalities," address the "market failure" of climate change, "pay for ecosystem services" (PES), and enable corporations and other institutions to claim carbon neutrality ("net zero"). Doing all this promises to change the relationship between capitalist economies and the environment, incorporating ecosystems as monetarily valuable components of those economies, rather than just "cheap nature" to be exploited.[17] Tropical forests have been key to this vision because they are important stores and sources of carbon and have long been imagined as sites of underutilized value more generally.

Belying liberal tropes about the "free market," green capitalism has not emerged on its own. Like other forms of capitalism, the state and law have been important to its variegated development.[18] I started learning about the entanglement of law, economics, and environmental markets while in law school around the same time as the 2007 UN conference. At the law school I attended, at least, studying environmental law often entailed studying economics. I learned about environmental harms as "externalities" that could be addressed through property rights and trading mechanisms.[19] I learned that pricing harm and cost-benefit analysis could create better environmental regulations. And I learned about market-mechanisms, offsets, and taxes as legal tools that could efficiently address environmental harm and overcome the inefficiencies of more traditional *command and control* regulations. Economic theory may shape the world, but it entails a lot of legal work and education to do so.[20]

My legal training was in keeping with the neoliberal environmentalism of that time. Even as economies reeled through the 2007–2008 financial crisis, the 2009 UN climate negotiations floundered, and the US Senate abandoned carbon market climate legislation in 2010, carbon market programs and their offset programs proliferated through the UN's Kyoto Protocol, voluntary offset purchases, and national and subnational programs. This latter category includes cap-and-trade programs, also known as compliance carbon markets. They are government programs that require specified types of polluters to obtain allowances or permits to emit GHGs, with the allowances adding up to an overall cap that is ramped down over time to an emissions level deemed to be acceptable.

Key to these programs is an understanding of the climate that I learned in law school: that it does not matter *where* GHGs are emitted or reduced because their climatic impact is the same. Emissions reductions and removals therefore should occur wherever they are cheapest. Carbon offsets and credits are meant to enable this economic efficiency.[21] In compliance carbon markets, regulated polluters can obtain allowances (essentially permits to pollute) from governments. If they reduce GHG emissions or remove

GHGs from the atmosphere, polluters can also generate and trade carbon credits within carbon markets. In addition, depending on a carbon market's rules, polluters can purchase generally lower-cost carbon offsets from outside of the regulatory cap to cover some percentage of their emissions.[22] These offsets represent GHG emissions reductions or removals, sometimes geographically far removed from the compliance carbon market itself. As I explore in chapter 1, to generate credits or offsets, an entity (for example, a business, government, NGO, or individual) sets a reference level rate of GHG emissions. The reference level is a seemingly technical counterfactual projection of the emissions that *would have happened* in the absence of a decision or effort to reduce emissions—often a projection of *business-as-usual* (BAU).[23] If the measured emissions rate is lower than the reference level (or a lower baseline rate), the difference between them might be sold as carbon credits or as offsets to outside polluters—quantifications and sales that are mediated by other institutions including regulators, auditors, and certifiers. Each generally represents a ton of carbon dioxide equivalent (tCO$_2$e). Such programs seem to make GHG emission reductions and removals standardized and fungible. In the process, carbon itself can seem like a new global commodity. As the UN explains about its emissions trading program, "A new commodity was created in the form of emission reductions and removals. Since carbon dioxide is the principal greenhouse gas, people speak simply of trading in carbon. Carbon is now tracked and traded like any other commodity."[24]

The alluring potential that forest carbon would soon be integrated into compliance markets as offsets—and the simple logic of their economic efficiency—spurred and shaped REDD+'s development in the late 2000s and early 2010s. In anticipation of market money flowing, many REDD+ initiatives were started in tropical forests around the world, creating what Rob Fletcher and coauthors call "an economy of expectations."[25] As they waited for that money to arrive, some in tropical forests sought to access funding by selling offsets in the so-called voluntary market, in which businesses and individuals buy offsets voluntarily in what they frame as compensation for their own emissions. New institutions, organizations, businesses, and research projects (including my own) proliferated, and old ones were repurposed to include this latest conservation trend. "Carbon cowboys" looked to cash in,[26] businesses saw the potential for profit, and many committed environmentalists—who had worked for years to protect forests with limited funding—were sold on carbon's commodification as a way to bring significant money into forest conservation. As with other tropical forest resources, the "myth of profit," as Bolivian sociologist René Zavaletta put it, was powerful; the "culture of miracles" described by Venezuelan playwright José Ignacio Cabrujas was strong.[27] The pursuit of forest carbon's value could be as tantalizing, and illusory, as El Dorado—the mythical Amazonian city of gold that seemed to be, somehow, always just around the next river bend.

There was also some nonmarket REDD+ funding, often portrayed as a stopgap until market funding arrived. This funding could also monetarily valorize forest carbon. Some of it came through familiar channels: foreign aid, development lending, and conservation funding. Often, well-known development and conservation institutions were powerful players. For example, the World Bank, through its new Forest Carbon Partnership Facility, offered funding to get tropical forest regions REDD+ "ready" for the carbon market envisioned to come.[28] Some money came through new programs that paid for emissions reductions or ecosystem services. In Brazil, the federal government established the Amazon Fund in 2008, into which other countries (Norway most prominently) paid for reductions in Amazonian forest emissions.[29] In Acre, the most important source of nonmarket funding during the early to mid-2010s came from the German development bank KfW. Through the REDD+ Early Movers Programme (REM) that was launched at Rio+20, KfW paid the Acrean government €25 million in "payments for emissions reductions" for the years 2011–2015—paid at US$5/ton of avoided emissions across two contracts.[30] Acre was REM's first recipient. Acrean government interlocutors characterized these payments to me as more akin to donations than market transactions, since the emissions reductions were not traded, nor were they were meant to compensate for German emissions. Yet the distinction between gift and commodity is not so definitive.[31] The payments monetized emissions reductions and signaled the quality of Acrean forest carbon, and the socioenvironmental relations keeping it sequestered, for potential carbon credit buyers to come. At a government meeting that I attended, one SISA administrator explained to her colleagues that she hoped that the KfW contracts would be "tools" to advance REDD+ nationally and internationally. "We want to show that this works," that the payments could be "channeled" well, benefiting those on the ground so they could, in turn, provide more emissions reductions. That was their "commitment." In this, the KfW payments made forest carbon valuable even in the absence of carbon trading. They were meant to prime Acre for the anticipated offset sales to come.

Green capitalism expanded as I was writing this book at the end of the 2010s and early 2020s. Concepts such as natural capital, ecosystem services, and nature-based solutions are increasingly mainstream and value-producing.[32] Renewable energy, ecotourism, and sustainability are big business. As the CEO of BlackRock, the world's largest investment firm, put it in his 2021 annual letter to other CEOs, "The climate transition presents a historic investment opportunity."[33] His 2022 letter was titled "The Power of Capitalism" and touts US$4 trillion in "sustainable investments" in the global economy.[34]

Tropical forests continued to be enmeshed in green capitalism, though not always in the ways imagined during my fieldwork. REDD+ initiatives did not become part of regional compliance carbon markets or a global one in the 2010s. Some scholars and conservation practitioners said that REDD+ was just the latest conservation "fad" to fade or that "REDD+ is dead," while others argued that it had never really been born.[35] At the same time, deforestation increased substantially in the Amazon in the late 2010s, around the presidential election of the right-wing politician Jair Bolsonaro and then with support from him and other agribusiness-linked politicians.[36] This occurred in Acre too, where deforestation increased substantially during that time. Acrean voters supported Bolsonaro and allied candidates in 2018 and 2022, and no state forest carbon offsets were sold.

Yet green capitalism, and the role of tropical forests in it, only grew in the late 2010s and early 2020s—a period of increased corporate commitments to "carbon neutrality" via voluntary forest carbon offset purchases,[37] new financial technologies such as forest bonds, prominent pledges of tens of billions of dollars to combat tropical deforestation, and the potential to access even more funding via new pathways for selling forest carbon credits and offsets.[38] This period saw the creation of new organizations like Emergent, an NGO described on its website as an "intermediary acting between tropical forest countries and the private sector" to "creat[e] a new marketplace in large-scale transactions of high-integrity carbon credits at the jurisdictional scale."[39] Echoing Stoltenberg's 2007 speech, Emergent posits that "reducing tropical deforestation is the largest near-term natural climate solution and one of the most cost-effective, gigaton-scale opportunities to reduce emissions over the coming decades." Even Bolsonaro tried to cash in, demanding a billion dollars to reduce Amazonian deforestation and "fair payment for ecosystem services."[40] His government positioned Brazil as a carbon offset exporter, committed to ending illegal deforestation by 2028, and adopted a bill to pay rural people for environmental services.[41] Meanwhile, related green capitalist efforts at *low-carbon rural development* continued in many tropical forest regions, including in Acre through SISA.[42]

Through green capitalism's expansion, then, carbon—usually only monetarily valued when extracted to create energy from fossil fuels—appeared to now have value generated from staying put in tropical forests. The latter value is meant to counter the climatic harm of the former. Monetarily valuing carbon in this way tantalizingly promises to "civilize" markets, as Michel Callon envisioned.[43] It promises the fulfillment of not only the usual goals of profit maximization and economic growth, but new ones too—climate mitigation among them.

There are good reasons to be critical of this promise—reasons that have been articulated by Indigenous and other forest activists and critical scholars, including some in Acre.[44] They include critiques about the veracity of forest emissions reductions,[45] the fungibility of these reductions with fossil fuel emissions, and the simplifications entailed in commodification—with concerns about forests being managed to maximize carbon sequestration above all else. More fundamentally, carbon credits and especially offsets have been understood as part of a neoliberal environmentalism and a wily iteration of capitalism that extracts value to address the very harms that it causes—harms which are unequally borne and perpetuate environmental injustices in frontline communities, including where fossil fuels are extracted, processed, transported, and burned.[46] In this, there is concern about REDD+ being premised upon, and reinforcing, global inequity—a kind of green colonialism or "green grabbing" in which wealthy polluters in the Global North pay poorer communities in the Global South to absorb their emissions, and at a low price that is based on ongoing colonial relations.[47] Forest carbon can thus appear as another exploited resource of the Global South, one that can facilitate "accumulation by dispossession," dividing and harming Indigenous and other forest communities.[48] Packaged as carbon offsets, reducing forest emissions in the Amazon promises to prop up the continued burning of fossil fuels elsewhere, primarily in the Global North. In this sense, forest carbon as commodity extends extractive, colonial, and racialized capitalist dynamics.[49]

This book does not dispute these accounts, but its emphasis is different. Green capitalism has expanded, I found, but incompletely and in ways that can consequentially diverge from neoliberal doctrine and some critical accounts of it. Approaching capitalism as constructed and yet potent, I explore the contingent and relational work that underpinned Acre's prominent effort to valorize the forest and its carbon via an environmentally premised welfare state, and how such efforts do not always proceed as planned. In doing so, I build on feminist anthropological work on capitalism as "produced," as Sylvia Yanagisako puts it—as powerful and yet not comprehensive, and as composed of often gendered and racialized processes that can shape the world but require a lot of work to do so.[50] In the case of green capitalism, at least, these processes are not always so all-encompassing as our stories (both critical and supportive) often make them out to be. Rather than coherent or standardized, green capitalism in Acre was fragile. It was an incomplete experiment whose "achievement" was elusive and contested,[51] and one that could rely upon and reproduce dynamics it sought to counteract. At times, green capitalism seemed to barely exist, and yet it also shaped lives and landscapes in the Acrean Amazon.

Following Forest Carbon and Producing Nothing

I started off my research by trying to study forest carbon offsets in a manner that critical scholars have often studied commodities—to "follow" them from production to purchase.[52] But I found forest carbon to be quite elusive. When I brought it up in discussions, interactions, and interviews, or when the term was introduced in workshops, meetings, or events that I attended, it tended to quickly recede from the conversation. I was in Acre to study what was arguably the world's most prominent effort to make forest carbon monetarily valuable, and yet forest carbon itself was difficult to hold in focus. It was difficult to follow.

This elusiveness points to the strangeness of forest carbon as a would-be commodity. It is worth dwelling on this strangeness for a moment—to be curious about how forest carbon is valorized and commodified, and what understanding the process can help us to fathom about this moment in early climate change. Unlike many of the things we buy, forest carbon is not something you can hold, wear, eat, sit on, listen to, or display. Other than documents, there is nothing to see—not to mention hear, smell, taste, or touch.[53] Moreover, its value comes not from being removed and traded but rather from reductions in its emissions, often referred to as emissions reductions. For example, through SISA and working with other institutions, the Acrean state could sell forest carbon offsets or otherwise receive payments for emissions reductions, compared to a BAU forest emissions rate (the reference level referred to earlier).[54] In other words, it is not forest carbon's extraction or circulation that generates monetary value, in contrast to many other commodities. Rather, forest carbon's value comes from carbon staying in place, from a tree not being cut down, as Stoltenberg put it. There is, then, no rupture into independent being, nothing tangible that readily appears to hold or create value. Instead, what creates value is a reduction, an absence, an action not taken. Offsets, payments for emissions reductions, and other carbon credits entail producing nothing to balance the too-many somethings made. Producing nothing was REDD+'s great value proposition in the effort to enlist capitalism to combat the climate crisis and other environmental harms of the Anthropocene.

But how do you produce nothing? It might seem that it would entail doing nothing, perhaps paying people not to deforest. Indeed, this is frequently how REDD+ has been discussed, as Stoltenberg's speech indicates. After all, trees and other plants need no help in absorbing carbon. They build their bodies with carbon via photosynthesis and absorb billions of tons of it from the atmosphere without any human say-so.[55] Stoltenberg's elegant logic follows from this: REDD+ should be an easy and cheap way to reduce carbon emissions because everyone knows how not to cut down a tree.[56] Leave it alone, and it will help to address the climate crisis itself.[57] Yet forests are not just trees, and when they are exploited or threatened—as they are in Acre and many other places—doing nothing seldom leads to their protection.[58] Instead, deforestation is increasingly the norm. Deforestation accounts for approximately 16 percent of yearly global GHG emissions, and some tropical forests may now emit more carbon than they store.[59]

There are different ways to understand this widespread and consequential deforestation. It is partly a consequence of the integration of forests into global capitalist supply chains and of their resulting monetary value when destroyed for timber or to make space for infrastructure, occupation, monocrops, and cattle. Yet there are also other connected dynamics at work: Since forests are notoriously hard to govern—serving, for example, as places to evade domination—deforestation can be a strategy of state control and colonization.[60] Relatedly, in places like Brazil, where deforestation has historically indexed progress, it can be a way to claim land, citizenship, power, and belonging.[61] Consider what the leader of a "land grabbing gang" told reporters in 2014, after he was fined for extensive illegal deforestation in the Amazonian state of Pará, far to Acre's east: "I regret doing something illegal, but I don't regret having deforested because if we didn't deforest, Brazil wouldn't exist, nothing would exist."[62] Modern existence itself, he asserted, depends upon felling the forest.

Not cutting down forests is, therefore, not so much a question of knowledge, as Stoltenberg positioned it. Rather, those who deforest may feel they have no other option if they are to survive. This was how smallholders in Acre frequently put it to me. They deforested for the money, security, status, and sense of belonging that doing so can bring in the Brazilian Amazon.[63] In this, deforestation has less to do with individual decision-making or knowledge than with large-scale colonial and capitalist processes in which smallholders are frequently enlisted—processes that value deforestation and shape things like knowledge, the perception of choice, and deeply felt desires.

In threatened forests, then, producing nothing requires a lot of work to avoid the deforestation that has been the main way to make forested places monetarily and culturally valuable. Significant critical scholarship has elucidated some of that work, focusing particularly on the measurement, calculation, and other forms of standardization that are essential to creating new carbon commodities and enabling them to move through markets.[64] As Donald MacKenzie puts it, this work makes different GHG emissions appear the "same" via the common metric of tCO_2e.[65] Through such standardization, carbon can come to appear like a fungible and valuable commodity—an offset, for example. Carbon's standardization can enable it to take on other roles as well: it can act as a "common denominator for thinking about the organization of social life in relation to the environment," a "metric of the human," and a "universal standard" that "enables comparability and even commensurability between different forms of life and different actions across spheres."[66] In this, carbon's standardization is important to making forest carbon valuable, monetarily and otherwise.[67]

Yet forest carbon's valorization also depends on producing nothing from places like Acrean forests. And producing nothing involves a different kind of work that is not about standardization: protecting forests. This second kind of work is necessary because it is carbon's state of being—rather than simply its existence—that is of climatic relevance. Carbon's existence, after all, is not the cause of the climate crisis. Rather, carbon—created as distant stars die out—is one of the most common and oldest elements in the universe.[68] It is, as Steffen Dalsgaard points out, a "constant in nature."[69] It is "the element of life," Primo Levi tells us, made available to us mere mammals primarily by the greenery of trees and other plants when they convert energy from the sun.[70] The problem with carbon, instead, comes when its state of being is changed: when it is released into the gaseous atmosphere, primarily when some humans burn very old trees and plants—a.k.a. fossil fuels—and, to a much lesser extent, when they burn living ones. Valorizing carbon also centers on its state of being: it requires not that carbon be extracted or disappear but instead be held in place.

When I began to focus on the ontological issue of carbon's state of being, instead of on carbon as a thing I could follow, I began to approach it relationally. Specifically, I began to focus on the diverse social and environmental relations that determine whether carbon stays in place or is released. Using a translation of the Brazilian term *socioambiental*—with its helpful amalgamation of nature and culture—I explore these as "socioenvironmental relations."[71] If, as Sarah Besky and Alex Blanchette write, drawing from feminist scholarship, "the ability of one person to stand on a factory floor is the product of a vast assemblage of contingent relations,"[72] something similar is true of the ability of one tree to stand, especially in a threatened landscape. After all, there is no standardized forest, even if forests may be represented as such in climate modeling, discourse, and negotiations. There are only living forests. And each living forest lives through particular, if patterned, socioenvironmental relations. Neither is there standardized forest carbon, even if it can be made to seem so.

While forest carbon offsets and other carbon credits may be immaterial in that they represent an absence, then, they are also dependent upon the diverse socioenvironmental relations of living forests and their protection. As Tracey Osborne and Elizabeth Shapiro-Garza describe in their analysis of forest carbon projects in Mexico, "Because these [forest carbon] offsets are actively produced when carbon is sequestered in trees, they have an unbreakable and continuous bond to living biomass and can therefore never be fully divorced from the place of production or the people who produce them."[73] That bond means that producing nothing necessitates the ongoing, contingent, and contested work of engendering and shaping socioenvironmental relations in order to protect forests. These relations constitute forest carbon—carbon held in place in forests rather than released—and each forest carbon offset or other monetized unit of emissions reduction represents some assemblage of them.

In Acre, forest carbon valorization entailed forest protection of a particular state-centric sort. There are a number of existing conservation policies that work through restricting and punishing deforestation. Foremost among them is Brazil's federal Forest Code, which requires Amazonian landholders to keep up to a striking 80 percent of their land as forest, with fines for those who violate this requirement. At least on paper, a significant portion of the Brazilian Amazon is also protected through being designated as an Indigenous territory or other kind of protected area.[74] Yet while the effort to produce nothing in Acre was linked to these laws and regulations, it was more directly pursued through a different approach: it entailed a state-driven effort to make forest carbon and the living forest itself valuable. The aim was, I came to understand, to shift the "diverse, intimate network of human and non-human relations" that Laura Bear, Karen Ho, Anna Lowenhaupt Tsing, and Sylvia Yanagisako argue comprise capitalism—to make those relations protect the living forest by making it into a source of economic value, and, in this sense, to make capitalism green.[75]

Instead of following forest carbon, then, this book traces efforts to valorize Acre's living forest and the carbon sequestered in it as processes of making and modifying socioenvironmental relations.[76] While they were meant to keep Acre's forests standing, these relations tended not to center on the forest itself. Often they were governmental, working through state governance of the forest and those smallholders who lived in and near it. Here the state did not seek to create a neoliberal market, corporate ownership, or private property within its borders, but instead sponsored environmentally focused redistribution that promoted certain socioenvironmental relations. The relations could involve multiple species and enlist different trees, plants, and animals—both within and outside of the forest. They could be infrastructural, bound up in the building of roads and water sources, as well as associated visions of developmental progress.[77] They sought to produce more from rural landscapes, rather than less. They were enabled and constrained by histories and presents characterized by colonizing resource extraction, deforestation, and property relations. They conjured a speculative green capitalist future in which the value of the living forest both kept climate-changing carbon in place and secured the well-being, belonging, and citizenship of poor rural people. Producing nothing, then, entailed efforts to produce many other things—different sorts of socioenvironmental relations as well as myriad types of products from forests and fields. And, unintentionally, these efforts to produce nothing produced other effects as well.

Some of the work to produce nothing takes place at international conferences, like Rio+20 and the 2007 UN conference at which Stoltenberg spoke. Some of it is undertaken by corporations, governments, and intermediaries (like carbon traders, NGOs, and auditors) based in the Global North's centers of GHG emissions. But much of the work occurs in and around tropical forests located in places like Acre, both in the rural areas that are often the focus of critical analysis and imagination of the Global South, and in urban government and NGO offices, museums and public spaces, universities, and businesses working to valorize the forest and its carbon. That work is the focus of this book.[78]

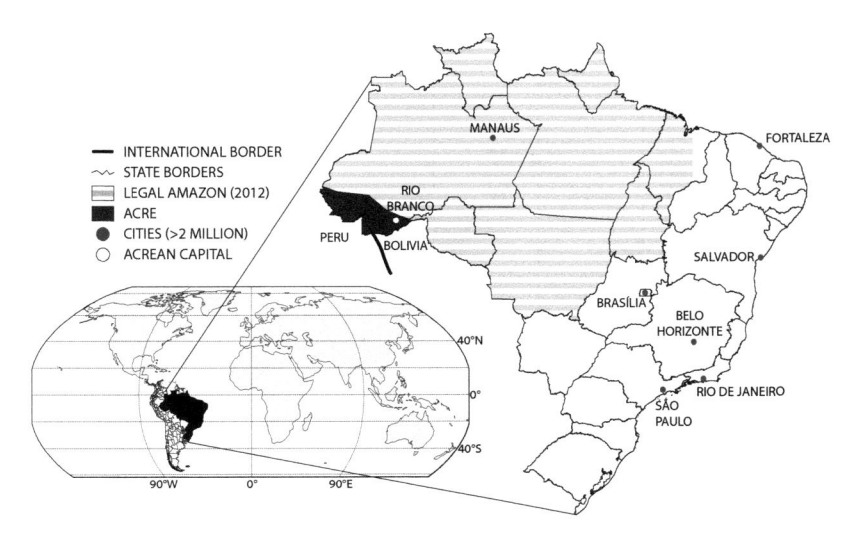

Map I.1 Map of Acre. Created by Justin Mankin.

O Acre Existe!

The question of how to produce nothing held particular resonance in Acre—a state small in terms of both human population (about 830,000 people) and size (about 164 thousand km²).[79] As in many other parts of the Brazilian Amazon, the vast majority of Acreans (about 96 percent) do not identify as Indigenous, though they may have Indigenous heritage.[80] The state is located far from Brazilian centers of cultural, economic, and political power (see map I.1), and it became a part of Brazil only at the beginning of the twentieth century and a state in 1962. To many who lived there, Acre could also feel peripheral to the Brazilian economy. Acrean friends and acquaintances alike frequently griped to me that nothing was produced there—*O Acre não produz nada*, they said. Everything they bought, it seemed, had to be trucked in from Brazilian states to the south or imported during periodic shopping trips to the Bolivian border town of Cobija. For some Acreans, this was a failing that seemed linked to the forest that covered over 85 percent of the state.[81] "There is no economy here," one Acrean taxi driver warned me when I told him I was studying the state's economy. "There is only forest." To study the economy, I should head to São Paulo, he advised.

The state's forest and lack of production could seem like evidence that a common Brazilian joke about Acre was, in fact, true: the state didn't really

Figure I.1 "Acre: do you believe?" T-shirt. According to the Acrean journalist Altino Machado, the shirt was made through a website called Jovem Nerd. Machado writes about it in his discussion of the jokey theory that Acre does not exist. Altino Machado, "Não dá pra levar a sério," November 6, 2008, http://www.altinomachado.com.br/2008/11/no-d-pra-levar-srio.html.

exist; "*O Acre não existe!*"[82] When I mentioned to people elsewhere in Brazil that I was living in Acre, they almost always told me some version of this joke. "Acre?!? Are you sure it even exists?!?" the woman behind an airport check-in counter in Rio de Janeiro teased. "Wasn't it just full of Indians (*índios*) and jaguars," others quipped. Wasn't it where Santa Claus and the tooth fairy lived? Did they even speak Portuguese there? One acquaintance silently crossed himself, pressed his hands together in prayer, and turned his eyes upward to God, as if pleading on my behalf. *O Acre não existe* is a hashtag on Twitter, the "conspiracy of Acre" has its own Wikipedia page, and there is a T-shirt that asks, "Acre: Do you believe?" (figure I.1). A 2014 documentary—*O Acre Existe*—made by São Paulo–based filmmakers is meant as a sort of rebuttal. The film poster advertises that "it's not known whether [Acre] is the beginning or the end of the world."

The joke points to something serious: the Amazon's long-standing position as both a frontier supplying global capitalist supply chains and a place of alterity from which global capitalism has been challenged and reworked.[83]

It is a place that might generate "ideas to postpone the end to the world," as Indigenous Brazilian thinker Ailton Krenak puts it.[84]

In Acre, the forest and peoples' relationships with it have been central to this dual role. Most directly, Acre's forests were a key source of rubber during the nineteenth- and twentieth-century rubber boom (Ciclo da Borracha). In the late nineteenth century, when Acre was predominantly Bolivian territory, the high price of rubber and Acre's high quality rubber trees attracted thousands of Brazilian migrants to the area, eventually leading to its incorporation into Brazil. Most Acreans I knew offered family histories linked to this and subsequent rubber-driven migrations. These included both government and NGO employees and the smallholders they enlisted in valorizing the forest. Acrean rubber enabled accumulation and industrial economies in far-off economic "centers," like so many frontiers.[85] Floated downriver, Acrean rubber formed the tires of the new bicycles and automobiles that multiplied in distant cities, and the United States colluded with Bolivia to secure a piece of the rubber-rich territory for itself.[86] For a time, Acre was "one of the most commercially desirable stretches of territory on earth."[87] So, while green capitalism was new to Acre, Acrean forests have long been enlisted in global capitalism through rubber.

However, after the rubber economy collapsed, starting in the late 1910s and then again after a temporary resurgence during the Second World War, Acre's forests made it a target within a national economy, culture, and governmental system that often valued deforestation as a tool of the entwined processes of colonization and extractive capitalism.[88] Recall the Amazonian land-grabber's assertion: "If we didn't deforest, Brazil wouldn't exist, nothing would exist." Even the country's colonial-given name comes from a tree—*pau-brasil* (brazilwood)—which the Portuguese cut and exported to Europe to make a valued red textile dye. The widespread extraction of the trees spurred further clearing in the Atlantic Rainforest, which was decimated—and violently colonized—to grow coffee, sugar, and other cash crops. More broadly, government officials have often sought to shore up control over the massive territory by encouraging deforestation.[89]

Starting in the 1970s, similar processes were underway across the Amazon. Brazil's dictatorship (1964–1985) sought to consolidate sovereignty there via deforestation-led "development."[90] Most of the deforestation in Acre has been carried out by large-scale cattle ranchers, many of whom immigrated there at the dictatorship's encouragement. That deforestation entailed not only killing trees but also the violent dispossession of rubber tappers and Indigenous peoples. Encouraged by policy and practice around property rights, infrastructure, taxes, and credit, as well as demand for beef, deforestation accelerated in Acre through the neoliberalization of the 1990s. Monocrop agribusiness loomed to Acre's south as an "arc of deforestation"—fire then soybeans—seeming to move inexorably northward.[91] Acre seemed like it might become just another extractive frontier.

Frontiers are sources of not only material resources, like rubber, beef, and soy, but also imaginative ones. Since its colonization, the Amazon has loomed in European and North American understandings of nature as a biodiverse and resource-rich paradise untrammeled by human hand. It has drawn white macho adventurers from the Global North like Teddy Roosevelt, as well as innumerable profit seekers pursuing visions of El Dorado.[92] At the same time, deforestation and biodiversity loss has meant that the Amazon has also symbolized environmental destruction, as a paradise lost.[93]

Most recently, the Amazon has been understood internationally as the imperiled "lungs of the Earth" and as both the cause of and cure for the climate crisis. When Amazonian deforestation increased dramatically before and during the presidency of Jair Bolsonaro (2019–2022), France's president, Emmanuel Macron, tweeted in August 2019, "Our house is burning. Literally. The Amazon rain forest—the lungs which produces 20% of our planet's oxygen—is on fire." A tweet from the actor and environmentalist Leonardo DiCaprio from the same time concurred: "The lungs of the Earth are in flames." Carbon markets can also position the Amazon and other tropical forests as "climate frontiers," where sequestered carbon keeps the planet habitable even as fossil fuel emissions are allowed to continue elsewhere.[94] In this, carbon markets are aligned with long-standing dynamics of extraction in the Global South that facilitate enrichment in the Global North, as well as related Brazilian nationalist concerns about their control over the Amazon.

Yet frontier spaces can also allow for modes of thinking and being that may differ from the dominant societies they supply, materially and imaginatively. Such is the case in the Amazon, which thinkers like Brazilian writer Eliane Brum have positioned as "the center of the world."[95] Exemplifying this dynamic, Acre has played an important role in environmentalist thought internationally. Acre's rubber tappers famously organized against deforesting ranchers in the 1970s and 80s, garnering significant international attention and support for reframing deforestation as a matter of socioenvironmental injustice. In the ruins of the rubber economy, rubber tappers built on their socioenvironmental forest practices to create new forms of land rights and forest conservation that shaped environmental organizing and governance around the world.

In the decades that followed, Acre also became known as a leader in forest-focused development in certain environmentalist circles. An environmental-technocratic middle class, comprising both Acreans and outsiders drawn to this project, developed in the state. They sought to reinvent Acre as a center of forest governance and culture. Particularly after the 1998 election of the self-proclaimed state "Government of the Forest" led by the left-of-center Workers' Party (PT), government officials and NGOs promoted a forest-linked Acrean identity, developed systems for participatory governance, and adopted state policies meant to simultaneously spur forest protection and economic development.[96] They promoted this as a new approach—my state interlocutors told me—that, in the words of former Acrean governor and senator Jorge Viana (brother of Tião Viana), "demonstrate[d] . . . that development does not depend on the destruction of the forest, but rather on its survival."[97] As Marianne Schmink describes it, "The goal was no less than to create—for the first time—an articulated, statewide model of sustainable development with interlinked ecological, economic, political, and cultural goals."[98] The forest was meant to be Acre's "passport to the future," as Carlos Antônio da Rocha, the Secretary of Forests and Extractivism under the Government of the Forest, put it.[99] In this, he invoked what Kregg Hetherington calls the "promissory nature" of infrastructural and development projects in frontier settings.[100] Hard work in the present would pay off in future prosperity. But, while usually in frontier settings that hard work entails deforestation, in Acre, it required refraining from it. We might understand this forest-centered development strategy as an assertion: *O Acre existe!*—not through the forest's destruction but rather through its valorization.

The last legislative act of the Government of the Forest was the 2010 passage of the law creating SISA.[101] SISA included a forest carbon program that was the lodestar of this new state policy. NGOs and government officials had initially considered creating a delineated REDD+ project in a few priority areas within the state but then decided to expand it to include the state's entire territory. "We started with a project and ended with a system," two of its developers marveled to me separately at the Rio+20 International Forest Day event. It was as if they had surprised themselves. In this system, the Acrean government measured deforestation across the state's entire territory—rather than focusing on a specific region within it—and compared it with the projected BAU rate. The government could then sell the difference as carbon offsets or otherwise receive monetary compensation for it, as it did, for example, through the €25 million in payments from KfW via the REM program. In a second phase of the program (2017–2022), agreements offered €30 million from Germany and a new contributor (the United Kingdom), among other funding sources.[102] Overall, a 2022 state government estimate stated that REM had raised more than R$175 million, with a former SISA official estimating that the program brought in closer to R$200 million.[103]

The idea of making Acrean forest carbon monetarily valuable was not simply an external imposition from the Global North, then—one that imagined the Amazon as a carbon reserve.[104] Rather, forest carbon's valorization was also entangled in Acrean environmental politics and the state's effort to enact a form of forest-dependent—rather than deforestation-dependent—development.[105] In this effort to valorize the living forest and its carbon, Acre's existence comes into view not just in the injustices of extractive capitalism and resistance to it—dynamics for which the Amazon is so often known—but also through efforts to create a different sort of development and, with it, greener capitalist practices. These efforts worked through a logic of inclusion and were funded in part by forest carbon's new monetary value.

Inclusion

Instead of pushing rural people or different species off land to maximize carbon sequestration, the Acrean state sought to include many of them in its efforts to make the forest and its carbon valuable. These efforts often involved the distribution of benefits to rural people and aimed to include them in the creation of a productive, low-carbon rural economy envisioned to benefit both them and the forest. These benefits were partially funded by forest carbon's new international value, through payments like KfW's for Acrean emissions reductions. Many of them aimed to get rural people to produce more in ways deemed to not require more deforestation, in contrast to the dominant form of deforestation-reliant cattle ranching (participating Indigenous communities tended to receive different kinds of benefits).[106] The benefits included a subsidy for collecting native rubber, a monetary "bonus" paid to those who committed to "sustainable" agricultural practices, fishponds constructed on "degraded" pastureland, açaí palm seedlings meant to be planted in deforested land, and mucuna seeds—a legume that fixes nitrogen, thereby negating the need to burn the forest to fertilize the soil.[107]

In other words, the effort to valorize the living forest and its carbon did not focus on getting rural people to produce nothing. Instead, it tried to entice them to produce more. This strategy was clear when I visited or stayed with smallholders in rural Acre. Sometimes I visited them with state agri-

Figure I.2 Mucuna drying in the sun. Photo by Maron E. Greenleaf. Feijó, Acre.

cultural extension technicians. One took me to see mucuna at work on the land of a smallholder family he was visiting. While the government usually distributed mucuna seeds, a few smallholders grew them themselves. Removing the seeds from their pods was labor intense work, I saw. A boy, perhaps eleven or twelve years old, wielded a heavy stick, repeatedly hitting the ground. His body moved rhythmically up and down with each swing. But his eyes did not leave his target: a layer of dark podded beans spread out to dry in the blazing sun (figure I.2). The pods cracked open under his effort. Under the shade of a thatched, wall-less structure nearby, the boy's mother separated out the cracked pods from their insides: small black mucuna seeds.

Mucuna was also the focus of the ongoing conversation between the woman's husband and the technician who had brought me. Mucuna, while toxic to eat, could be useful for growing other crops because of how it fertilized the soil. As the boy's father walked us around, he told me that this family had been growing and using mucuna in their small, rainforest-backed fields for six years. With pride, he gestured to pineapples growing fat in "dead mucuna." He held up a sprawling, multipronged manioc root—a staple food (figure I.3). The technician explained to me that because of mucuna, the

Figure I.3 Manioc grown in "dead mucuna." Photo by Maron E. Greenleaf. Feijó, Acre.

man no longer had to cut and burn the forest to create fertile soil. Instead, the family could simply keep growing crops on the land they had already cleared. He compared the mucuna-fertilized fields around us to a well-kept house, contrasting them with what he described as the mess (*bagunça*) of the neighbors' land.

I later learned that, around this part of the settlement, the family was known for their mucuna. Others sometimes called them "mucuna" as a nickname. I picked up the habit too, coming to think of them as the mucuna family. Months after my visit, I recognized the patriarch of the mucuna family showcased in a photo on the website of a state-linked environmental NGO. He was the picture of success.

I came to think of such state programs as inclusive, and particularly as promoting a kind of inclusive productivism. Rather than simply pushing poor rural people out in favor of large landholders or corporate interests, these programs enlisted them in green capitalism. The state programs provided benefits (in the form of inputs like mucuna seeds, for example), training about how to use the benefits, or simply monetary rewards. Mucuna and other forest benefits funded by forest carbon's new value were meant to bring smallholders, like the mucuna family, into a forest-protective version of the Brazilian productivism that had long linked rural belonging to land use and yield. By producing more through mucuna and less through deforestation, the mucuna family would help to create, and be part of, green capitalism. In so doing, this approach sought to inclusively combat the nagging anxiety that Acre—where nothing was produced—might also not actually exist. Forest carbon was the latest resource, the latest crop, whose production was meant to secure poor rural people's welfare and citizenship and the state's sovereignty.[108] As Carlos Edegard de Deus, the Acrean state environment secretary at the time, said at a SISA-focused event in 2017, the state government's goal was the "inclusion of the neglected: Indians [*índios*], rubber tappers, riverside peoples [*riberinhos*], and land reform settlers. We have always fought for the inclusion of everyone, provided they adhere to the principles of sustainability."[109]

There were several reasons that the Acrean state took an overtly inclusive approach to making the forest and its carbon valuable, I came to understand. For one, forest protection can be difficult without local support. Some resources depend on minimal local labor or can be produced or extracted under compulsion, generating significant wealth that states can capture as rent.[110] Keeping carbon sequestered in tropical forests is quite different. Forests can be difficult for often distant city-based bureaucrats to effectively govern; there is often much they do not know.[111] As Tania Li writes, "The forest edge is a site of struggle, but it is difficult to control by coercive means."[112]

In Acre, while past deforestation had been predominantly executed by large-scale ranchers, an increasing portion of it was being undertaken by smallholders, sometimes at the directive of larger ranchers or because of smallholders' increasing imbrication in the ranching economy. Large-scale deforestation was, in some ways, easier to police. It was easier to detect via state-run remote sensing monitoring, and it was easier to villainize and fine large landholders. Poor smallholders were a more sympathetic group, one which the PT saw themselves as helping. Moreover, their smaller-scale deforestation was harder to detect via remote sensing, and they were harder to punish, both politically and administratively. And small-scale deforestation could be part of legally permissible traditional forms of swidden agriculture, in which clearing a hectare or two to grow crops is followed by decades of encouraged or permitted forest regrowth. The need to address smallholder deforestation therefore militated for a more inclusive approach to forest protection—a form of "environmentality" in which rural people were enlisted in making the living forest valuable.[113] Inclusion was, in this sense, a governance strategy for an under-resourced and city-based government in the context of smallholder deforestation.

"Social inclusion" (*inclusão social*) was also an explicit priority and strategy of the PT, which governed at the federal level from 2002 to 2016 and in Acre from 1999 to 2018.[114] Through social inclusion, the PT aimed to, in party discourse, "give to the poorest people in Brazil the right to fully realize their citizenship and be respected as people."[115] The PT's conception and practice of social inclusion relied in part on distributing material benefits. In Acre, I will explore, those included benefits linked to forest protection. Funded by forest carbon's new international monetary value, forest benefits can be considered part of the natural resource–funded redistribution typical not only of Brazil's PT governments but also of many so-called Pink Tide leftist and left-of-center governments elected in the region in the early twenty-first century.

In Acre, the living forest seemed to bolster the PT's hold on state power and its approach to social inclusion, like "petro-states" whose legitimacy and authority is bound up with their access to oil.[116] The PT administrations promised they would be able to access the forest's new international value. At the same time, keeping forest carbon sequestered also became a way for the state to pursue smallholder inclusion. In this, forest carbon's new monetary value funded, justified, and shaped forest-focused redistribution. In other words, that value enabled an environmental version of the distribution-centered inclusion that has developed in articulation with neoliberalism in parts of the Global South. As James Ferguson writes, "While many influential accounts of neoliberalism have seen only ever-growing social exclusion, we here must also take stock of a new kind of *in*clusion as millions of poorer citizens previously ignored or worse by the state have been direct beneficiaries of cash payments."[117]

As part of market-based efforts to combat the climate crisis, the Acrean forest's new value facilitated and was facilitated by state redistribution and the attendant development of a type of welfare state that governed and sought to benefit—to "make live," to use Foucauldian terms—certain types of subjects, multispecies relations, and landscapes.[118] In distributing forest carbon's new value as state benefits, this environmentally premised redistribution shaped the state and its relations with those it sought to govern. It engendered a kind of nascent "carbon democracy," to use Timothy Mitchell's term, that impacted political practice, governance, and citizenship.[119] It was a carbon democracy, though, centered not on carbon's extraction for fossil fuel–based energy production—as in the carbon democracies that Mitchell analyzes—but rather on the effort to keep carbon sequestered in the earth and the trees that grow from it.[120] Green capitalism, in other words, developed in Acre not through private property, markets, or other hallmarks of neoliberal capitalism, but rather through the elaboration of a state that sought to include and benefit some marginalized people and landscapes on environmental terms.

The reasons for the prominence of social inclusion in Acrean forest valorization are in some ways particular to Acre, and yet also speak to a common dynamic in green capitalism. Inclusivity does not deterministically inhere in forest carbon itself and yet it is widespread.[121] While there are many of examples of forest carbon programs that have excluded and otherwise dispossessed rural people, inclusive approaches have been common to forest carbon programs beyond Acre, in part because of Indigenous and forest community organizing.[122] Inclusion can also be seen in some other green capitalist initiatives, such as Green New Deal proposals and aspects of the United States' 2022 Inflation Reduction Act. Such programs seek to include those seen as excluded from traditional forms of capitalism, through state-sponsored integration into reformed but recognizable systems of work, politics, and culture. This inclusion is part of the purported win-win-win of green capitalism—profit, environmental protection, and improved human well-being. In this, inclusion is one of green capitalism's constitutive logics.[123]

Yet, as I explore, this kind of inclusion can also reinforce existing or engender new forms of marginalization. Take the conditionality of the second part of the statement by Carlos Edegard de Deus, the Acrean environment secretary I quoted earlier: "We have always fought for the inclusion of everyone, *provided they adhere to the principles of sustainability.*" Those who do not adhere to these principles remain excluded. This implicit threat is the underbelly of Acre's forest valorization strategy, which relies on incentivizing forest protection. Inclusion, then, does not necessarily replace exclusion so much as rework it. Intentionally or not, inclusion can engender, enable, or fail to effectively constrain processes of dispossession. In particular, those who are neglected, as de Deus put it, can be included in ways that are materially and culturally meaningful, but that still maintain their marginality. Take the mucuna family. Using mucuna enabled them to produce and earn more, deforest less, and be part of culturally valued productivism. Through their efforts, Acrean rural production increased, if only slightly. Yet in other ways, they remained on the margins: poor; without ready access to running water, health care, or education; and looked down on by cattle ranchers and urban residents. Inclusion can thereby perpetuate aspects of the status quo, even through efforts to transform it. This contradiction may be less an indication of green capitalism's failure than of its efficacy in engendering capitalist practices for the Anthropocene that redraw lines of inclusion and exclusion, but maintain much of the economic, political, and cultural status quo.

The Amazon at Eye Level

Outsider ideas about the Amazon (my own included) and Amazonian ideas about outsiders (myself included) shaped my research, analysis, and writing—what I noticed, who spoke with me or invited me into their homes, and what we said to each other there. I was part of a long history of outsiders from the Global North coming to Amazonia for adventure, for profit, to "save the rainforest," or to work with the rubber tapper movement and Government of the Forest.[124] Many of those most critical of REDD+ and green capitalism suspected that I was in Acre to further them. Others who favored more deforestation thought I was another foreigner there to tell them to stop, even as my own country's wealth was predicated on it. "You've already cut down all your trees," one acquaintance teased me in what became a familiar refrain over lunch at his house on my second day in Acre. Now I had come to tell Acreans not to cut down theirs.

    I worked to see and think beyond these established narratives, starting with how I studied the deforestation that has made the region of international environmental concern. Like many outsiders, before my research, I primarily knew the Amazon from above, through shocking images in popular media and land-use change science of apparently untouched rainforest turned to monocropped soy or cattle pasture. These images are important: remote sensing and other analysis reveal socially and ecologically consequential deforestation and forest degradation, which can direct resources, shape policy, and illuminate essential socioenvironmental dynamics. Yet aerial imagery and analysis can also be used to perform a kind of "god trick" that can conjure a clear distinction between "natural" forests untouched by human hands and fields fully converted to human use.[125] The reality is more complex. Ancient geoglyphs unearthed by recent deforestation in Acre and research on continued Indigenous and other traditional forest cultivation practices, for example, challenge the firm distinction between forest and field, and through them, nature and culture.[126] Moreover, the label *deforestation* can homogenize smallholder forest clearing—which is important to their precarious livelihoods and traditional swidden practices—and the violence of large-scale forest clearing.

    Approaching deforestation at eye level, as I did in this project, can reveal its heterogeneity and, with it, the socioenvironmental relations that shape

Amazonian landscapes and either hold carbon in them or release carbon from them. The forest often appeared to me as neither unbroken nor fully destroyed. Rather, the rural landscapes I encountered were inhabited and reworked. There were, of course, the large cattle ranchers' fields—mostly empty, a monochrome and blanched green-brown color, and lined by distant forests. Yet next to them were patchy landscapes that smallholders narrated to me in ways that blurred firm divisions between natural forest and decimated human landscapes. They were what Hugh Raffles calls "aspirational landscape[s]"—the outcome of hard and uncertain work, enmeshed in global capitalist processes, and tended with an eye toward making a better future.[127] The mucuna family had, for example, just burned a small patch of forest to grow more of the bean, the man explained to me proudly as we walked around his land. It was a testament to the entanglement of forest and field within the state effort to make the living forest valuable. My account of this effort reveals some of this complexity, which can be obscured in aerial analysis.

This book, though, is not primarily an ethnography of rural Amazonian life. For one, the forest I studied was not just the rural arboreal one. With more than 70 percent of the population living in cities and over 85 percent of the state considered forested, Acrean relationships with the forest were also urban.[128] And the Amazon is also more than its forest. It is also composed of roads, factories, fields, and the people who build them, work in them, and dream of them. All of them were also entangled in the effort to make the forest and its carbon valuable. This book is about them too.

Most of the time, I lived in Acre's capital city Rio Branco, where I conducted about eighty in-depth interviews with government officials, NGO workers, academics, and businesspeople about the forest, its destruction, and its protection. Most seemed happy to talk, or at least were willing to do so. Many had worked with foreign environmentalists or researchers, some of whom were from elsewhere in Brazil. I also attended workshops, community meetings, and forest-related events and visited government institutions involved in valorizing the forest and forest carbon. In particular, I spent a lot of time at the Institute for Climate Change and Environmental Services (o Instituto de Mudanças Climáticas e Serviços Ambientais—called the IMC), the new state institution created to administer SISA. At the IMC, I interviewed staff, attended meetings and workshops, provided comments on draft documents, once filled in as a last-minute translator, and often sat at a desk I was generously provided. I also spent a significant amount of time at the Secretariat for Agroforestry Extension and Family Production (a Secretaria de Extensão Agroflorestal e Produção Familiar—SEAPROF), which administered some of the SISA-funded forest benefit programs. There, I attended meetings, interviewed staff, and accompanied some of them on visits to the field (*o campo*).

The rural area in which I spent the most time was a municipality called Feijó, which also has a small city by the same name. That was where I first worked as the coleader of a twelve-person research team. Comprised primarily of Acrean university students and recent graduates, the team conducted a survey for the Center for International Forestry Research (CIFOR) study. This prominent forest-focused research organization, based in Bogor, Indonesia, was studying REDD+ efforts at twenty-six sites in seven countries around the world.[129] Our team administered the survey, asking over two hundred smallholders about their land uses, possessions, agricultural production, and the impact of SISA-linked interventions, among other questions. The following year, I hired one of the CIFOR team members, whom I call Fernanda, as a research assistant to return to Feijó with me.[130] We stayed and talked with some of the smallholders whom CIFOR had surveyed, interviewing about thirty of them and spending time with a smaller subset.

The mixed-heritage rural smallholders with whom I spoke and spent time often referred to themselves, and were often referred to by government officials, as rural producers (*produtores rurais*)—a term I also sometimes use in this book. This term differs from many others used to describe rural people in the Brazilian Amazon. Accounts of themselves as former rubber tappers or descendants of them emphasized stories of their forebears' migration to Acre from the parched Brazilian northeast to tap rubber in Acre's forest, eliding Indigenous ancestry.[131] Yet this rubber tapper heritage did not mean that they described themselves to me as extractivists (*extractivistas*)— the local term used to describe those who collect forest products like rubber. Neither did I hear peasant (*camponês*) or the derogatory term *caboclo*. I heard the term *rural worker* (*trabalhador rural*) only occasionally; it had been important to the rubber tapper movement and helped to position rubber tappers as workers in a society in which that category carries legal rights and sociocultural status.[132] Yet I heard it infrequently and mostly in relationship to rural workers' unions, which smallholders in Feijó described to me either as now being corrupt or irrelevant, except to access certain government benefits.[133] Another term—*fazendeiro* (literally *farmer*)—implied high social and often outsider status.[134] It was usually reserved for cattle ranchers who lived in the city or another state. *Rural producer*, in contrast, was a term associated with many rural people's diverse livelihood strategies, including small-scale agriculture (primarily manioc), paid daily labor (for *fazendeiros* or for other rural producers for short-term work, such as forest clearing), cattle ranching, hunting, and the collection of forest products (e.g., Brazil nuts and açaí).[135]

The prominence and acceptance of the term *rural producer* points to the dominant cultural valorization of certain forms of productivism.[136] That it was used by smallholders themselves in many communications (that I witnessed) with state representatives and separately with me indicates that inclusion in that productivism was something which some of them aspired to or used to access resources and cultivate relationships with the state. In other words, calling oneself or someone else a rural producer could itself be a claim to inclusion and belonging in a society that valued production.

The Road Ahead

The rest of this book consists of five chapters interspersed with interludes and an afterword. A road called the BR-364 meandered through my research, and it meanders through this book as well.[137] In the Amazon, roads are the classic harbingers of deforestation.[138] Before coming to Acre, I had read about them in articles on deforestation and watched Google Earth and NASA time-lapse videos of aerial imagery that showed deforestation seeming to inexorably seep from roads in a fishbone pattern. I had also learned about what was implied but could not be seen in these images: the movement of people—like colonists (*colonos*), miners, and loggers—and attendant violence against Indigenous and other forest communities that is so often part of deforestation.

In particular, I had heard about a road called the BR-364, which stretches from Limeira in the state of São Paulo to Mâncio Lima, close to the Peruvian border in Acre, where it essentially dead-ends (see map I.2).[139] In Acre, much of it was a two-lane highway that, in places, was still in the process of being paved. And it was the only paved road linking the state to the rest of Brazil. I had first heard about the road because of how its paving had facilitated massive, rapid, and violent deforestation in the neighboring state of Rondônia in the 1980s.[140] Seeking to avoid a similar outcome in Acre, the famed Acrean rubber tapper and community organizer Chico Mendes had, along with many others, worked against the BR-364's paving into the state. This activism, along with transnational environmental organizing, led the World Bank to temporarily suspend funding for the project and to the recognition of some Acrean rubber tappers' land rights.[141] Decades later,

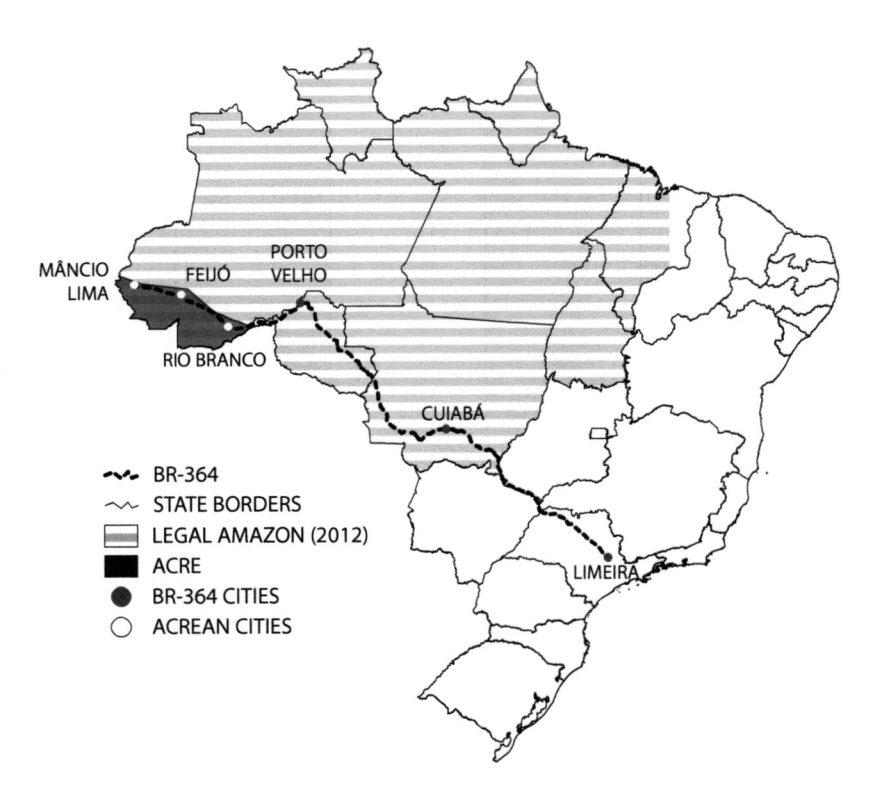

Map I.2 Map of the BR-364. Created by Justin Mankin.

when the last stretch of the BR-364 was being paved in Acre in the early 2010s, the Acrean government sought to avoid the extensive road-linked deforestation that 1980s Rondônia had exemplified. Making the forest and its carbon valuable was central to this effort.[142]

I use the BR-364 to connect and elucidate some of the socioenvironmental relations enlisted in the effort to make the living forest and its carbon valuable, as well as to reveal the tensions within this effort. This approach builds on my experience of the road. I rode certain stretches of it a lot since it connected Rio Branco, where I lived most of the time, and the municipality of Feijó, located about five to six hours northwest, where I conducted most of my rural research. I always traveled the road in other people's vehicles. I rode on it with the CIFOR research team. Sometimes I squeezed into crowded trucks next to agricultural extension technicians who were trying to get smallholders on the BR-364 to stop deforesting. Once, I traveled it in a spacious luxury van with two forest carbon project developers and the two

American auditors whom the developers had hired to visit one of their forest carbon projects. And I rode in the truck of a traveling salesman named Carlos, whom I hired to drive a research assistant and me on the BR-364 and its dirt side roads (*ramais*) in Feijó. Most of the smallholders I write about lived directly on the road or on a side road or trail off it. We sometimes drove on the road together. And the road also moved through many of our conversations. The smallholders talked about how much it had improved their lives and how it was already falling apart, just after being paved.

In Acre, in other words, the BR-364 was more than a harbinger of deforestation. It was also integral to both daily life and its governance. Smallholders in Feijó told me that the BR-364 facilitated a form of inclusion via improved access to education, health care, and the stuff of consumer life. And some urban residents traveled it frequently for work and pleasure. In its integration in everyday life, the road aggravated inequality, emphasized rural peoples' continued marginality, and indexed governmental ineptitude and corruption, even as most residents also valued it.

The BR-364 was also important to green capitalism. It connected state officials with those they sought to govern, enabling the distribution of forest benefits to smallholders. It was how agricultural extension technicians originally brought mucuna to the mucuna family and then subsequently visited them to promote its use. And it connected smallholders with buyers for the products that the state wanted them to produce. In other words, the BR-364 was simultaneously the thing from which forest carbon needed to be protected and a means of protecting it. In this, it elucidates both the effort to value the living forest and the tensions within this iteration of green capitalism.

In the book's first chapter, the BR-364 appears only briefly, to note the deforestation its paving precipitated in Rondônia. Chapter 1 centers on a key component of the effort to make Acrean forest carbon valuable to potential carbon offset buyers in places like California: Acrean rubber history and its retelling. I call this retelling the *rubber narrative* and explore how it both differentiated and standardized Acrean forest carbon in ways that elucidate green capitalism. The chapter also examines forest carbon's materiality and temporality. Chapter 2 explores how Acrean forest protection centered on increasing, rather than decreasing, production—of many things other than carbon. Specifically, the chapter focuses on efforts to increase açaí berry production in forests and fields as part of the effort to make the forest valuable. It explicates what I call *inclusive productivism* to show how green capitalist inclusion can reinforce the marginalization it purportedly seeks to combat, in ways that can both modify and reinforce the status quo.

Chapters 3 and 4 examine forest benefits distributed by the state and NGOs along the BR-364, centering on the socioenvironmental relations and politics this distribution entailed. Chapter 3 traces how the Acrean state made forest carbon's new international value into a kind of public wealth it then redistributed to some rural people. Examining this as a form of statecraft, it also argues that this approach engendered an environmentally premised welfare state that, while inchoate and not always effective, differed substantively from the private property–making and –enforcing state envisioned in both supportive and critical discussions of forest carbon and neoliberal capitalism. Yet, in avoiding property, this approach also skirted the powerful forms of belonging that rights to land can engender. Chapter 4 examines the forest beneficiary as a figure of environmentally mediated and negotiated citizenship, in conversation with the Acrean state's understanding of the concept of *florestania*—a term often translated as forest citizenship. It traces negotiations between agricultural technicians and smallholders over what it should mean to be a beneficiary, pointing to the mutual dependence in the Anthropocene that the term ultimately reveals.

Chapter 5 explores forest valorization, and green capitalism more widely, as a cultural project. Efforts to make the forest culturally valuable were deeply linked to efforts to make it monetarily valuable in ways that reshaped the Acrean capital Rio Branco and the lives of some of its residents. This urban forest, as I explore it, sought to include the forest and forest people in dominant culture, transforming it in the process. Many urban residents benefited from this cultural valorization through its forest-themed public space and culture, as well as secure middle-class employment governing the forest from the city. Rural poverty and marginalization, though, continued in ways that the crumbling BR-364 could seem to embody. The chapter elucidates how the limits and contradictions of the forest's cultural valorization undermined green capitalism in Acre.

In between these chapters are four interludes centered on the BR-364. They ethnographically elucidate deforestation, production, inequality, and aspiration in daily Acrean life along the BR-364 in Feijó and beyond. Finally, in the afterword, I look at how Acrean forest carbon valorization helps us to understand both the changes that were to come—namely, the rise of Jair Bolsonaro, allied politicians in Acre, and deforestation there—as well as the continued expansion of green capitalism around the world.

Carbon Boom

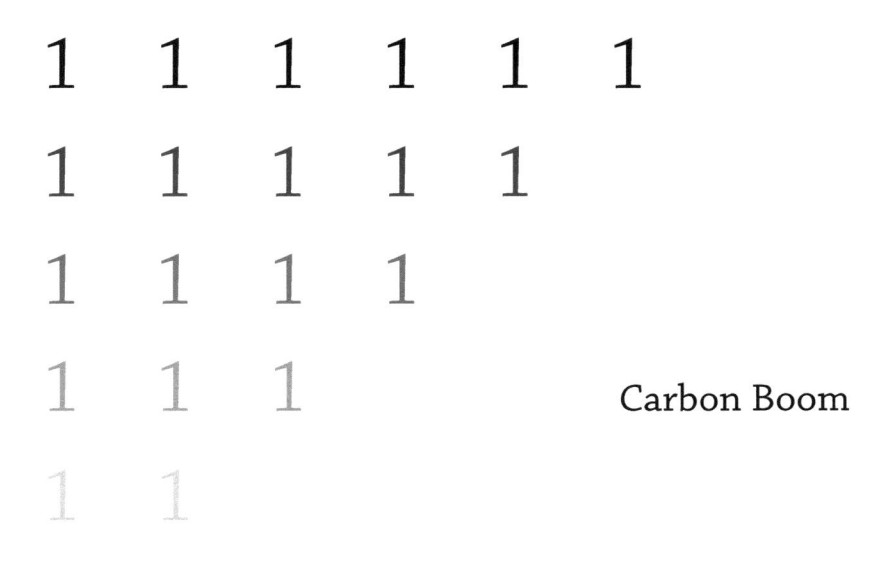

In the 2010s, the California Air Resources Board (CARB) deliberated for years about whether to admit carbon offsets from tropical forests into the state's cap-and-trade program.[1] Doing so would allow regulated polluters—power plants, oil refineries, and the like—in California and linked jurisdictions to buy carbon offsets from some tropical forests to compensate for a small portion of their own climate-changing GHG emissions.[2] This kind of carbon offset had yet to be admitted into any cap-and-trade program, and the prominence of California's program meant that CARB's decision would be a momentous one. Acre's renowned forest carbon program featured in the deliberations.[3] In 2010, the Californian and Acrean state governments signed a memorandum of understanding (MOU) to cooperate on climate mitigation, and in subsequent years, CARB staff and others suggested Acre as the most likely source of tropical forest offsets for California's carbon market.⁴

As a result, when I was in Acre in the early to mid-2010s, there was anticipation that polluters in California might soon be buying millions of dollars' worth of Acrean offsets. Other potential buyers, like the airline industry, postured that they too were interested in Acrean offsets, or in otherwise paying for forest conservation there.⁵ It was during that time, for example, that KfW paid the Acrean government €25 million for some of the state's forest emissions reductions. An Acrean government interlocutor likened the KfW payments to an act of God: it both floated the internal Acrean effort to valorize forest carbon and seemed to bless that carbon as of international worth. Acre—a small state that Brazilians liked to joke did not even exist—seemed to be on the brink of a forest carbon boom, one that would confirm its position at the cutting edge of green capitalist efforts to address the climate crisis.

Understanding green capitalism, and Acre's one-time prominence in it, necessitates looking beyond the thing being valorized—reductions in carbon emissions—to the broader set of socioenvironmental relations involved. Among those relations in Acre were those surrounding another resource: rubber. During the Amazonian rubber boom of the late nineteenth and early twentieth centuries, native *Hevea brasiliensis* trees were a key source of rubber for tires and other components of the fossil fuel economies developing around the world. Most of the Acreans I knew recounted stories of ancestors who had tapped the trees. In the 1970s and 80s, some of them organized against cattle ranching–linked deforestation and dispossession in what would become an internationally famous socioenvironmental movement that innovatively integrated social justice and environmental concerns. Accounts of the rubber tapper movement—and particularly its slain leader, Chico Mendes—circulated worldwide, helping to redefine conservation and recast forest communities as leaders of environmental conservation.⁶ A savvy state government subsequently recirculated this history of rubber in its effort to valorize the living forest, both within Acre (as I explore in chapter 5) and internationally, in its marketing of Acrean forest carbon to potential outside carbon offset buyers, other forest protection funders, and government regulators—like those in California. In other words, Acre's living forest gained monetary value at least twice—once because of its rubber trees and once because of its forest carbon. The value of the latter was partially based on the value of the former.

Green capitalism is not a coherent global system created in or imposed from centers in the Global North.[7] Rather, it is also created by people and negotiations in places like Acre.[8] There, as this chapter traces, state and nonstate actors recruited rubber as part of their effort to make forest carbon internationally valuable, helping to—almost—spur a carbon boom in the state. This recruitment indicates how green capitalist value relies not only on how "people assert the value of their cultural capital," as Lisa Rofel and Sylvia Yanagisako explore, but also on how people assert the cultural capital of specific places' multispecies and other socioenvironmental relations, such as those surrounding Acrean rubber.[9]

These relations include the stories told about them. In Acre, I came to think of those stories as "the rubber narrative" after repeatedly hearing versions of them in settings like a 2016 workshop that was part of CARB's offset deliberation process.[10] The workshop included a presentation entitled "Linkage Process and Acre, Brazil," referring to the possibility of admitting Acrean forest carbon offsets into California's cap-and-trade program.[11] The presentation offered information on Acre—its size, biodiversity, and forest cover. Then came a key component: a description of the rubber tapper–led socioenvironmental movement, how this "grassroots environmental justice movement bec[ame a] dominant political force" in the state, and how the subsequent Government of the Forest made the living forest a focus of rural development.[12] This rubber-linked story was not just contextual background. Rather, the story and its articulation were paramount to California's and other potential offset purchasers' interest in Acrean forest carbon and therefore to its value. They were part of the multifaceted labor undertaken to produce Acrean forest carbon offsets and to otherwise valorize the living forest.

Such stories can create value for many commodities. They may be particularly important for "future-oriented" commodities characterized by speculation, including offsets.[13] In "speculative enterprises," Anna Lowenhaupt Tsing tells us, "profit must be imagined before it can be extracted."[14] Stories and their telling are part of that imaginative work. They attract needed investment now by conjuring future profit. But, crucially, for offsets, this conjuring centers not on the future extraction that is the mainstay of extractive economies, but rather on forest carbon's continued sequestration.[15] The rubber narrative projected that continued sequestration, positioning Acre's rubber-focused past as enabling a forest carbon–focused future.[16]

Doing so entailed differentiating the Acrean Amazon from other forests. This differentiation contrasts with the standardization that has been the focus of much of the critical scholarship on carbon offsets. Adam Bumpus and Diana Liverman, for example, show how "once a tonne of reduced carbon becomes a [carbon] credit [or carbon offset], it is largely assumed to mean the same thing as other tonnes of reduced carbon, despite the potentially different material circumstances . . . in which they were produced."[17] Such standardization is essential to carbon's commodification, yet those "different material circumstances" can matter significantly too. Stories about them can create value or diminish it. Acrean forest carbon had to be standardized to be sold into carbon markets like California's, but to even gain admission into that market, it also had to be made different.[18] Such "singularization" is important for many commodities in what Michel Callon, Cécile Méadel, and Vololona Rabeharisoa call "the economy of qualities."[19] Yet singularizing stories may be particularly salient for new environmental commodities, like offsets, that rely on ongoing socioenvironmental relations to keep carbon in place or to otherwise ensure environmental quality into the future. Those working to produce offsets and otherwise get payments for emissions reductions in Acre in the 2010s used the rubber narrative to highlight the specific relations that kept more than 85 percent of the state forested and had seemed to dramatically reduce its deforestation rate starting in the mid-2000s. The rubber narrative made a commodity premised on absence (of carbon emissions), rather than extraction, seem real.

As part of its valorizing work, the narrative showcased inclusivity. It highlighted how rubber's socioenvironmental relations not only protected the forest but also benefited rural people. Rubber, in other words, made Acrean forest carbon offsets seem like a component not of an extractive capitalism that harmed rural people and the environment, but rather of a different extractivism (*extrativismo*)—one conceptualized in Acre as a means of producing economic value from the forest in ways that benefited both it and those who lived there. Through rubber, forest carbon offsets could seem like part of Acre's "grassroots environmental justice movement," as the CARB presentation put it. Without rubber, Acre might have been the site of just another threatened tropical forest, of international concern but no special value. Or it might have been the site of just another REDD+ project, with dubious environmental and social credentials to be barred from reputable carbon markets like California's. Instead, rubber helped make the Acrean Amazon appear like a socially inclusive and effectively governed forest, and therefore California's most likely source of tropical forest carbon offsets.

The Value of the Living Forest

To understand the rubber narrative as a value-adding story of green capitalism necessitates exploring a bit of rubber's history. Much of it was highly exploitative, not inclusive. By the late nineteenth century, Acre, with its quality *Hevea brasiliensis* rubber trees, was booming. Rubber processing technology had improved, and buyers in industrialized economies were eager to purchase Amazonian rubber for car and bicycle tires. Tens of thousands of Brazilians (mostly men) moved to Acre—then primarily claimed by Bolivia—many of them looking to escape a brutal drought in the country's northeast.[20] Some of the new arrivals revolted against Bolivian rule and briefly formed the independent Republic of Acre—romantically named the "Republic of Poets"—before being quite willingly subsumed into Brazil in 1903.[21] The area was governed as a federal protectorate until 1962, when Acre was granted statehood.[22]

Brazilian governance, though, did not improve most locals' brutal living conditions, which were largely an outcome of the exploitative rubber economy. That economy included the routine murder of Indigenous people, including the Huni kuĩ, Apurinã, and Yawanawa. It also introduced new diseases and pressured Indigenous people to assimilate via forced marriage and other tactics.[23] Migrant rubber tappers (*seringeiros*) and their descendants, meanwhile, were extensively exploited. A small number of rubber barons (*seringalistas*) employed large numbers of rubber tappers under conditions of exploitative debt peonage (*aviamento*). This system kept tappers poor and isolated, barring them from growing their own food and, often, from seeking other forms of livelihood or even work on different "rubber estates" (*seringais*).[24] Their labor produced significant wealth that was largely accumulated by rubber barons, trading houses, and companies, most based in distant Amazonian cities and the Global North. The rubber economy then declined precipitously starting in the late 1910s, with a brief revival during the Second World War. This decline was due in part to colonial biopiracy: British agents stole Amazonian *Hevea* seeds and grew profitable rubber plantations in their Southeast Asian colonies. The price of rubber fell, and Acre's rubber economy collapsed.[25] Many rubber tappers left for cities and other states, ushering in an era of population decline and governmental neglect referred to in Acre as the *decadência*.

It all seemed a typical story of how colonial resource extraction harmed workers and Indigenous communities and made local economies vulnerable to the vicissitudes of global commodity markets and powerful Global North interests. Yet when I encountered rubber in Acre in the early 2010s, it was being used and talked about very differently. The Acrean government was enlisting rubber and its history in the inclusive and forest-protective development they promoted, of which forest carbon was a key part.

This strategy became clear to me, among other places, at a factory that manufactured the world's only condoms made from native rubber. There, I watched workers quietly and methodically sheath shiny, rotating metal phalluses in newly minted condoms—a process that, the guide of the factory tour explained, ensured quality control. Trying to overcome the juvenile humor I found in this phallic assembly-line work, I reminded myself of the import of what I was witnessing: this factory, called Natex, was a celebrated hub of the state's effort to make the living forest a basis of the contemporary economy. It used the forest to create jobs and a valued modern commodity.

Later that afternoon, the tour group I was with visited that forest, which was managed by a rubber tapper cooperative. Our guide, a co-op member, described how the state supported the cooperative and paid tappers a subsidy for the rubber they collected. This enabled cooperative members to live better than people in the city, he told us. Located in Xapuri, the historic center of Acrean rubber production, Natex was doing well too: its condoms were purchased and distributed by Brazil's Ministry of Health as part of its successful HIV reduction programs.[26] Along with the rubber subsidy, the demand supported the area's historic rubber economy and tappers' forest-based livelihoods.

Walking through the forest with the tour guide, I recalled the first living rubber trees I had seen a few months before. Previously tapped for rubber, the trees were now part of a small outdoor memorial in a Rio Branco park. An acquaintance of mine named Angelo had taken me to the park and explained the tapping process to me. He knew all about it because he had grown up tapping rubber near Natex in Xapuri. The trees in the Rio Branco park had grooves carved diagonally into their bark, one on top of the other (see figure 1.1). While I had never seen a living rubber tree, it was a pattern I recognized from pictures at the state Forest Library's exhibit celebrating Acrean rubber. Angelo softly traced the grooves with his finger as he described a long trail that he made through the forest in Xapuri connecting some 150 rubber trees. He walked it twice a day, he told me. In the morning, he would gently scrape open the grooves on each tree so that white, milky latex would slowly ooze down and drip into a tin cup fastened

Figure 1.1 Rubber tree in park memorial (with Angelo's hand, *at right*). Photo by Maron E. Greenleaf. Rio Branco, Acre.

just so below. In the evening, he would walk the long path again to collect the latex from the cups.

This method scars the trees, but it does not kill them. Practiced in much of the Brazilian Amazon, it was different from methods used elsewhere, which often entailed felling the tree. The difference was due, in part, to the species of tree involved. In Acre, rubber is tapped from living *Hevea brasiliensis* trees during an eight-month growing season. Traditionally, tappers rotated daily among trails that snaked through the forest (*estradas*), each connecting 120 to 180 rubber trees across over a dozen or so kilometers.[27] In contrast, during the rubber boom, another type of rubber, *caucho*, was commonly collected in other places by felling and draining the species of rubber trees that grew there: *Castilla ulei* or *Castilla elastica* trees. In portions of the Upper Amazon (e.g., what is now Panama), this approach entailed both the enslavement of Indigenous people forced to fell the trees and the decimation of forests.[28] Indigenous people and tappers suffered significantly in Acre as well, but rubber came from the trees' continued life, not their death. As Susanna Hecht explores, Brazilians justified their country's 1903 annexation of Acre in part based on claims about the superiority of Brazilian migrants' rubber extraction, which "tam[ed] the wilderness" but did not destroy it, as the Brazilian journalist and writer Euclides da Cunha wrote at the time.[29]

Acrean rubber also relied on *Hevea* trees' relationships with other species around them. In the Amazon, when grown in plantations, *Hevea* trees tended to die from the South American leaf blight that spreads easily among them. This was a lesson that Henry Ford learned in his failed Fordlândia rubber tree plantations to Acre's east. Created to source rubber for Ford car tires, they were decimated by leaf blight.[30] In contrast, dispersion throughout the famously biodiverse Amazonian forest offers some protection from the fungal disease. Rubber tappers had to walk long *estrada* trails through the forest because rubber trees were so spread out within it. In this sense, the Natex condom factory's rubber supply relied not just on the *Hevea* trees from which the latex was tapped but also on all the other trees from which it was not. Acrean rubber production relied upon the socioenvironmental relations constituting Amazonian biodiversity.

After the rubber cooperative tour, I hung back with a few people I knew. Among them was Breno, a lawyer in his thirties who had recently been hired by the state government to work on its forest carbon program. We took turns taking pictures of each other with the rubber trees that the guide had pointed out. Breno had just moved to Acre from the state of

Rondônia to Acre's southeast, where his father had tapped rubber decades earlier. But Breno himself had never seen a rubber tree before the tour, he told me. Rondônia's rubber trees were mostly gone, destroyed as part of the notoriously rapid deforestation that took place in the state in the 1980s and 90s, linked to the paving of the BR-364 there. Breno's eyes welled up as he kissed a rubber tree for a picture. He could not wait to show it to his young children, he told me. This was why they had moved to Acre—for moments like this. These trees were why his work valorizing the living forest and its carbon was important. It was work enabled by the trees that would help protect them into the future.

The living forest was central to Acre's recent history. The idea that it could be profitable was, then, not an import of a global green capitalism. Rather, Acrean rubber—an export product without much direct local use—showed that the living forest could generate great wealth. That wealth was a product of socioenvironmental relationships that included rubber tappers, *Hevea* rubber trees, non-*Hevea* trees and plants, rubber barons and foreign rubber purchasers in the past, and governmental subsidies and purchases in the present. In keeping the forest standing, these relations also helped to hold forest carbon in place.

Inclusive Extractivism

"SONG OF THE RUBBER TAPPER"
Let's give value to the rubber tappers.
Let's give value to this nation.
For it is through their work
That car and airplane tires are made.
Bicycle tires are not made of cheese.
It is not cattle leather that truck tires are made of.
It is not cattle horns that erase letters, no.
These are all products of rubber, made by our hands.[31]

J. S. Araújo, first President of the Rubber Tapper National Council (first sung October 1985)

Yet it was not only the socioenvironmental relations of rubber's extraction that helped to hold forest carbon in place. It was also the way that Acrean activists, researchers, politicians, and people in allied organizations reframed and reworked these relations. They sought not only to give value to the living forest but also to "give value to the rubber tappers" themselves, in the words of J. S. Araújo's "Song of the Rubber Tapper." They too were part of the socioenvironmental relations of the forest that helped to hold forest carbon in place.

Their valorization began with a famed socioenvironmental movement. While many tappers left the forest as the price of rubber fell, some stayed through the decades of the twentieth century, cobbling together diverse forest-based livelihoods through tapping rubber, growing subsistence crops, hunting, and foraging. In the 1970s and 80s, southern Brazilian cattle ranchers began to move in at the behest of the Brazilian dictatorship and the Acrean government. Rubber tappers, particularly those near the Natex factory in Xapuri, organized to resist the ranchers' violent land claiming and clearing.[32] They nonviolently occupied and dismantled rancher-financed deforestation camps and created a new tenurial form that would become very influential, both within and outside of Brazil: extractive reserves. Extractive reserves are protected areas in which rubber tappers, among others, receive formal usufruct rights to land within an area with more stringent forest protection requirements than the surrounding region. The rubber tapper movement developed the idea of extractive reserves in the 1980s, including in response to the widespread violence and disease linked to the paving of the BR-364 in Rondônia and planned paving through to Rio Branco. Institutions such as the World Bank and Inter-American Development Bank, which had funded the road's paving, eventually embraced and pushed for extractive reserves in response to extensive protests and organizing by rubber tappers, Indigenous organizations, and US environmentalists focused on the BR-364. Extractive reserves have since become a mainstay of Brazilian conservation governance and a model for protected areas that include human use.[33]

One of the rubber tappers' essential insights was that their fate was entangled with that of the forest: ranchers' deforestation meant both forest destruction and rubber tapper dispossession. Their movement sought to redefine their entanglement with the forest from being exploitative, immiserating, and a sign of stagnation—as it had been during the rubber boom and decline—into something that could enable access to land and work. In so doing, they articulated a socially inclusive vision of environmentalism in which forest protection was also a means to protect rural peoples' rights. It was a vision that resonated far beyond Acre too, with regional and international grassroots movements as well as with powerful external organizations, some of whom became valued allies to organizers in Acre.[34] The tappers worked with liberation theologians, rural workers' unions, and the left-leaning PT, which was organizing against the military dictatorship.[35]

Rubber tapper leader Chico Mendes, who helped establish the PT's Acrean branch and was a rural union organizer, received international awards from the United Nations and the National Wildlife Federation and spoke with members of the US Congress. His 1988 assassination by a cattle rancher's hired hand was covered on the front page of the *New York Times* (a stark contrast to the lack of media attention paid to most other murdered Amazonian activists).[36] The attention indicated not only Mendes's own skill and charisma but also how the rubber tapper movement's articulation of tapper-forest relations resonated internationally as a story uniting social and environmental concerns.

In the years following Mendes's assassination, the rubber tapper movement changed, and some within it gained political power. The 1985 ending of the dictatorship created political openings, and some of those who were part of or connected to the rubber tapper movement gained support among voters. Marina Silva, a former rubber tapper and social movement leader, was elected to the Brazilian Senate in 1994 and was later appointed Minister of the Environment by President Lula da Silva (2003–2008), known as Lula. Jorge Viana, a charismatic forest engineer turned politician with links to the rubber tapper movement, became mayor of Rio Branco in 1992 and governor of Acre in 1998.[37] He served for two terms, and his administrations and the allied one after it were self-titled the Government of the Forest (o Governo da Floresta, 1999–2010).[38]

These administrations and their NGO partners sought to elevate forest use and protection as the basis of rural development. Building on the rubber tapper movement, the Government of the Forest redefined extractivism itself as socially inclusive, environmentally protective, and economically efficient; a valuable form of heritage; and an effective basis of resistance against agribusiness and cattle-dominant culture and land use.[39] In Acre, the term *extractivistas* does not refer to miners, loggers, or ranchers depleting the environment, as it often does elsewhere. Rather, it refers to those who collect from the forest—rubber, Brazil nuts, or açaí berries, for example—without destroying the forest itself. This vision of extractivism was invoked in state-promoted concepts of forest-based identity like Acreanidade (Acrean-ness) and *florestania* (often translated as forest citizenship), which I take up in later chapters. Extractivism was, in the words of José Fernandes do Rêgo, the State Secretary of Production under Jorge Viana, "a way of relating to the forest, to people, and to spiritual things."[40] And it was a centerpiece, economically and conceptually, of the Government of the Forest's vision of a socially inclusive and environmentally protective development.

The government's first major legislative victory was the Chico Mendes Act, passed within two weeks of Jorge Viana's inauguration as governor. It created a rubber subsidy for tappers—payments that, for example, the Xapuri rubber cooperative members received.[41] The government also supported the formation of rural extractivist cooperatives, not only to tap rubber but also to collect other forest products like Brazil nuts, and helped to form Natex.[42] In 2008, the government created the State Policy for the Valorization of Forest Environmental Assets, which, as its name implies, framed the forest as containing assets (*ativos*) and laid the groundwork for SISA. The Government of the Forest also created a state Secretariat of Forests and Extractivism, restructured state agricultural extension to focus on agroforestry and agroecology, and turned an existing agricultural school built on an old rubber estate into a new educational institution called the Forest School. There, select extractivists and smallholders came to live for extended periods to be trained as agroforestry and agroecology extension agents, with the idea that they would return to their communities to work. In this model of inclusive development, rubber offered a model for other forest products—both traditional ones, such as Brazil nuts and açaí berries, and new ones, like forest carbon offsets.

Adopted by the state legislature at the end of the Government of the Forest in 2010, SISA built on and sought to support this inclusive forest-based approach to development. Sequestered forest carbon's new international monetary value would fund forest protection and low-carbon rural development in ways meant to benefit rubber tappers and others. Through the KfW payments for emissions reductions, for example, SISA funded the Chico Mendes Law's rubber subsidy, supporting Natex and rubber tapping in Xapuri and elsewhere. This kind of state distribution can be understood as extractivist in the Acrean sense articulated by Acre's do Rêgo and others: as promoting socially inclusive and environmentally protective development that, to return to Araújo's song, gives value to rubber tappers and other rural people based on the new value of the forest.

The Rubber Narrative

Or so the story went. For it was not just this history of rubber and its socioenvironmental relations that enabled forest carbon's valorization in Acre. It was also the retelling of that history in settings such as the 2016 CARB workshop in California. The rubber narrative, as I came to understand it, was a story of environmental protection and social inclusion that helped make

Acrean forest carbon valuable internationally.[43] As such, it speaks to how value is constituted in green capitalism, as well as to Acrean politics of the late twentieth and early twenty-first century, when one kind of oppositional leftist politics was partially subsumed by another more state-based kind.

I was accustomed to hearing versions of the rubber narrative both in Acre and outside of it. I heard the narrative, for example, in Rio Branco in 2013 during a research team training ahead of the CIFOR study on SISA on which I worked. During the training, a government agricultural economist recounted rubber's boom and bust, the arrival of cattle ranchers and the violent expulsion of rubber tappers from their land, the struggle (*luta*) that Chico Mendes led, and the current work to valorize the forest in response, of which SISA was a part. "Everything comes from this history," he told the research team, composed primarily of Acrean college students and recent graduates. "We are the counterexample, the counter model," he said, to the deforestation that happened in the neighboring state of Rondônia, in southern Pará, and that had happened all over the world.

I also heard the rubber narrative in interviews when I asked about Acre's position as a REDD+ leader and as a likely source of tropical forest carbon credits and offsets for places like California. As one government official put it, after describing Acre's rubber-based colonization and rubber tappers who "needed the living forest," "because of our history, it is natural for us to do REDD+." And I read versions of the rubber narrative in supportive NGO and funder reports, which referenced Chico Mendes, the rubber tapper movement, and the Government of the Forest in describing their support for SISA.[44] At a government meeting on the KfW payments I attended, a SISA administrator asked a small group of environmental bureaucrats whether it was clear to them why Acre was receiving the payments. In response to the ensuing silence, she offered that it was "because we have [carbon] credits"—millions of tons' worth of them registered with the Markit Environmental Registry. Her colleague then supplemented her explanation with a version of the rubber narrative: Acre was also receiving the money because of its history with rubber, its existing "good policies," and its "functioning system of forest governance." Without this, he said, there would be too high a risk that deforestation in Acre would increase substantially again.

These iterations of the rubber narrative were not its first. A different version of it had already been circulating internationally for decades. The *New York Times*'s Andy Revkin wrote a book about it, for example.[45] And it helped to inform Indigenous and other rural organizing in other places with tropical forests like Kalimantan, Indonesia, where the story of Chico

Mendes and the rubber tappers was mobilized as "an inspirational one about the possibility of effective mobilization."[46] The rubber narrative's power lay in its articulation of a form of conservation that was led by poor rural people, in contrast, for example, to the fortress conservation of protected areas that so often pushed them out.[47] The later retelling of the rubber narrative to make Acrean forest carbon valuable in places like California was possible in part because of this earlier circulation, which both gave Acre an international reputation for community-led conservation and helped link social inclusion with forest conservation. Previous iterations and circulations of the rubber narrative, in other words, enabled and became part of the contemporary one.

The rubber narrative simultaneously helped to explain Acre's early success in valorizing its forest carbon and helped to create that value through its retelling. Take the 2014 annual Governors' Climate and Forest Task Force (GCF Task Force) meeting, hosted by Acre in the capital, Rio Branco. Leaders in forest conservation and rural development from the GCF Task Force's dozens of member jurisdictions gathered to discuss strategies of forest protection and rural development—REDD+ among them—and to sign the Rio Branco Declaration "reaffirm[ing]" signatories' "commitment to reduce tropical deforestation, protect the global climate system, improve rural livelihoods, and reduce poverty."[48] The GCF Task Force is collaborative, meant to support officials and civil servants working on forest protection from around the world.[49] Yet events like the annual meeting could also contain an element of competitive spectacle.[50] At the 2014 meeting, Acrean officials displayed their state's prized capacity to protect and value the forest, and some attendees praised their host's successes and leadership. The stakes felt high. The audience included representatives from CARB, which was still deliberating about whether to admit tropical forest carbon offsets into the California carbon market. The event also included an optional day of tours through which attendees could see the Acrean forest economy in action, including the tour of the Natex condom factory and the rubber tapper cooperative that I attended. Before dinner one night, we all watched young Acreans perform a folk dance wearing traditional garb—the girls in long skirts, the boys in white shirts. As we watched, a lawyer who had helped to write SISA's 2010 law whispered to me that no other Brazilian state had this kind of performance, this kind of pride. Acre was special.

The rubber narrative featured in some of the gathering's events. For example, Senator Jorge Viana, former Governor of the Forest, offered a version of it in his speech that I attended along with many people I knew: Acrean students who had been part of the CIFOR research team, CIFOR staff from other parts of the world, GCF Task Force staff from the United States, an Acrean real estate agent working on private forest carbon projects, and many members of Acre's environmental middle class who worked for the government and allied NGOs. Viana hit upon all the rubber narrative's key elements. He described how Acre had the "the largest production of the most important product of the Industrial Revolution"—rubber—and how this "connected [Acre] with the world without destroying the forest." He then offered an account of the rubber tapper movement and how Chico Mendes had taught that "the best way to defend life" and to "defend people" was to "defend the forest."

Viana turned next to issues of governance, discussing the outside alliances the rubber movement forged and the Acrean political leaders it helped create—himself included. "We are all," he said, "children of this movement." And, he went on, "I am not an heir, I am a son; there is a big difference in that," he said, implying, as I took it, kinship and continuity. When he and Mendes's other "children" were elected to office in 1998, he reminded the audience, they created the Government of the Forest. At a gathering of Brazilian governors at that time, he said, a prominent economist publicly referred to him as the administrator of "that NGO that is Acre." This was only partly a joke, Viana said. After all, his was a purposefully "different" kind of government, one that prioritized the forest and those connected to it from the start. He offered an example—the Chico Mendes Law subsidizing rubber—which he positioned as Brazil's first policy to offer "remuneration for environmental services." It "created the basis" for SISA. The Government of the Forest—from the Chico Mendes Law that began it to the SISA legislation passed in its final months—positioned the forest as an "environmental asset."

The importance of Acrean rubber's socioenvironmental relations to making forest carbon valuable was not just my insight, in other words. It did not come into view only through my anthropological training to see how the particular—here, Acrean rubber relations—modifies and constructs the general—here, green capitalism. Instead, it was an assertion by those working to valorize Acrean forest carbon that this particular history mattered, that rubber had helped to hold forest carbon in place in the past and would into the future. In Acre, the rubber narrative suggested, the economic value of the living forest was not the import of a foreign green capitalism. Rather, the narrative positioned forest carbon's monetary valorization as a homegrown outcome of rubber's political ecology and the socioenvironmental movement and socially inclusive approach to economic development that came from it.[51]

Value-generating stories are not uniform, and their composition and performance are indicative of the commodities and economies they help to constitute. Commodities that are, for example, incompletely commodified or "unfinished," as Heather Paxson writes in her discussion of artisanal cheese, may be made in part by "bring[ing] select elements of the social and material backstory . . . to the foreground."[52] The rubber narrative did just that: it foregrounded a part of forest carbon's backstory to make it into an internationally valuable commodity.

Forest Carbon as Commodity

The differentiation and standardization entailed in commodification can be synergistic, rather than at odds. In this, the rubber narrative's utility in differentiating Acrean forest carbon from that of other tropical forests also contributed to the standardizing and speculative work necessary to make that forest carbon monetarily valuable. It did so in at least two ways.

Reference Levels and Baselines

Selling offsets or credits usually involves creating a reference emissions level and crediting baseline.[53] The reference level is a projection of the rate of deforestation or deforestation emissions that would have happened without interventions to stop it. The baseline is the rate below which credits can be issued or other monetary compensation may be awarded. The two can be identical, but the baseline can also be set lower than the reference level (see figure 1.2).

For example, for SISA, officials adopted a deforestation reference level that was originally developed as part of the state's forest governance strategy (not its forest carbon valorization strategy) in its 2010 Plan for the Prevention and Control of Deforestation (PPCD).[54] Officials also had to make SISA's approach compatible with the less conservative baseline adopted in the Brazilian Federal Government's 2009 National Policy on Climate Change, and then with later federal REDD+ policy.[55] Following federal guidelines, Acrean officials set the baseline equal to that reference level. The baseline and reference level were compared with the measured deforestation rate, determined

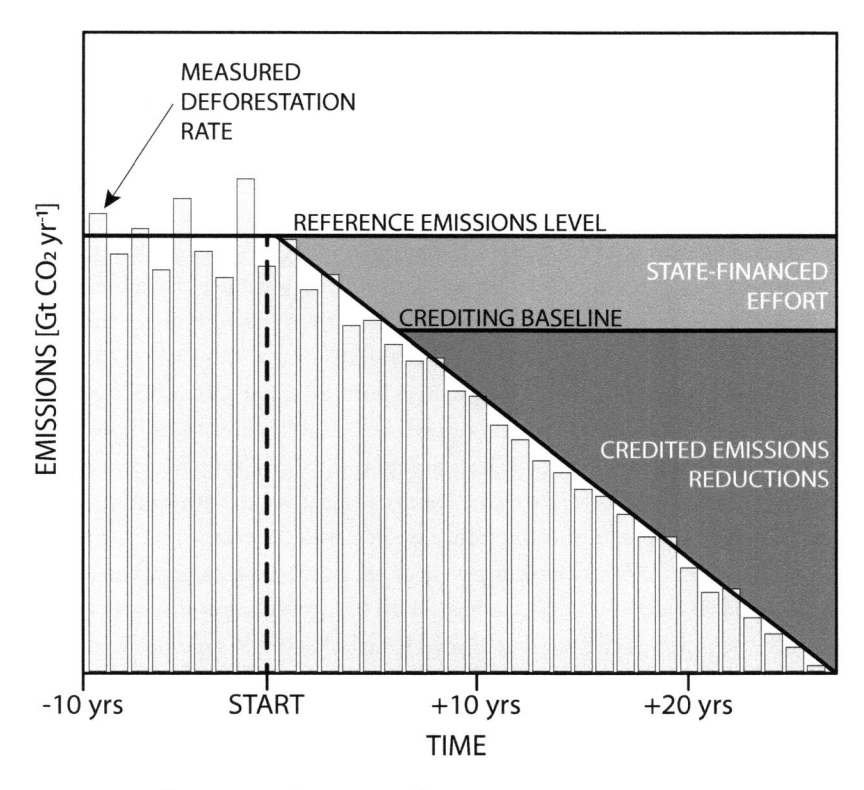

Figure 1.2 Illustrative schematic of forest emissions accounting over time. Credited emissions reductions can be compensated for, for example, via carbon offset or credit sales or payments for emissions reductions. Setting the crediting baseline below the reference emissions level allows for a forest carbon project, program, or locality's own efforts to reduce deforestation to be excluded from the credited emissions reductions. A lower crediting baseline can also create a buffer to avoid overcrediting. The baseline and reference level can also be made the same. Adapted by Justin Mankin from Evan Johnson, "California, Acre and Chiapas: Partnering to Reduce Emissions from Tropical Deforestation," REDD Offset Working Group, 2013, 25, accessed November 5, 2023, https://ww2.arb.ca.gov/sites/default/files/cap-and-trade/sectorbasedoffsets/row-final-recommendations.pdf.

by the Brazilian National Institute for Space Research (INPE)'s PRODES system, which is internationally respected for its robust remote sensing–based deforestation monitoring.[56] These measures of deforestation could be translated into emissions via two conversion rates, adopted by SISA's Scientific Committee based on existing state and federal policies and independent scientific research: each hectare of forested land was understood to hold an average of 123 tons of carbon—a conversion rate considered conservative because it includes only above-ground carbon and was lower than the federally adopted average—and each ton of stored carbon was recognized to create 3.67 of tCO_2e in the atmosphere if released.[57] The REDD Offset Working Group—a group of scientific and legal experts charged with advising CARB on tropical forest carbon offsets—described Acre's methods as "consistent with . . . internationally recognized best practices."[58] Comparing the actual deforestation emissions rate to the baseline rate gives the amount of creditable emissions reductions and hence determines the amount of monetary compensation that can be received via offset or credit sales or payments for emissions reductions (see figure 1.2). With their baseline and reference levels set, the Acrean state was positioned to sell offsets and credits and receive payments for emissions reductions into the future.

As projections that enable forest carbon's monetization and commodification, baselines and reference levels can be considered "technologies" of speculation, as Laura Bear puts it, which are "deployed to anticipate the future; to stimulate its emergence; and to control it."[59] But they raise tricky epistemological issues. Calculating them requires summarizing place-specific and unstable socioenvironmental relations into standardized metrics of deforestation-linked emissions per year, and then projecting them into the future. Following accepted best practices, these calculations are often based on past deforestation, a so-called BAU approach.[60] In Acre, for example, the state's average 1996–2005 deforestation rate was used to set the reference level and baseline through 2015, with the 2016–2020 reference level set based on the lower 2001–2010 deforestation rate.[61] Changing socioenvironmental dynamics—like the changing climate or the construction of highways—can also be included via modeling of the future.

Such calculations necessarily rely on counterfactuals: once an intervention (to use the common climate lingo) to reduce deforestation occurs, the reference level and baseline are simply projections of what would have happened without it. Their validity is ultimately unknowable (though certain calculation techniques are considered more or less valid): how can we know if any reductions in emissions are actually *additional* to reductions that would have happened anyway? In other words, maybe deforestation would have declined even without the intervention to stop it. Maybe the future would not have been *business as usual*. Maybe it would have differed from the past. Moreover, these counterfactual projections can also provide cover for increases in deforestation. Since it remained below the state's reference level, for example, some government interlocutors told me that the increase in Acrean deforestation that occurred during my fieldwork and after would have been worse without their efforts to limit it. The answer was, then, to double down on existing strategies to valorize the living forest and its carbon.[62] Baselines and reference levels enable such would-have-been-worse reasoning.

Stories can be helpful to salve such epistemological disquietude, shoring up forest carbon's value. These stories can be quantitative. For example, quantitative model projections often associated with REDD+, as Andrew Mathews shows, enact a form "storytelling, of looking back to measure several past movements of the world and then of turning to project this past into a range of credible futures."[63] Yet to effectively generate value, these quantitative projections also benefit from other kinds of storytelling that interpret and narrate them. These supplemental stories can make reference levels and baselines seem credible and compelling, and therefore able to create value. Such supplemental storytelling speaks to what reference levels and baselines really do: condense a complex assemblage of socioenvironmental relations linked to deforestation into simple lines. While such lines are necessary, they are not always sufficient to generate value. Stories about the underlying socioenvironmental relations can be essential too.

This is what the rubber narrative did. On the one hand, its account of the villainous cattle ranchers' violent deforestation made Acre's high reference level legible and compelling, offering a morally charged explanation of why deforestation had been so high and would be high in the absence of efforts to reduce it. On the other hand, its description of the rubber tapper movement and Government of the Forest's socially inclusive and environmentally protective approach to development explained the departure from that reference level—the reduction in deforestation and related emissions in recent years. And, importantly, this account made it seem plausible that the reduction in emissions would continue into the future, because of Acre's socioenvironmental relations.[64] In other words, the standardized baseline and reference level projections did not speak for themselves. To make forest carbon monetarily valuable—especially amidst rising deforestation—other stories were needed as well.

Third World Forest Imaginary

The rubber narrative, though, was different from the stories surrounding many other forest resources. After all, forest carbon is a different kind of resource. First, there is the matter of its materiality. Carbon offsets, credits, and payments for emissions reductions work through absence: they represent emissions *not* emitted. Yet this apparent immateriality is misleading; in fact, they rely on carbon's arboreal sequestration—rather than extraction—and, therefore, on socioenvironmental relations that keep forests living and carbon sequestered in them. Second and relatedly, forest carbon's materiality also points to its temporality: the future is important to forest carbon's new value, not because of its imagined future extraction, but instead just the opposite. Forest carbon needs to stay in place. And since holding carbon in place in living forests requires place-specific socioenvironmental relations, making offsets entails the speculative promise and belief that these or similar relations will endure.[65] Rather than "making value uncertain," then, forest carbon's speculative value relies on conjuring certainty against the reality of an uncertain future, made more so by the changing climate.[66]

Forest carbon's materiality and temporality shape the stories told about it. Extractive resources often rely on classic "resource frontier" stories, in which these regions are depicted as "'vacant,' 'un-governed,' 'natural,' or 'uninhabited,'" as Mattias Borg Rasmussen and Christian Lund put it.[67] This positioning can happen through what they call "frontier moments—when existing regimes of resource control are suspended," such that resources seem ready for the taking in a globally legible and homogeneous way. As Tsing describes, "frontier culture . . . conjures a self-conscious translocalism, committed to the obliteration of local places."[68] Resource extraction can rely on such standardized depictions of empty and ungoverned frontier spaces.[69] To be a forest carbon frontier, though, necessitates *not* being seen as a frontier of the usual sort. To appeal to California carbon market regulators, Acre had to seem not devoid of regulation, empty, or equivalent to other tropical forests, but rather as effectively governed by a state at the forefront of forest protection. It had to appear inhabited by rural people who were not poor and primitive, but rather who were innovators of that forest protection and who benefited from it.

This differentiation, though, also involved a form of standardization. To draw on Julie Guthman and Sarah Besky, the rubber narrative conjured a kind of "Third World forest imaginary" that depicted Acrean "forest people" (like rubber tappers) as the historical vanguard of socioenvironmental organizing and forest governance.[70] In it, the socioenvironmental relations of Acrean offsets also appeared as inclusive—benefiting diverse rural communities and forest ecosystems. In green capitalism, such place-specific but globally legible stories of people and the environment jointly thriving can be of particular worth, enabling commodification and creating value. Those studying designations such as fair-trade and organic have shown, for example, how stories of happy family farmers and cage-free chickens or those implied through "fair trade" labels on bars of chocolate make these products valuable.[71] Something similar was true for those seeking to get Acrean offsets admitted into carbon markets like California's.

Making Acrean forest carbon valuable internationally entailed a different kind of standardization than that of making greenhouse gases and diverse forests the "same."[72] Instead it was a standardization that emphasized stories of forest-protective inclusion. This standardization was itself not just a figment of a globally standard green capitalism, but was also partly the outcome of earlier circulations of Acrean stories about Chico Mendes and the rubber tapper movement that had positioned forest communities as the potential vanguard of forest protection in the first place.[73] The rubber narrative emphasized some of rubber's socioenvironmental relations in ways that both drew from and reinforced ideas about such communities as innovative pioneers of green capitalism who helped to hold forest carbon in place. In doing so, the narrative facilitated Acrean forest carbon's value in places like California.

Conclusion: Rubber's Past, Forest Carbon's Future

Acre's rubber-connected past made a future forest carbon boom seem possible. But that possibility depended on a common tension within commodification: it necessitated that Acrean forest carbon be both standardized and singularized. To become a valuable environmental commodity, Acrean forest carbon needed to be made into fungible and purchasable units. But it simultaneously needed to be differentiated from and of higher value than forest carbon elsewhere. In green capitalism, being seen as well governed, socially inclusive, and environmentally committed was essential to this value-generating differentiation.

The rubber narrative served this purpose. It distinguished Acrean forest carbon by linking it to rubber and the Acrean rubber tapper movement's charismatic history in a way that made sense to an audience of potential offset buyers and those otherwise interested in paying for emissions reductions. Through it, Acre appeared as a different kind of resource frontier than the extractive ones typical of many tropical forests, and that Acre itself had been during the rubber boom and subsequent bust. Acre appeared as socially inclusive, environmentally protective, and managed by an effective government connected to a pioneering social movement in ways that aligned with outsider views of what offsets should be—views that were themselves partially formed by previous international circulations of the rubber tapper narrative. These views were articulated, for example, in carbon standards like the Climate, Community, and Biodiversity Alliance (CCBA) Standard and the Tropical Forest Standard (TFS) that CARB eventually adopted in 2019.[74] Such standards make the ability to create and sell offsets contingent on socially inclusive socioenvironmental relations that keep forest carbon in place. The rubber narrative was a story of place-specific inclusive relations that protected the forest and made Acrean forest carbon offsets seem worthy of admission into the California carbon market. Offset sales promised to fund, and thereby maintain, those socioenvironmental relations, keeping forest carbon in place into the future.

Yet, like all stories, the rubber narrative was incomplete. Rather than "see[ing] the future while looking at the present," as the southern African planners whom David McDermott Hughes writes about do, the rubber narrative largely focused on the past.[75] Rubber served as a template of a form of rural development that would bring rural prosperity through forest protection. Yet the Acrean rubber economy itself continued to decline.[76] It is telling that I primarily encountered rubber in museum exhibits, guided tours, and park memorials. Angelo, who had fingered the grooves in the rubber tree at the Rio Branco park memorial, had abandoned rubber tapping long ago. He told me that he was happy to escape its drudgery and dangers, to just visit the trees in city parks. Many other Acrean rubber tappers had also stopped tapping rubber, despite the SISA-funded rubber subsidy and other governmental support. Those who continued on could be seen as "heroes," as one government official called them at a 2014 workshop to promote Acrean rubber. But tappers at the workshop asserted in response that rubber's price was just too low.[77] It was not worth it, even for those who lived in the extractive reserves that the rubber tapper movement had helped to create and with payments from the state rubber subsidy.[78] The price did not justify the hard work and exposure to the forest's dangers.[79] Moreover, as Jeff Hoelle found, rubber tapping continued to be associated with poverty and low social status in Acre.[80] Some rural Acreans still identified as rubber tappers, but many moved to cities or sent their children to attend school there to make a life away from the forest.[81] Other rubber tappers started accumulating cattle, including within the state's flagship Chico Mendes Extractive Reserve, where I attended the rubber tapper cooperative tour.[82] Rubber tapping became a story of the familial past—something parents or grandparents did, something to tell anthropologists when asked about family origins. Rubber trees were something to look at in parks and on tours, something to kiss for a photo on your cell phone so you could show your kids in the city.

The state government also seemed to be de-emphasizing rubber, at least in its extractivist form. The governorship of Jorge's brother Tião Viana (2010–2018), for example, eliminated the word *extractivism* from the name of the state secretariat. Using new species and planting techniques to counter the South American leaf blight, the state also supported new rubber tree plantations and planted rubber trees to supply the Natex condom factory.[83] I saw one of the new plantations in Xapuri, not far from the Chico Mendes Extractive Reserve. The trees were planted in long rows, neat and straight. My friend—the daughter of a rubber tapper—pointed them out as we were driving to Rio Branco. "Look how good they look," she said. "I'll bet it's nice and cool in there; a great place to hang a hammock for the day." Others lamented the change. Another friend, who worked at the local office of an environmental NGO, mourned the shift away from extractivism as we walked out of a late afternoon meeting with government officials. She asked me if I noticed how the officials were not interested in extracting rubber from within the forest and how they only wanted to talk about planting rubber trees in plantations.

With rubber's decline, the rubber narrative could appear as a story increasingly divorced from reality.[84] Some Acrean scholars and activists have called out this disconnect. Dercy Teles, for example, the first female president of the Rural Workers Union in Xapuri and a leader in the rubber tapper movement, argued that "rubber extractivism has died and the government, to date, has not presented an alternative form of income generation to replace the one that rubber extractivism provided."[85] Yet the rubber narrative also indicates that the imbrication of rubber and forest carbon was more material and, often, more meaningful than straightforward co-option. It was an effort by state officials to use forest carbon's new value to do what rubber had not been able to do: make the living forest monetarily valuable, not to benefit bygone rubber barons but rather rubber tappers and other poor rural people. In this vision, forest carbon's new value would fund inclusive green development. As Jorge Viana wrote in a public Facebook post just before the GCF Task Force meeting in August 2014 (and a few months before a crucial election in which he was running),

> I want to see Acre prosper and become a good place to live, but with the face of Acre [*a cara do Acre*]. With the wealth of its forest exploited sparingly, without destruction and shared by all. Anyone who doubts that this is possible should try to follow the international GCF [Task Force] meeting that will take place in Rio Branco from the 11th to the 14th of this month. Specialists from 7 countries (United States, Brazil, Indonesia, Mexico, Nigeria, Peru and Spain) will present strategies and propose negotiations that guarantee payment to those who do not deforest and, therefore, do not release carbon dioxide into the atmosphere, generating the greenhouse effect. Whoever does this provides an environmental service: they preserve Nature, save the Planet and, of course, need to be paid well.[86]

Rubber's decline, though, indicates that forest carbon might not easily stay put, undermining its value and disrupting the vision of inclusive green capitalism that Jorge anticipated. Rubber's socioenvironmental relations could not be counted on to keep carbon sequestered in forests since fewer people were tapping it and the government was not sufficiently supporting it. The past would not be the future. For a time, the rubber narrative was effective in giving Acrean forest carbon potential value in places like California. But it also obscured many other socioenvironmental relations shaping the Acrean Amazon, which entailed a much wider array of livelihoods, landscapes, species, and attendant socioenvironmental relations than just those surrounding rubber. I could see some of these along the BR-364.

INTERLUDE I

Highway Landscapes

When I rode fast on the BR-364 between the cities of Rio Branco and Feijó, it often looked to me like nothing was being grown or made alongside of it. That was not true. In fact, people were growing food, raising animals, and collecting various forest products. But this extraction and production tended to be extensive and, in this sense, different than the monocrops and feedlots I was familiar with from the United States. Their relatively low yield per hectare was hard for me to see when traveling at high speeds.

Rubber

Rubber was the form of extraction I most expected to see because of its showcasing in the rubber narrative. Yet it was mostly invisible from the BR-364.

Rubber trees in Feijó tended to grow far from the road since they were dispersed in the forest that usually stood at a significant distance from it. The forest often appeared only as a thin green line in the distance, as if receding, pushed back from valuable roadside real estate. Many of the smallholders who lived on the BR-364 in Feijó used to spend a lot of time in the forest tapping rubber, but most I knew no longer did.

Among them were Luis and Flávia, a couple in their fifties, and one of their daughters and her husband. They were distant neighbors on the BR-364. All were born in rural parts of Feijó, where Flávia's parents had collected rubber. They tapped the trees without felling them, clearing the surrounding forest, or seeking to increase their yield per hectare. She was thirteen when the family, seeing the price of rubber decline, moved to the small city of Feijó so that she and her siblings could attend school. Many rubber tappers in the area followed a similar path. Flávia remembered life before the move to the city as difficult (*uma vida sofrida*), with her closest neighbor living over an hour away. But it has also been really good (*tão boa*)—simple, beautiful.

Luis's father, in contrast, had been part of rubber's upper class. One of Feijó's founders, he had been a *seringalista* rubber baron. He had bought latex from tappers who lived and worked on rubber estates to which he had lease rights, and the family had thus accumulated some wealth. But he had died when Luis was only two, and the family fortune subsequently dwindled, especially as the price of rubber fell. When Luis was a teenager, he spent seven years tapping rubber himself to make a living. It was hard work, he told me, but what else could he do? There had been few other options. Now, though, Luis and Flávia's focus was different. Their lives no longer revolved around rubber. Rather, they centered on their fields, other forest products, and the road on which they lived.

Swidden

Luis and Flávia described themselves to me as rural producers, though Flávia was also a teacher at a new school a short distance from their house on the BR-364. Yet, like their neighbors, they did not seem to grow—to produce—much, if anything, for sale. They were not the owners of intensive monocultured soy, as "producers" might be in other parts of Amazonia.[1] Theirs was, rather, primarily a subsistence agriculture that did not emphasize increasing production from each hectare.

Instead, they often cobbled together different strategies for making life livable. Rubber barons, who had wanted to keep workers dependent on them

Figure InterI.1 Small-scale burning off of the BR-364. Photo by Maron E. Greenleaf. Feijó, Acre.

for sustenance, had often forbade subsistence farming. They required tappers to trade rubber for the food and supplies they needed, often at a substantial markup—a company town model without a town. But as the rubber economy declined and the rubber barons mostly left, former tappers who stayed in rural areas began growing food for themselves. Along the BR-364, some had small gardens of a few square feet close to their houses. In them, women might plant collard greens, green onions, and cilantro for home use. They sometimes had small clusters of banana trees or neat rows of planted fruit trees (at least some of which were distributed through recent SISA-linked government programs to stop deforestation). Farther from the house, they might have a field of a hectare or two, likely planted with manioc—the staple food and the crop that they were most likely to sell—and perhaps beans or corn. These fields were traditionally part of swidden agriculture. Residents might clear and burn a hectare or so of land approximately every year, as they were legally allowed to do (figure InterI.1).[2] Doing so would fertilize relatively infertile Amazonian soils to enable a few years of cultivation. Traditionally, they would then allow the forest to regrow for several decades.

Many supplemented swidden agriculture with other pursuits. A few still tapped rubber. Some hunted or collected other things of use from the

Figure InterI.2 Cow hauling cooking gas. Photo by Maron E. Greenleaf. Feijó, Acre.

forest—resins, fruits, medicinal plants, açaí. But many described being afraid of the forest now. Even if they spent a lot of time in it during their youth, they no longer knew it well and some feared getting lost or being stalked by jaguars. The long days collecting rubber from hundreds of trees to eke out a living seemed to inspire little nostalgia for many of those I spoke with. They might have a cow or two, which were useful in transporting other goods, like cooking gas (figure InterI.2), on long dirt paths and roads (there were few cars).[3] They also described the cows as a form of "savings" (*pupança*) that could readily be sold for cash. The cows even walked themselves to the deal. Men, especially, might also do some paid day labor for larger landowners in the area, and there were a few formal jobs available at schools that had been recently built along the road, like the one where Flávia worked.

A rural producer named José, who lived about an hour southeast of Luis and Flávia on the BR-364, was a revealing exception. When a few members of the CIFOR team visited him, they found a very different landscape from that of the over two hundred other respondents we had talked with previously. When the CIFOR team members returned to the group, they excitedly described how densely cultivated it was with grapes, sweet *amara* berries, and other crops we had not seen before. Though we were all exhausted and eager to finish administering our survey so we could return to Rio Branco, a large group of us decided to make a special visit to spend more time with José on his land. He took us on a long meandering walk through valuable teak and mahogany trees he had planted in neat rows, past his small shaded coffee plot, and through the grapevines and other crops he grew. He walked fast, his two young daughters running behind him, but paused to describe each crop, how long he had grown it, and the trials and victories surrounding its cultivation and sale. He seemed to appreciate our amazement.

Unlike the others whom the CIFOR research team surveyed, and those whom I interviewed myself the following year, José had moved to Acre from outside of Amazonia. He came from Paraná, a southern Brazilian state known as a powerhouse of agricultural production, including by smallholders. His Acrean neighbors called him by the state's name, which he told me was not complimentary. It symbolized his difference, a difference that manifested in his use of his land. I asked José how he was able to grow so much, a question to which he gave two answers: good soil and hard work. His neighbors called him crazy (*douda*) for working so much, but he insisted that the land was very good in Acre (an unpopular opinion). It would "produce whatever you planted, as long as you worked it." He told me that his neighbors were simply lazy. The area where we were between Rio Branco and Feijó was called the "ass" of Acre because the people who lived there did not produce anything, he quipped, recalling the common description of Acre as, itself, the "ass" of Brazil. But José said that he and his family lived well, mostly eating fruit. You could eat as much fruit as you want and never get fat, he explained, patting his taut shirtless stomach, his eyes bright and smiling.

I heard and read various explanations as to why there were not more Acrean smallholders like José. Some explanations, like José's, invoked tropes of laziness. Environmental technocrats often faulted poor soil quality or gave cultural explanations: smallholders would only practice the low-yield agriculture that their parents had taught them. In contrast, José grew more because it was his culture to do so, my research assistant Fernanda—who came from the same state as he did—told me. That was what small farmers from southern Brazil did. A man who had worked in rural development in other parts of Brazil offered a similar analysis: as the descendants of rubber tappers, Acrean rural producers were "workers, not farmers." This is how they had organized during the rubber tapper movement—via rural workers unions and the Workers' Party—and this was how they remained.

Smallholders themselves, however, tended to emphasize another factor: the lack of reliable roads to bring out (*escoar*) their crops. There just were not enough of them, and those that existed were too muddy when the time came to harvest. It was not worth it to grow crops if you could not get them out to sell. Sometimes it was just better to let them fall to the ground and rot.

Cattle

Many rural producers along the BR-364 were, however, producing something: cows. Unlike the smallholders and rubber tappers who kept a cow or two for savings and transport, they undertook extensive ranching, with only one to two head of cattle per hectare of pasture, created by clearing forest. This large-scale cattle ranching has been blamed for much of the deforestation in Acre and the Amazon overall.[4] Dominant thinking has held that Amazonian pastures degrade quickly and cannot sustain more intensive ranching, though some in Acre have been working to change this.[5] No wonder, then, that when I drove past the parched, monochrome fields alongside the BR-364, they struck me as empty, save for the occasional lone Brazil nut tree.[6] The trees seemed to stand defiantly tall, growing branchlessly upward from the field and then unfolding at the top into a tuft of lush green foliage. Every now and again I saw a few cows—white bovine reminders of how demand for beef underpinned much of the deforestation in Acre.

Extensive cattle ranching does not work through increasing yields—in fact, just the opposite. I sometimes heard Acrean cattle referred to as *boi verde* (green cattle)—a laudatory term meant to indicate their "natural" quality.[7] The cows were for beef (milk was trucked in on the BR-364 from Rondônia). They were green not because they were sustainably produced or linked to forest preservation, but because they had large swaths of spacious pasture to consume. All that pasture had been forest, and when the cattle needed more grass, ranchers cleared more of it. That could happen quite fast, as pasture rapidly dried out and became degraded (*degradada*). This extensive cattle ranching was cheaper—at least in the short term—than investing in new and unfamiliar methods that might allow ranchers to raise cattle more intensively, like planting new grass species or adopting rotational grazing. And even if cattle could not eat from it for long, degraded pasture had its uses. It could be a way for cattle ranchers to claim land, access tax breaks or other benefits, or gain resource rights.[8] And it carried cultural power as well. As Jeff Hoelle has shown, Acreans often equated both extensive cattle ranching and the cleared land it entailed with development and prosperity.[9]

For decades, starting with the military dictatorship, the government had encouraged large-scale ranching *fazendeiros* from Brazil's south to move to the Amazon as part of a larger effort to colonize and govern the region. By the time I was doing my fieldwork, similar practices had been adopted by some former rubber tappers, like Luis and Flávia. Although they laughed in seeming surprise at my query as to whether they were *fazendeiros* (a class-linked term they associated with wealthy outsiders), they owned a comparatively well-appointed house on the BR-364 with an unspecified but large number of cows in the fields surrounding it (I could never get an exact number).[10] Many neighbors were integrated into cattle ranching in other ways—as wage laborers on *fazendeiros*' land or by fattening *fazendeiros*' cattle on their own.[11] It was part of what scholars have called a "*pecuarização*" ("cattle-ization") taking place as extensive cattle ranching and cattle culture spread in Acre and elsewhere in Brazil.[12] As cattle ranching came to dominate, more people along the BR-364 and elsewhere in Acre were becoming part of this third highway landscape. All three valued expansion over increased yields.

2

Producing the Forest

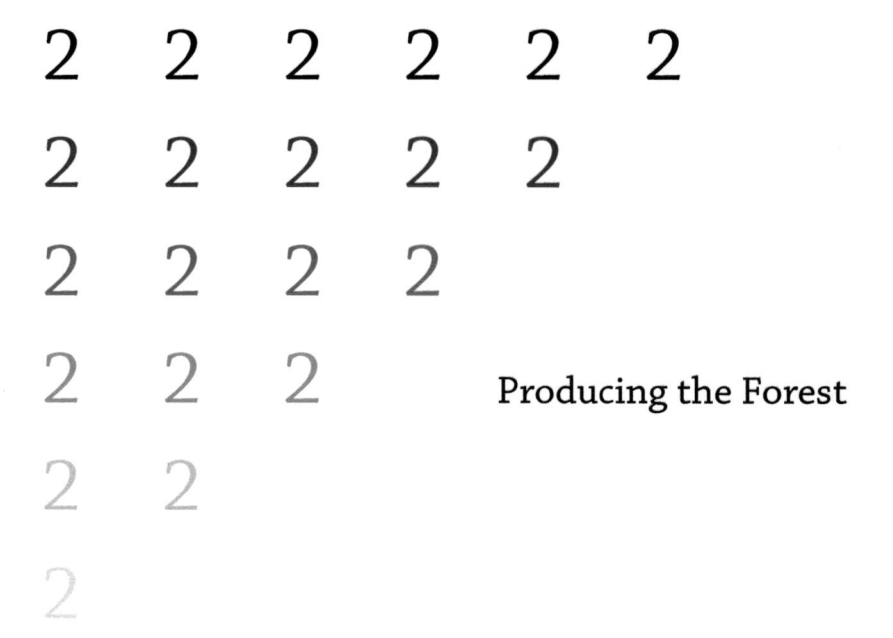

On the side of the BR-364 in Feijó, there was a sign. *Nosso Açaí*, or Our Açaí, it read. Below that: "Made with care for you. Açaí from Feijó." The sign was barely noticeable if you were driving fast. Situated in an apparently vacant field, the mint-green building behind it did not draw much attention either (figure 2.1). This was how buildings and fields often looked along the road. Yet the sign not only advertised the specific business housed in the building, but also signaled an economy expanding from within the famed Amazon rainforest behind it—one linked to the Acrean effort to make the living forest valuable.

The building housed a new business that processed and sold açaí—a berry that is part of Indigenous Amazonian cuisine and is used in savory dishes in eastern Amazonia, eaten as a chilled and sweet snack throughout

Figure 2.1 Nosso Açaí on the BR-364. Photo by Maron E. Greenleaf. Feijó, Acre.

Brazil (*açaí cremoso*), and, more recently, sold as a healthy international "superfood."[1] The business was owned by a woman named Aline who had recently moved back to Feijó, where she was raised. As we talked in her small, immaculately clean processing room, she told me that she was drawn home by açaí and the business opportunity it now presented. Tall native açaí palms grew well in watery parts of the forest in Feijó, and the recent paving of the BR-364 and construction of reliable bridges along it finally made it seem feasible to process and sell the fast-spoiling berries outside of the municipality. Five to six hours down the road in urban Rio Branco, homemade cardboard signs announced small stalls promising the good stuff: açaí from Feijó.

Aline bought açaí from former rubber tappers and their descendants who lived nearby. Young men and children had long scaled the palms to collect berries for their families to eat. Along with many other practices of cultivation and collection, it helped them make do. Now though, many were collecting more of the berries and selling them to people like Aline. Rubber was on the decline, but açaí was on the rise: between 2004 and 2015, the quantity sold in Acre went up an estimated 735 percent, and its monetary value increased by 2,531 percent.[2] By 2013, açaí had overtaken rubber in the state's economy.[3] Much of that increase occurred in Feijó, which had a relatively large number of native açaí palms (*açaízeiros*) growing along forest streams and rivers.[4] Along the BR-364, those collecting and selling açaí to buyers like Aline expressed excitement. It felt like another Amazonian commodity boom.

Some state officials, NGOs, businesspeople, and academics were excited too, and they encouraged açaí's expansion. As I explore in this chapter, encouraging more production—of açaí and a host of other products grown in and collected from forests and fields—was key to Acre's state-led effort to reduce deforestation by valorizing the forest and the carbon it sequestered. In other words, the state did not try to "carbonize" the landscape by creating carbon-sucking monoculture tree plantations or otherwise seeking to maximize carbon sequestration.[5] Nor did it pay smallholders to forgo deforestation or stop using their land altogether. Instead, it sought to enlist smallholders in the effort to make the living forest valuable by incentivizing them not only to produce differently—without using deforestation and fire but also to produce more. This inclusive productivism, as I explore here, may be typical of green capitalism: it seeks to incorporate people viewed as excluded from traditional capitalism's advantages in ways meant to benefit both them and the environment. Tracing açaí in Acre is, therefore, a way to explore how inclusion can help to expand green capitalism, despite or perhaps through its contradictions.

Inclusive productivism embraced a somewhat diverse array of multispecies relations and landscapes. Forest carbon was to be kept sequestered in the forest via socioenvironmental relations between people (like açaí-collecting former rubber tappers and açaí-buying entrepreneurs like Aline) and other species (like açaí palms). These relations were part of landscapes composed of fields, where buildings and businesses like Aline's stood, and forests, where groves of açaí palms grew in watery soils among multiple other species. This diversity contrasted with the uniform, dispossessive sameness of monocrop agriculture and tree plantations. Instead, Acrean forest carbon was a composite created through diverse relations—a multicrop, not a monocrop. The key was that these relations had to yield monetarily valuable products, like the sweet *açaí cremoso* that Aline sold from her roadside store. This increased rural production would supplement the comparatively paltry new international monetary value of carbon and the even more limited amount accessible in Acre. Açaí would help the forest live, securing that carbon. The Amazon would be saved by the productivity of its socioenvironmental relations.

In some ways, this approach to valorizing forest carbon is familiar. Its emphasis on individual productivity and entrepreneurship seems neoliberal.[6] It also echoes efforts dating to the 1980s and 90s to bring communities into conservation (such as *community-based natural resource management* [CBNRM]) and to protect the forest via the expansion of *non-timber forest products* (NTFPS) based on the theory, as Michael Dove puts it, that "the forest is being cleared because its riches have been overlooked."[7]

In other ways, though, protecting the forest by increasing production broke from entrenched patterns. In Brazil and elsewhere, increased production is typically linked to forest destruction via monocrop agriculture, mining, ranching, and infrastructure construction. Intensifying production—as in increasing yield per hectare—is sometimes posited to "spare" forests, but largely in an exclusionary and often violent way linked to racialized forms of smallholder and Indigenous dispossession, labor exploitation, and land grabbing.[8] Forms of forest conservation premised on people- and production-free protected areas can be similarly problematic.[9] Meanwhile, conservation efforts that have aimed to empower or benefit rural people, such as those linked to CBNRM and NTFPS, have often fallen short, including when they have tried to pay people to stop using their land.[10] In contrast, Acrean officials and environmentalists tried to create what some of them described to me as a "new economic model" in which diverse forms of increased production by broad sectors of the rural population would lead to forest protection, carbon sequestration, and rural development.[11]

This inclusive approach allowed for agreement across divergent swaths of Acrean society, from environmentalists to businesspeople—a version of the "politics of agreement" that Susanna Hecht writes about as facilitating Amazonian forest conservation in the 2000s.[12] Importantly, it aligned forest protection with the dominant Brazilian valorization of intensive agriculture and promised to address the anxiety that Acre did not produce enough— that, in a society that so values rural productivism, it does not exist. Yet this was an agreement facilitated by the misunderstanding it accommodated.[13] Within it lie some consequential tensions—tensions that point to the ways that efforts at green capitalist inclusion can be both a promising improvement on the status quo and effectively perpetuate exclusion.

Incentivizing Inclusive Production

In Acre, extensive cattle ranching and rubber tapping have traditionally been opposed in their relationship to the forest: the former is premised on forest destruction, and the latter on forest protection. Often these landscapes and the people in them have clashed, as in the deforestation and rural violence in Acre in the 1970s and 80s and beyond. Yet these landscapes can also enable each other. Many of those who lived along the BR-364 engaged in both, sometimes at different times in their lives and sometimes concurrently. A shared spatial logic facilitated this compatibility: rather than seeking to maximize yield within a given plot of land, practitioners instead expand outward. Ranchers clear more forest for more cattle. Tappers move farther into the forest to tap more trees. Swidden agriculture, too, works through a form of expansion: clear forest to grow food. This spatial logic has seemed suited to the Amazon's apparent vastness, and to the related view of land and forest as limitless resources.

I saw traces of these practices in overlapping highway landscapes: the seemingly empty fields lining the BR-364; lengthy, winding, and sometimes abandoned forest trails that connected dispersed rubber trees in the forest beyond those fields; and small swidden patches cut from that forest. All were made through hard work: smallholder deforestation undertaken for extensive cattle ranching on their own land or as hired labor, the opening of new rubber trails and long daily treks on them, and the forest clearing and cultivation of swidden agriculture—all are laborious.[14] Yet none seek to intensify output. The smallholders who practice them are enmeshed in landscapes where maximizing productivity is not valued in this sense. This extensive approach to production and extraction shaped the Acrean effort to valorize the living forest and its carbon. In that context, it appeared as opportunity.

It is worth teasing out the logic at work. Acrean state officials made increasing production in these expansive landscapes into a forest protection strategy. The BR-364's and other rural Acrean landscapes' low productivity meant that they appeared ripe for low-cost climate mitigation and, therefore, investment. To quote an influential report by McKinsey & Company, REDD+ provided "a large, low-cost amount of [greenhouse gas] potential abatement" because the land use practices in tropical forest areas "yield little economic value."[15] In this context, making those landscapes more productive became a strategy to avoid forest emissions.

Yet the Acrean state's emphasis on increasing production was consequentially different from the proposals from McKinsey and many others. The latter tended to center on payments to rural people for forgoing production. The McKinsey consultants, for example, suggest small "compensation payments and income support to the rural poor and forest people."[16] Such payments aim to compensate these people for using the land less: for abandoning the "slash-and-burn" agriculture that, in the McKinsey consultants' and many others' view, not only clears forests but makes the resulting cleared land insufficiently productive—an understanding that builds on long-standing, racialized tropes of the environmentally destructive and/or "lazy native."[17] As Tania Li writes, such compensation has been "given to forest communities and indigenous people so that they will not cultivate or indeed do anything at all with their land except conserve and restore forest."[18]

This form of REDD+ echoes earlier efforts at including communities in conservation schemes. As David McDermott Hughes writes about a prominent conservation program in Zimbabwe, its aim to get Black smallholders to forgo agricultural production and engage in ecotourism "put . . . [them] in an exceedingly difficult position. Their success as agricultural producers constitutes a failure as environmental stewards. Nor can they *work* their way out of this problem. Investing more labor only confounds their ecological folly."[19] Consequentially, in this approach to environmental conservation, such visions of reduced production tend to be accompanied by intensified forms of environmentally destructive production elsewhere, perhaps quite close by. Purportedly efficient monocrop plantations are often meant to exist alongside areas where smallholder and Indigenous people are enticed or compelled to forgo production in the name of environmental protection.[20] Either formally (via offsets, for example) or informally, the latter can seem to compensate for the former. In this view, smallholder production (and often Indigenous production) is environmentally destructive and unproductive; their rightful place is to exist and subsist apart from global commodity markets.[21]

In apparent contrast, Acrean officials sought to entice rural people to deforest less by producing more. Luana, a SISA administrator, explained this logic to me over bowls of sweet, cold *açaí cremoso* at Pinguim—a spacious upscale ice cream and açaí shop in Rio Branco—after work one day. She explained to me that "any activity" could address environmental destruction if "done well." All that was needed was "just a little help" to produce "sustainably." As another government interlocutor explained to me, "You have to help the person to produce, help the person to work." In SISA, this help came in the form of what, using a prominent neoliberal concept, were called *incentives* (*incentivos*), as in SISA—the System of Incentives for Environmental Services. When distributed to smallholders, these incentives were often called *benefits* (*benefícios,* a term signaling state-citizen relations, as I analyze in the chapters that follow). Benefits included some cash payments—like the state rubber subsidy—but were primarily in-kind materials, services, and trainings, including fishponds, land tilling services, mucuna seeds, agricultural assistance, and açaí seedlings.[22] Instead of payments for smallholders to produce less, these were benefits to entice them to produce more (see figure 2.2), in ways that were understood to reduce deforestation.

Figure 2.2 *MAIS PRODUÇÃO* (MORE PRODUCTION). Logo on an Acrean government (SEAPROF) car. Photo by Maron E. Greenleaf.

Forest carbon's new monetary value both enabled and relied on this vision of increased production. For example, much of KfW's €25 million payment to Acre's state government for 2011–2015 emissions reductions was distributed to rural people to entice them to produce more while deforesting less. Theoretically, subsequent reductions in deforestation-linked emissions would then enable the state to receive more international monetary compensation—a kind of green capitalist positive feedback loop in which profit, smallholder benefits, and environmental sustainability reinforce each other. A win-win-win.

My interlocutors in the state government and allied NGOs framed this strategy as being socially inclusive because smallholders were positioned as central to the low-carbon green economy they were trying to create. They were usefully not too productive, such that they could now be enticed to cultivate and collect more in ways that could grow the economy and protect the forest. Moreover, focusing on increasing rural production also meant that they did not have to "stop earning" because of the state effort to value the living forest, Luana explained to me. Nor would SISA have to cover their "opportunity costs," as envisioned in some scholarship on environmental valuation—a good thing, considering how little money forest carbon was actually bringing in.[23] In fact, smallholders could earn more than they would otherwise by growing, collecting, and selling more crops and forest products. These increased profits would make new low-carbon forms of production self-supporting after about fifteen years of international investment in forest carbon, a SISA administrator estimated in an internationally broadcast English language webinar, when she was asked how long payments would be needed. After that, low-carbon rural production would become "so natural" that external funding would no longer be necessary, she explained.

Luana told me that this approach was better than the popular idea of paying smallholders to forgo deforestation. In part because of the small amount of REDD+ funding that was actually available, the payment size would be too trivial to significantly impact smallholder lives or induce them to stay on their land instead of migrating to the cities' poor peripheries (where they were seen to contribute to urban crime).[24] Besides, paying smallholders not to use land could undermine their ability to make usage-based claims to it, and could also further exclude them from the culturally dominant rural productivism.[25] In contrast, inclusive productivism promised to incorporate smallholders into the dominant economy and culture by supporting intensified agriculture and market engagement.

The aspirational landscape of governmental imagination, then, was not composed of carbon sequestration–maximizing tree plantations or forest-sparing monocropped fields—anticipated standardized landscapes of the Anthropocene.[26] Rather, it was a landscape of an inclusive green capitalism populated by productive smallholder farmers of a sort that was uncommon in Acre—people like José, the smallholder who amazed the CIFOR team. In the case of açaí, at least, many smallholders were eager to participate. They wanted to be included in this vision of increased production.[27]

Açaí on the BR-364

When I spent time on the BR-364 in fall 2013 and spring 2014, I ate açaí both at Aline's small factory and down the road with Luis and Flávia, where I stayed. Their freezer had bags of it, collected by their son-in-law Francisco, who lived on land adjacent to theirs. The family would sometimes thaw the berries for a sweet after-dinner treat that would leave the grandchildren's faces briefly tinged a dark, earthy purple.[28] But Francisco sold most of the açaí he collected to buyers, including Aline. In recent years, he had started collecting more and more.

He was not the only one. Many people in Feijó were collecting açaí during that spring harvest, which ran from April through June (a smaller harvest occurred in the fall). Açaí was a frequent topic of conversation: its price, where middlemen were buying up bags of it and for how much, whether this year's crop was as good as it was in 2012 or at least better than it had been in 2013. Like many young men in the area, Francisco spent hours climbing the tall, skinny, branchless açaí palms to collect (*tirar*) their berries. These native palms—often called *nativo*—tended to grow along the banks of rivers and streams and other low-lying, watery places where they were shaded by a diversity of other trees and watered by Amazonian hydrology.

Climbing the palms could be dangerous, and collectors tended to use minimal equipment, Francisco included. They put both feet through a small loop, often made of discarded plastic sacks, and used it as leverage. With a machete tucked in their shorts and without helmets, they pushed the loop against the palm's trunk, little more than a half a foot in diameter, and hoisted their way upward, grasping the trunk first by their hands, then pushing it with their feet, then hands, then feet again. At the top, perhaps twenty meters up, they cut large bunches of berries and then slid down the trunk to bring the fragile harvest to the ground. There, they gently coaxed the berries into awaiting cans or baskets. And then up the tree they went again.

Usually, Francisco sold to two local buyers. Aline was one of them. But sometimes he sold to middlemen (*atravessadores*), who paid less but bought at a dependable frequency. On Wednesdays and Saturdays during the harvest, the middlemen drove the BR-364 to buy berries—placed in sacks for transport—from collectors, who waited with their sacks along the road's hot, newly poured asphalt. Since they did not drive as far as Francisco's land, he sometimes hitched a ride to the place where the BR-364 crossed over the Jurupari River. Though the bridge was small and unremarkable, it was central to the area's growing açaí economy. The best palms were apparently a few hours up and down the river—their berries brought to the bridge by boat. Standing with full sacks, smallholders from the BR-364 and from the river waited for the middlemen to arrive in open-air flatbed trucks (see figure 2.3). In the meantime, women sold long homemade *dindin* popsicles from portable coolers. Teenagers gathered under the shade of tarps erected earlier in the day, seeking respite from the hot sun and a place to flirt. With the sale of açaí, this place where the BR-364 crossed the Jurupari became a rare rural gathering place, replete with smiles, chatter, and moneymaking.[29] The middlemen inspected and weighed the sacks—for which they paid R$32 each when I visited—and stacked them into the back of

Figure 2.3 Bags of açaí being sold on the BR-364. Photo by Maron E. Greenleaf. Feijó, Acre.

their truck. Then they drove on, to factories in the towns of Boca do Acre or Plácido de Castro, where the açaí was processed before being exported. All that had to happen within three days of picking, before the berries rotted. Some of the berries didn't make it. Rotten açaí, dark and swollen, was cast off on the roadside, a testament to the berries' fragility, bounty, and the ongoing rush to collect more.

Açaí was reshaping life in this part of Feijó. Business was good. And Luis and Francisco wanted to talk about it when Fernanda and I showed up to stay. Until three years prior, Francisco had only collected açaí occasionally for his family to eat. He and Luis told me that so much had changed since then. They never imagined açaí could be like this. The selling price had almost tripled in three years. And it did not even fall during harvest time. Francisco marveled about how easy it was to make money.[30] In three hours, he and two others could collect enough berries to fill fifteen sacks. Calculating aloud, Fernanda exclaimed, "You could make over R$2000 in a week!" Francisco and Luis nodded in affirmation. They said that they and many of their neighbors had earned a lot of money in the past few years. Fernanda and I just needed to look around to see it. Houses were being renovated, additions built, and flatscreen TVs, refrigerators, and generators bought along the BR-364 and up and down the Jurupari River. Their neighbors already had plans for what they would do with money from this year's harvest. Luis was calculating the number of bricks he would need to pave their steep and mud-prone driveway from the BR-364 for a car he and Flávia had just bought. Their first, it was set to arrive in a few months. Such were the açaí-fueled futures being envisioned along the road.

Luis told me that açaí reminded him of the rubber of his youth. Collecting, buying, and selling it—that was all anyone did. He said that açaí could feel like that, or at least it seemed like it might be like that in the future; it was still so new. Given the growth in açaí in Acre, Luis was not the only one making such comparisons. There was talk on the BR-364 that—as with rubber—Acre's açaí trees were better than those in Pará, that distant state that was still the source of most açaí, and that Acre would soon surpass its rival to meet the growing global demand.

Açaí in the Forest

There was also excitement about açaí in Rio Branco. There, NGO employees, academics, businesspeople, and government officials met as part of an açaí working group I frequently attended. The group had meetings most weeks at the government secretariat SEAPROF. Some participants from the local federal university and NGOs were working to help smallholders like Francisco collect and sell more açaí from the *nativo* palms that grew within Feijó's forests. This type of açaí was understood as a non-timber forest product, like rubber. Collecting more of it from within the forest—extraction in the Acrean sense of the word—would make the living forest monetarily valuable. In doing so, açaí would protect the forest while also benefiting those smallholders and extractivists who collected it.

This was how some government officials understood it. Among them was Raimundo, a former community organizer turned state agricultural economist for SEAPROF. Raimundo offered açaí to me as an example of how SISA protected the forest through increasing production. As we sat by his desk, he quickly sketched a map of Acre onto which he wrote two numbers: 87, for the percentage of the state covered by forest at the time, and 13, for the percentage covered by cleared land. He kept referring to these two numbers, circling and underlining them repeatedly in emphasis. Land uses that increased the value of both forests and fields would keep these percentages stable. The forest and its carbon would be protected. By building on existing methods and spaces of collection, increasing açaí production was a way to valorize the 87 percent of the state that was forested.

For example, the local office of an international NGO and a group of university researchers and students I knew in Rio Branco were working with açaí collectors in Feijó, including Francisco, to make their collecting safer and sufficiently hygienic to meet commercial standards, as well as to teach them to waste fewer of the quickly spoiling berries and to plant and select for açaí palms within the forest. Staff at the NGO were also encouraging collectors along the BR-364 in Feijó to organize into a cooperative so that they could sell the berries at higher prices. This support could make forest-based açaí more profitable and productive, they believed, yielding more marketable berries per hectare.[31] It was a forest-based, extractivist form of intensification.

It was a vision that many açaí collectors in Feijó supported. At a workshop that the NGO organized in Feijó's city, the president of one of the local rural producers' associations—a longtime rural workers union organizer—advocated for collective organizing to increase their security and the prices they received. As it was, he lamented, middlemen did not help them if they fell and broke their leg while scaling a tall açaí palm. One by one, he pointed to each of the other association presidents in the room. When they started to organize, their income would grow, he projected. The men nodded in assent. They included Francisco, who led the association on a stretch of the BR-364 nearby. With the recent paving of the road, Feijó's açaí "market opened for the whole world"—another association head proclaimed. Aline was there too and offered that she would buy açaí from the planned cooperative. An energized NGO staffer remarked supportively, "*This is how we protect the forest.*"

This form of increased açaí extraction projected an inclusive Amazonian landscape. In it, smallholders could earn more from the living forest, to the benefit of both. Rather than seeing the forest, extraction from it, and resident smallholders as outside of the productive economy, they were to be brought into the productivist fold.[32] In combination with other forms of forest-protective rural production, collecting more *nativo* açaí from the forest would enable a vibrant and diverse low-carbon rural economy. This vision included some degree of standardization via the promotion of certain practices of açaí hygiene, safety, and forest-based cultivation. So too did the larger suite of forest benefits meant to intensify smallholder production, including the fishponds and mucuna seeds I explore in the chapters that follow. Yet the benefits also centered smallholders, like Francisco, and forest-based extractivism, like his collection of açaí berries, as productive parts of a low-carbon economy that were to be integrated into global capitalist supply chains. In the Acrean Amazon, this green capitalist vision seemed to conjure more diverse and inclusive Anthropocene landscapes than other more standardized and exploitative ones.[33]

Açaí in the Field

Yet, even as açaí collection boomed in the forest, something different was happening in Feijó's already-cleared fields. Those fields promised another approach to açaí intensification, often referred to as *plantado*. It too was seen as a way to create an inclusive low-carbon rural economy, but by increasing the productivity of cleared land while sparing the forest. This would be a very different açaí landscape and political ecology than that in the forests off the BR-364. It entailed new species, capable of growing outside the forest in drier fields, and new practices of cultivation.[34] In this, it resembled other forms of intensified agricultural production, including that of plantations.

Many of the participants of Rio Branco's açaí working group were enthusiastic about field-based açaí. They included researchers at Embrapa, the Brazilian Agricultural Research Corporation. Embrapa had been key to the successful intensification of agriculture elsewhere in Brazil, and it had started to put that experience to work in developing new varieties of açaí meant to enable cultivation on dry soil (*terra firme*) outside of the state of Pará's wet *várzea* landscapes, where smallholders dominated the decentralized açaí economy.[35] In Acre, Embrapa was an important institutional player in the effort to make forest carbon valuable, and at the Embrapa campus close to Rio Branco on the BR-364, researchers grew experimental plots of intensively cultivated açaí palms in fields.[36] The aim was to increase yield. Embrapa was also reportedly providing seeds for large-scale açaí plantations—for example, to a businessman who planted a million açaí palms on his cattle ranch on the BR-364.[37] Other açaí working group attendees were also focused on field-based açaí. Two of them, Rio Branco–based açaí processors who currently bought from smallholders like Francisco, described planning to grow commercial "plantations" (*plantações*) of açaí in the future.[38]

Some state officials were also working for more açaí to be grown in fields. Acre's governor at the time, Tião Viana, was consulting with a prominent açaí businessman (based elsewhere in the Amazon) named Eloy Luiz Vaccaro, the so-called king of irrigated açaí, who grew large fields of irrigated palms in Pará and had also been accused of extensive deforestation there.[39]

Vaccaro had declared that Acre "possesse[d] good soil for [açaí] plantations," and he was in the final stages of developing his own açaí variety that he believed would be better for growing large-scale plantations on dry land than Embrapa's.[40] "Planted açaí will be the big business of the future when we have more technology," he said.[41] Some interlocutors in and outside of the state government told me that this focus on field-based açaí reflected a more general governmental shift—in terms of budgets, institutional organization, and administrators' attention—away from forest extraction and toward field-based production.

Although field-based açaí was new in Acre, it recalled long-standing efforts to make the Amazon into a space for producing "a cornucopia of respectable crops."[42] Amazonian rubber's forest-based extraction had always been out of keeping with the intensive monocrop plantation-based production innovated by colonizers and enslavers elsewhere in Brazil. Particularly as the price of rubber declined after World War II, the Brazilian federal government aimed to "develop" the region's agriculture with the same techniques used elsewhere in the country.[43] In part through Embrapa's research on intensifying tropical agriculture, Brazil went from being a food importer to one of the world's leading agricultural exporters by the end of the twentieth century.[44] "Produce in Acre, invest in Acre, export via the Pacific," Acre's governor (linked to the federal dictatorship) in the 1970s advertised, as he sought to attract other Brazilians to move to the state.[45] Extensive cattle ranches were, in this vision, a prelude to this productivity to come, the forest-clearing vanguard for fields of soy, corn, and other monocultures of industrial agriculture.[46] In a field, açaí offered a legible form of production. It looked like a "respectable crop."

Local supporters of field-based açaí in the state government and burgeoning açaí businesses rooted their approach in Acre's history, just as Luis had in his enthusiasm for forest-based açaí. They too recalled the history of rubber production, but their remembrance was different from his. Where he recalled rubber's boom, they recalled its bust—how Acre lost out to the rubber plantations of colonized Southeast Asia. One of the head administrators at SEAPROF, for example, explained to me how much Acre had lost because its economy had relied on extractivist rubber collection from native trees. Being spread sparsely and unevenly throughout the forest offered protection against the easy spread of the endemic South American leaf blight. But it also meant that rubber trees had to be encountered. Rubber tappers had to cut their way through thick old-growth forest, creating long trails to connect distant trees that they then had to traverse to collect rubber. Something similar was true for native açaí palms, which grew in a greater concentration than rubber trees but still within forests. It was all very inefficient.

Some açaí collectors shared the SEAPROF administrator's concern. At the açaí workshop I attended in Feijó, one collector recalled the centrality of rubber when he was young. "Then they took the seeds, and it was all over," he said, referring to the British theft and replanting in Southeast Asian plantations. Acrean rubber tappers' long daily collecting treks were no match. He worried that the same could happen with açaí soon. Then others would be able to produce more cheaply than they could from within Feijó's forests. His comment was met with a brief silence, seeming to dampen the enthusiasm in the room. As fantasy, at least, large açaí plantations promised efficiency, productivity, and competitiveness within a global economy that valued all three, in contrast to the inefficiencies of extractivist açaí and other low-yield BR-364 landscapes.[47]

Yet it is important to note that this type of field-based açaí was also meant to be inclusive. The Acrean state sought to include smallholders in different parts of the field-based açaí supply chain (*cadeia*). Through SISA and related programs in the early 2010s, the state grew açaí palm seedlings at a nursery on the BR-364 in Feijó and then distributed them to smallholders in Feijó and elsewhere in the state, including Francisco. The recipients were supposed to plant the seedlings not in forests, where they already collected açaí *nativo*, but in fields they had already cleared of trees. The goal, in the words of the state Forest Secretary João Paulo Mastrangelo, was to "recover[] degraded areas" and "avoid[] . . . itinerant agriculture based on deforestation and fire."[48] Through this field-based production in smallholders' fields Feijó could soon become one of the Amazon's largest sources of açaí. Support for this kind of field-based açaí continued under the administration of the next governor, who was from an opposing political party. His Secretary for Production and Agribusiness described açaí as a way for already "open" areas (meaning deforested) to produce more in ways that did not increase deforestation and would also produce jobs (*empregos*). He said, "All this is agribusiness and it is linked to the small producer."[49]

Conclusion: Açaí Futures

Acrean officials sought to keep the forest standing and its carbon sequestered by enticing smallholders to produce more in ways sanctioned as sustainable. In turn, this increased rural production promised to decrease deforestation, generating internationally valuable carbon offsets and other credits. In other words, while offsets and other credits are based on an absence—a reduction in emissions—they are made through and rely upon the production of many other things. Carbon was not monocropped but rather multicropped, produced through the making of many other forest and nonforest products, including açaí.

This effort at forest carbon valorization also entailed cultivating a form of inclusive productivism within green capitalism. It sought to bring smallholders and forests into economically and culturally dominant productivism in ways also meant to transform that productivism into something that would protect the forest. Native forests and cleared fields were both treated as spaces for increasing açaí yields, an intention usually reserved for the latter.[50] Both smallholders like Francisco and species like açaí palms were enlisted in the effort, which sought not only to valorize the forest and its carbon but also to make rural Acre more productive than the extensive cattle ranching, extractivist rubber, and subsistence agriculture that dominated BR-364 landscapes in Feijó and elsewhere. As an increasingly valuable domestic and international food, açaí berries seemed like they might help to counter the anxiety that Acre did not exist because it produced so little. Whether in forests or fields, açaí cultivation and collection could be intensified in ways that promised to protect the forest, benefit smallholders like Francisco, and increase Acrean production.

Yet in forest and field, açaí projected different landscapes, different actors, and different types of investments, inputs, and species—in other words, it projected different futures. On the one hand, açaí promised to increase the monetary value of the living forest since the palms depended on taller tree species' shade and forest hydrology's natural irrigation. Forest-based açaí would be collected by those, like Francisco, who were already collecting more and more of it along the BR-364 in Feijó. On the other hand, field-based açaí promised to protect the forest by increasing the productivity of cleared land such that more clearing would be unnecessary—sparing what forest remained. Relying on different species and irrigation practices, field-based açaí would be led by those able to harness and fund new agricultural technologies, species, and methods.

These different açaí futures could be consequential. Planted in fields, açaí had the potential for higher yields than when collected from palms that grew in the forest, with its many other species and transport costs.[51] Field-based açaí therefore promised to increase the value of cleared land more than intensified forest-based production would increase the value of forested land. This imbalance meant that the intensification of açaí—or other crops—in fields could increase the pressure to deforest, as forested land came into view as potential space for fields of açaí palms.[52] In fields, açaí could potentially be integrated into smallholder agroforestry systems. That was what many supporting it within the government envisioned. Yet it could also potentially be made into another monocropped commodity, whose industrialized management could be destructive to existing forest ecosystems, agroforestry practices, and other socioenvironmental relations. Even within the forest, intensively managed açaí can be "converted to plantation-like stands" of the palms.[53] The allure of increased production looms large in green capitalism, as in its antecedents.

As agribusiness monoculture, or even as a few trees in a former pasture or in lowland forest converted to an açaí stand, açaí can lose its forest-ness, its diverse socioenvironmental relations. In this sense, forest-ness and productivity exist in tension. The effort to protect the forest and the carbon in it via inclusive productivism threatened—or promised, depending on your perspective—to transform the Acrean Amazon, with its history of forest-based extractivism, into something more familiar: planted agroforestry systems managed by smallholders or the intensive agriculture and patches of spared forest that dominate an increasing portion of rural Brazil. In some ways, the latter possibility seems more likely.

Field-based açaí not only entailed replacing existing land use practices with those developed elsewhere, but it also threatened to exclude small-scale açaí collectors including Francisco, like many other efforts to intensify agriculture.[54] Following the instructions he was given, Francisco had planted the *plantado* seedlings that the Acrean state gave him in one of his already-cleared fields. The seedlings were meant to do well outside of the forest's irrigation and shade. Yet, once planted, they browned. They shriveled. They died. Other recipients of the seedlings reported similar outcomes. To me, they often did so with frustration, offering up the açaí seedling program as an example of the government's failure to provide them with what they needed to stop deforesting. By the time I arrived, the state açaí nursery on the side of the BR-364 had shuttered. Francisco was not surprised at the death of his seedlings. He knew that the palms needed the shade of the other trees in the forest and the water that flowed through it. That was where he always collected the berries. Cultivating açaí palms in fields was another matter altogether, one for which neither his experience in the forest nor the minimal government assistance he received offered sufficient preparation. Growing açaí in fields necessitated a different sort of expertise—the expertise of agricultural scientists, specialists, and the plantation owners who could pay them.

Moreover, this form of açaí cultivation could be expensive, difficult to access, and ecologically harmful. To grow açaí in fields could require costly irrigation, for example, estimated in Pará to be R$12–14 thousand per hectare, with total costs reaching R$25 thousand per hectare, with any returns taking significant time to appear.[55] In this, field-based açaí threatened to exclude people like Francisco, or to include them not as independent collectors but rather as something else—perhaps as workers collecting a monocropped commodity in a field owned by someone else. This is what was happening in parts of the açaí powerhouse of Pará, where the Amazonian Environmental Research Institute (IPAM), an environmental NGO, reported that açaí monocultures create "exclusionary spaces and spaces of wealth and land accumulation," as açaí's increasing value drives up land prices in a place where access to land is already insecure.[56] Like other monocultures, fields of açaí can also exclude other species, with negative effects on biodiversity and, potentially, hydrology, since açaí palms are part of not only landscapes but waterscapes as well.[57]

There were many people working within and outside of the Acrean state to promote a different and more inclusive vision of productivity, one that centered forests and smallholders. But the deaths of the açaí seedlings and the unpopularity of the state program distributing them attest to the durability of intensification's exclusions and the attendant difficulties of creating an effectively inclusive green capitalism. For residents of the Acrean Amazon, human and otherwise, the future açaí boom might not be so inclusive after all.

With some state support, perhaps some smallholders like Francisco would try to adopt the new technologies of açaí production, with their enticing possibility of big rewards. That is what some in Pará were doing. For example, a 2021 newspaper article describes how a smallholder in Pará named Nazareno Alves—who was born in an Indigenous reserve and had worked with açaí since he was twelve—recently decided that he needed to become a "producer" of açaí, as he put it.[58] If he was going to stay in the game, he had to stop just harvesting from native trees and needed to start planting other varietals himself. "The market is on fire," he said. "If I hadn't started planting, I would already be out of it." He experienced a loss of R$500 thousand in his first attempt at planting but projected that he would break even in 2022 and planned to double or triple the size of his fields in the coming years. Inclusion could be risky and rocky, as well as tantalizing.

For all its novelty, inclusive productivism and the green capitalism of which it is a key part attempted what many governments have long tried to do: make land and people more productive. Intensification efforts—whether under the mantle of the Green Revolution or green capitalism—recall a long history of governments, corporations, and NGOs characterizing swidden, subsistence agriculture, nomadic forms of existence, and extractivism as unproductive, environmentally destructive, and in need of "improvement," as Tania Murray Li writes.[59] That improvement, in green capitalism as well as in some of its forerunners, often entails trying to make their practitioners into more productive farmers in ways that rely on capital and expertise.[60] Through this form of inclusion, green capitalism can expand despite, or even through, smallholders' continued marginality.

Yet in some instances, smallholders are already producing for commodity markets in ways that may be more efficient than plantation-style agriculture, as scholars like Michael Dove, Tania Murray Li, and Pujo Semedi have shown.[61] Despite the growth of and excitement around large-scale irrigated açaí plantations on Pará's *terra firme*, the vast majority of the state's açaí still comes from small-scale production in the watery *várzea*.[62] In Acre, it was the same: it was in the forests near the BR-364, up and down the Jurupari River, and in some other watery and shady parts of the state—not in Rio Branco meeting rooms or parched graveyards of government-issued açaí seedlings—where açaí collection was increasing and açaí futures were being forged.

INTERLUDE II

The Flood

Since being paved in the 1990s, the BR-364 had become an essential part of consumer life in Acre. This was a goal of its construction. Begun in 1961, the BR-364 was a component of President Juscelino Kubitschek's attempt to integrate southern Brazilian territory via large-scale infrastructure. In connecting strongholds of industrial and agricultural production with Amazonian states, the BR-364 brings both the goods and promise of modern Brazilian life to places like Acre, deemed so remote as to be of dubious existence.

The road's importance to urban life in Acre became clear to me when it was temporarily inaccessible. When the Rio Madeira flooded in Rondônia in early 2014, the number of trucks that could get through to Acre from the rest of Brazil on the BR-364 dwindled. The media and government officials celebrated the drivers as heroes. They were brave men who risked their lives

Figure InterII.1 Trucks driving on water. Photo by Agência de Notícias do Acre, published in Fábio Pontes, "Isolado há 20 dias, Acre vive cotidiano de incerteza e preços altos," BBC, March 14, 2014, https://www.bbc.com/portuguese/noticias/2014/03/140313_acre_isolamento_pai.

to bring food to Acre. They spent weeks sleeping in their trucks, driving slowly through the meters of water that covered the highway or placing their trucks on the beds of even bigger ones or on makeshift ferries to forge the flooded river.[1] In some newspaper photos, they appeared to drive on top of the water (figure InterII.1). A miracle.

Finally, for over a month, no trucks were able to make it through to Acre. The water was too deep. Essential supplies—rice, beans, milk, oxygen for hospitals—were flown in by air, with the help of the Brazilian Air Force. Sometimes they arrived from the distant city of Manaus via boat. There were rumors that the water was rising close to the electrical wires that brought energy hundreds of kilometers from Rondônia to Rio Branco (from the dams whose construction was blamed for the flood). Maybe Acre would be left in the dark. Lines for gas grew hours long, with people sleeping in their cars overnight for a chance to access scarce deliveries. On Facebook and WhatsApp, friends alerted each other when a gas station received a shipment, and soon a line would appear of cars, motorcycles, and men on bikes carrying empty soda bottles to fill with gas. Construction slowed or stopped as materials failed to arrive. Computer parts were waylaid, and restaurants closed for lack of supplies.

In Rio Branco's supermarkets, the perishables went first. Yogurt soon ran out, and the cheese and butter dwindled quickly too. Vegetables and fruits also largely vanished; tomatoes, broccoli, and apples were soon gone. Rationing signs appeared—Five Bags of Rice per Person, one read. The supermarkets lined empty shelves with large bottles of soda that would never spoil. Shoppers filled their carts with whatever was available, whatever other shoppers were buying. They loaded their carts straight from any boxes that did arrive. They eulogized the taste of potatoes or reported, on Facebook, sightings of eggs for sale. They recalled the time before the BR-364 was paved through to Acre in the 1990s, a time when food had been limited too. Left on the shelves, besides the bottles of soda, were the few products made in Acre: collard greens, lettuce, manioc products, and beef. They were staples but did not fulfill the varied tastes of urban Amazonian diets. Everything else was imported from other parts of Brazil, trucked in on the BR-364, ferried across the bridgeless, flooding Rio Madeira.

The flooding revealed how reliant many Acreans were on the one paved connection to the rest of the country, a connection made more tenuous with more frequent flooding in a changing climate. When the flood diminished, things seemed to return to normal, as they do. But the flood had aggravated the common and linked anxieties about the state's limited production and ambiguous existence.

3

Robin Hood in the Untenured Forest

On a dirt side road off the BR-364 in Feijó, a middle-aged smallholder named Neldo lived in a new brick house painted a cheerfully bright green. Constructed and paid for by the government, for Neldo the house was both a house and more than a house. It was also an indication of his and his neighbor's changing fortunes. Talking to him about it helped me to understand how the living forest's new monetary value relied upon rural Acrean lives, landscapes, the state governing them, and the relations between them—and how this reliance both reshaped those relations and kept them the same.

As we sat in the shade of his house, Neldo responded to my queries about his family history. His responses echoed those I heard from other smallholders near the BR-364 in Feijó. Neldo's forebears had tapped rubber from native *Hevea* trees on an old, now-defunct rubber estate nearby since they had moved to Acre from the parched northeastern state of Ceará during the late nineteenth century or early twentieth century (he was not sure which). Rubber was booming then, but the lives they had lived were hard and isolating. Neighbors had lived far away, as was custom in the old rubber economy. His family had hung on through its collapse and the associated dissolution of its political and social structure during the early to mid-twentieth century.[1] Neldo and his family were largely on their own then. They received little, if any, government support, and many of the now cash-poor rubber barons—who had been the obligatory source of needed supplies—decamped to distant locales. When Neldo grew up, his family had gotten by through growing food (mostly manioc), hunting in and collecting from the forest, and continuing to tap rubber, as long as the price justified it. When the price finally fell too low, Neldo had given it up altogether.

In the 1990s, he moved to this land in the government settlement. There, he enjoyed the closer proximity of his new neighbors, a number of whose homes we could see across the small fields surrounding his house. They too were former rubber tappers who had moved from nearby rubber estates. For many years, the group had remained largely isolated from the outside world. When they arrived in the settlement, there had been no roads, houses, schools, clinics, or stores. There was only the thick forest and a narrow path through it. They were largely cut off from medical care, education, and supplies, particularly during the many muddy months of the long rainy season. A river merchant—a former rubber baron who stuck around—periodically came through on his boat, plying goods at wildly inflated prices. Sometimes there was no other choice but to buy what they needed from him. But mostly, they made do. They cleared small patches of the forest to grow manioc, usually letting it regrow after a few years, and then they cleared another patch. By their kitchens, they kept small rectangular gardens with cilantro, collard greens, and green onions. Those who could afford it often kept a few cows, to carry in what they needed to buy from town and to sell if money was tight. More recently, some of them sought to accumulate more cattle or grazed those of large-scale ranchers, and so they began to keep their fields cleared of the regrowing forest. Sometimes, they worked as day laborers on nearby cattle ranches on the BR-364, where they were also hired to clear the forest. Just as before Neldo moved to the settlement, the government gave them almost nothing. He and his neighbors were expected to deforest the area themselves—part of the history of Brazilian colonization via deforestation that predicated rural people's livelihoods and belonging on clearing the forest. He noted all of this without apparent nostalgia.[2]

Figure 3.1 Old house, new house. Photo by Maron E. Greenleaf. Feijó, Acre.

But things had begun to change in recent years, Neldo told me approvingly. He gestured around us to his and his neighbors' new homes (figure 3.1), which included the "dignity" of indoor bathrooms. They replaced older wooden houses made of roughly hewn wood from the forest that could now serve as a shed or other extra space. The new houses were linked by new electricity lines carrying state-subsidized energy and by a new road—unpaved, but still a significant improvement from the old narrow footpath through the forest. Around many of the houses were new fishponds, chicken coops, and fruit trees. These were among the benefits that the state government gave smallholders like Neldo as part of the effort to make the living forest valuable by getting them to produce more. Additionally, Neldo's wife, like women in many other families, received conditional cash transfers as part of the federal Bolsa Família program, with payments linked to children's school attendance, as well as some other forms of government support.[3] In other words, in his narration of what he saw as improvements in his home, landscape, and well-being, Neldo was also describing something else: a recent expansion of government benefits—those linked to forest protection and forest carbon sequestration among them.

Carbon sequestration's new international monetary value raises contentious questions about who should receive or otherwise benefit from it, and how—questions that were hotly debated in some environmental policy and academic circles.[4] In privately run forest carbon projects at the time of my fieldwork (including within Acre), the forest's new value often accrued to the large private landowners who threatened to deforest the most.[5] In contrast, the Acrean government, and some others, pursued a different approach. Property rights in the state were often messy and difficult for poor people to access in ways that incentivized deforestation. Attaching forest carbon's new value to property rights would threaten to further entrench these exclusions. Instead, the state sidestepped private property (to land and carbon) and turned forest carbon's value into a form of public property and wealth that it then redistributed to some rural people. These forest benefits were part of the state's approach to making the living forest and its carbon valuable by including smallholders and others in rural areas. The broad strategy was to increase the production of rural products, like açaí. But, at a more day-to-day level, many of the relations between the state, rural people (like Neldo) and the landscapes in which they lived centered on the distribution and use of forest benefits. These were socioenvironmental relations cultivated to keep the forest standing and the carbon in it sequestered. In these relations, state officials tried to act like "Robin Hood," as one of SISA's developers told me: they took money from the wealthy—who would have received it in a private forest carbon project—and distributed it to the propertyless poor.

In this, forest carbon was imbricated in the governing PT's strategy of resource-funded redistribution. This "green distribution," as I call it elsewhere, was aligned with new forms of governmental distribution in the Global South that James Ferguson analyzes as sometimes developing through and in the wake of neoliberalism.[6] This chapter and the one that follows examine some of the socioenvironmental relations of this redistribution and the inclusion it entailed—the politics and citizenship that this distributive approach engendered (chapter 4), and the state that it formed and approach to property and value through which it governed (this chapter). Doing so elucidates some of the key socioenvironmental relations enlisted and cultivated to keep Acrean forest carbon sequestered.

Forest carbon–funded distribution entailed a type of statecraft, but one different than that often anticipated to develop in tandem with markets. In both supportive and critical accounts, liberal and especially neoliberal states are understood to create and enforce property rights so that individuals and businesses can effectively and efficiently trade. In contrast, the Acrean state did not create private property rights, but rather used forest carbon's new value, and the apparently neoliberal concepts of *ecosystem services* and *ecosystem service providers*, to cultivate neoliberalism's apparent opposite: an environmentally premised welfare state in a form of "tropical Keynesianism."[7] States are always important to creating and sustaining capitalism, but that role may be particularly salient in efforts to make capitalism environmentally protective.[8] This is because developing green capitalism necessitates countering, modifying, or minimizing entrenched socioenvironmental relations that have long militated for carbon emissions and other forms of pollution and environmental harm. Brazilian property rights were among those relations. The Acrean state enlisted inclusive redistribution to counter them in its effort to valorize the living forest and its carbon. Yet this approach to forest valorization also reinforced some poor rural people's exclusion from the property regimes that have been so important, not only in enabling Amazonian deforestation but also in shaping power and belonging.

Visions of Property

When I started researching forest carbon, I began by asking questions about ownership: Who owns the land where forest carbon is being monetized? Who owns the trees that grow from that land? And who owns the carbon stored in them? After all, how could forest carbon offsets be bought and sold if it was not clear who owned the forest and the carbon in it? My legal and anthropological training—and a lifetime spent in the private property–centered United States—led me to think that private property would be essential. I thought that because carbon offsets entailed commodification, they must also require private property. I was wrong.

   Studying law and anthropology had helpfully reshaped my understanding of property from a thing that is owned to a set of social relations mediated by things.[9] These relations can be understood as socioenvironmental—both human and more-than-human, material and social. Private property's socioenvironmental relations are particular, not natural or universal, and they enable particular kinds of other relations. These include exchange relations within market systems, where individually owned things are bought and sold. Exchange value is created in the process, with profit accruing primarily to property and capital owners.

Policy discussions and scholarship about forest carbon offsets have often assumed or asserted that these are the socioenvironmental relations needed to commodify forest carbon and otherwise make it valuable. Forested land or the carbon in it must either be clearly owned already or its ownership needs to be clarified. Julia Dehm details, for example, the many prominent policy and NGO reports that have foregrounded property rights as necessary for REDD+ to work.[10] The Eliasch Review, commissioned by the British government, for example, asserts that "only when property rights are secure, on paper and in practice, do longer-term investments in sustainable management become worthwhile."[11] The law firms Covington and Burling and Baker McKenzie similarly argue that "uncertainty surrounding land title is the single most significant impediment to effective preconditions for a REDD+ scheme."[12] The World Bank required that those applying for money from its Forest Carbon Partnership Facility explain "the status of rights to carbon and relevant lands to establish a basis for successful implementation of the ER [emissions reductions] Program."[13] Climate, Community and Biodiversity Alliance (CCBA), a third-party carbon credit certifier, requires that a REDD+ project "respect[] and support[] rights to lands, territories and resources" to be certified.[14] Scholarship on REDD+ and PES often contains similar pronouncements. An article in the journal *Ecological Economics*, for example, argues that "the exclusiveness of rights to the land" is a "fundamental precondition" for payments for ecosystem services programs and characterizes "land tenure chaos" as "the single largest impediment . . . to REDD implementation."[15]

This emphasis on private property derives from several lineages within Euro-American culture that both predate and are part of neoliberalism. Since the early days of European colonization, for example, creating or formalizing private property rights in frontier regions has been envisioned as a way to both incorporate them into capitalist economies and dominant political systems and to ensure "proper" land and resource use.[16] As popularized by the American ecologist (and sometimes champion of anti-immigration eugenics) Garret Hardin in a 1968 article, the ecological concept of the "tragedy of the commons" posits the lack of private property rights as a primary cause of environmental degradation and their creation as its perhaps best solution.[17] The metaphor still shapes much Euro-American environmental thought and practice. Also in the 1960s, economics scholarship coming out of the United States and Canada helped to frame pollution in terms of negotiable property rights.[18] This laid the theoretical groundwork for the pollution markets—including carbon markets—developed in the decades since.

In neoliberal theory, assigning and enforcing property rights, for any of these reasons, constitutes a properly limited scope of state action. As Paige West writes, "Neoliberals propose privatizing and commodifying everything through the assigning of property rights . . . having the state withdraw from the realms of social and ecological life unless it is needed to enforce private property rights."[19] If property rights are properly assigned, the thinking goes, private actors' transactions in the market will lead to optimal outcomes—including environmental ones. In the case of forest carbon, this means that by assigning private property rights to land and attaching carbon to them, or assigning rights to carbon itself, offset proceeds or payments for emissions reductions should reach the right person: the one who otherwise would have an incentive and ability to deforest.[20]

Critical scholarship in the Marxist tradition also often focuses on the essential role of privatization in commodification and market creation, but it makes very different arguments. For example, David Harvey writes that "commodification presumes the existence of property rights over processes, things, and social relations," and Noel Castree highlights privatization as a key component of commodification.[21] Those critically examining carbon markets often also focus on privatization and private property.[22] In these analyses, privatization is often key to the violent dispossession and other harms wrought by nature's new commodification. Thus, while they may vehemently disagree as to its effects, both private property's proponents and critics tend to highlight its importance.

In practice, though, forest carbon programs operate through a range of property relations, of which private rights to land or carbon are not necessarily a central component.[23] In Acre, other socioenvironmental relations were enlisted to make forest carbon valuable. One reason for taking a different approach to forest carbon's valorization is that land ownership is often unclear in tropical forests, including those in Acre. From this perspective, private property can look more like a dream—whether a fantasy or a nightmare—than a viable tool to make forest carbon monetarily valuable.

Irregular Property

Privatizing forests can be difficult, thanks in part to "their unbounded spaces and diverse ecologies," as Anna Lowenhaupt Tsing puts it, as well as complex histories of land use, ownership, and power.[24] This has certainly been the case in Acre, where almost 30 percent of land is not formally owned, but is often occupied and claimed by the smallholders and others who live on it.[25] Moreover, even areas that are formally owned can also be home to many other claimants without formal land rights. They are often known as *posseiros*, and their occupancy and use of land are often legal but not formalized. This land and claims to it are not, in the terms employed by those working on land rights in Acre, "regularized."[26]

This informality is partly the outcome of government policy and practice. So is associated deforestation and often violent conflicts over land. In Brazil, confusion over land rights has been, James Holston writes, a "strategy of rule" of often distant and under-resourced governments.[27] After slavery's 1888 abolition, state policies sought to create a landless class of cheap rural workers to replace enslaved people and colonize the country's expansive *interior*, including the Amazon, through smallholder settlement. Settlers (often called colonists or *colonos*) had the formal possibility, but often practical impossibility, of earning land title through land use and occupation. This was true in Acre as well, where, as Keith Bakx describes, "colonization . . . was carried out under the principal of *uti possidetis*, that is, effective control, rather than juridical title."[28] In the absence of effective title, use, occupation, and, crucially, deforestation could facilitate land claims and control since they could be used as evidence of the prized standard of "productive use."[29]

In this chaotic context, Amazonian landholders, large and small, have often sought out title. As Jeremy Campbell writes, they "conjured property" through land use practices, such as deforestation and the cutting of boundary trails (*picadas*), and collecting, creating, and using forged documents meant to gain state recognition of property claims.[30] "These conjurings," he writes, "are made with the belief that they might be recognized [by the state] and thereby become the basis of individual wealth, a shared economy, and a rural way of life."[31] Through these dynamics, more land has been bought and sold than exists. As early as the 1980s, for example, more than 100 percent of Acre's land had been sold.[32] The fact that Acre itself has had so many claimants and governments since colonization—it was part of Spain, then Bolivia, then was briefly independent, then was a Brazilian federal territory, and then finally became a state—gave those seeking title a complex historical repertoire of documents to draw from and forge in making their claims.[33]

The political ecology of rubber also contributed to property's irregularity, social conflicts, and deforestation in Acre. During the rubber boom, much of Acre was divided up into rubber estates. Each was controlled by a rubber baron with usufruct lease rights to the rubber collected from its rubber trees. The rubber barons assigned rubber tappers land on which to live and work based on the trees' idiosyncratic spread and the barons' desire to keep labor disaggregated and unorganized. They assigned each tapper a *colocação*: an area composed of three *estrada* trails, each a dozen or so kilometers.[34] As Euclides da Cunha, the Brazilian writer who visited Acre in the early 1900s, described it, "the value of land is relegated to such insignificance compared with the unparalleled value of the tree that it has engendered a novel agricultural measurement—the *estrada*, the rubber route or trail—which alone summarizes the most varied aspects of the new society haphazardly perched on the banks of the great rivers."[35] Land was plentiful; it was rubber trees that were of real value.

Rubber baron land rights and their use of them added further complexity to property ownership in Acre. They were usufruct rights, formally inalienable with ownership meant to revert to the state. Yet with the collapse of the rubber economy in the decades after World War II, many cash-poor rubber barons sold what they saw as their land to southern Brazilian ranchers, who started to move to Acre and many other parts of the Amazon. But these pastures-to-be were, in fact, inhabited forests: rubber tappers and Indigenous people lived in them. The result was the violence, expulsions, deforestation, and rubber tapper social movement that characterized much of the state starting in the 1970s and beyond. Many of the resident tappers were *posseiros*. When falling rubber prices prompted many rubber barons to abandon control of their estates, the tappers often stayed, claiming the land on which they and their forebears had tapped rubber. Yet those claims did not lend themselves to easy formalization into regularized plots because they were based on the uneven spread of rubber trees throughout the forest and on the *estradas* assigned by the rubber barons. In addition, boundaries between claims were not always clear, since trees, not land, had been what mattered during rubber's reign.

In this context, access to land title was, not surprisingly, unequal. Efforts by two Acreans to gain land title illustrate this point.[36] Both were seeking formal land rights in part to start private forest carbon projects, but that is where their similarity ended. One was a wealthy Acrean landholder named Diego. Like many others, he had moved from the south of Brazil in the 1970s to raise cattle, deforesting significant swaths of rainforest toward this end. In recent years, however, Diego decided to instead keep the forest standing on one of his parcels of about one hundred thousand hectares off the BR-364. The forest was meant to be a tribute to a son who had recently passed away. But he wanted to earn money by doing so—by selling carbon offsets to outside polluters. He worked with a small US-based company that had partnered with a few other large landholders in Acre to form private forest carbon projects. These projects included one that sold offsets to FIFA, soccer's international governing body, to compensate for emissions from the 2014 World Cup, held in Brazil.[37] I reached out to this company and met two of the people who worked for it (one from the United States and one a US-based Brazilian citizen) when they were in Acre. They introduced me to Diego, whom they considered an effective partner in creating a high quality private forest carbon project on his land.[38] But to create the project, Diego first had to get title to the land, which involved an expensive and lengthy process that took nearly a decade. Diego and one of his partners told me how it involved hiring lawyers, real estate specialists, and companies to conduct land surveys and record GPS points, as well as cultivating "friendships" with people in private registration companies (*cartários*) and the government. Diego also had to compile many documents that demonstrated land occupation and use over the decades, such as deeds of sale, tax receipts, and bank documents. With enough money and connections, he eventually obtained formal title and started a private forest carbon project with the US carbon offset company.

Poorer landholders tended to be less successful in obtaining formal land tenure on their own. One couple I knew, Angelo—a former rubber tapper I introduced in chapter 1—and his wife Lucia, were trying to obtain title to twelve hectares that they had occupied for decades. They wanted to start a private forest carbon project based on what Angelo described to me as their "pioneering" work reforesting the land—former pasture on which they had planted fruit trees within forest they let regrow. When I visited, the forest surrounded a small field that they planted with manioc and other crops, some of which they sold in the nearby city of Rio Branco. Angelo estimated that he had visited Acre's land agency, Iteracre, some forty times in all to get title. Yet even with access to free legal counsel and the legal merit of their claim, he and Lucia had failed to get it. Officials denied them access to the GPS points the government had recorded—coordinates demarcating the extent of their claim—and the couple could not afford to hire a private company to take new ones. Moreover, neighbors disputed the boundaries of their claim, and the Iteracre president refused to issue the title without all boundaries in the area (whose boundaries itself were unclear) being agreed upon, as a "matter of public policy," he explained in a meeting with Angelo and his lawyer that I attended. Angelo and Lucia were using the land in ways that Acre's environmental officials were trying to get rural people to do: producing crops from it while simultaneously protecting the forest. But without land title, their efforts to start a private forest carbon project were stalled.

Their experience with land tenure echoed that of many smallholders I knew in Feijó. In the few years before my fieldwork, Iteracre technicians had visited many of them along the BR-364 to take GPS points. On those visits, the technicians had told them that the survey was the first step toward giving them definitive land title. But the technicians had never returned, multiple smallholders told me, leading some of them to feel betrayed, disgusted, or both. When someone raised this issue at a government meeting I attended, an Iteracre staff member explained to the gathered officials and NGO representatives that the agency had, temporarily, abandoned its work in that part of Feijó. They had decided instead to focus their land titling efforts in urban areas, where most people now lived.[39] Smallholders on the BR-364 were on their own.

Theoretically, at least, Indigenous groups and some others considered to be "traditional people and communities" (*povos e comunidades tradicionais*) could make place-based and culturally specific territorial claims that could be translated into forms of state-defined property.[40] Both Brazil's 1988 Constitution and some subsequent environmental legislation created mechanisms for some of these groups to gain land rights. The process is imperfect: it can be difficult to access, provide an insufficient amount of land in locations that are out of keeping with territorial claims, and offer inadequate protection from invasions by ranchers, loggers, and others.[41] Yet some communities have gained meaningful protections and rights. For example, some Indigenous groups have had land demarcated, some *quilombola* communities have gained rights, and some rubber tappers and other extractivists have accessed usufruct land rights through extractive reserves—the tenurial form created by the rubber tapper movement—and other types of conserved land, such as sustainable development reserves.

Those who identify or are identified as rural producers, however, generally cannot claim land based on identity or traditional land use practices. I saw this exclusion at work at a 2014 meeting off the BR-364 attended by about two dozen Acrean rural producers, state officials and agricultural extension agents, and a Brazilian World Bank official visiting from Brasília. The president of the rural producers' association to which those present belonged talked about how they were all former rubber tappers without land rights.[42] Their efforts to form an extractive reserve in the area had come to naught, he complained, describing five fruitless meetings with government officials over the past few years. The World Bank official responded bluntly:

they needed to actually practice extractivism to form an extractive reserve. Identifying as a former rubber tapper was insufficient.[43]

Smallholders might access formal land rights through moving to or being incorporated into government settlements, as Neldo did.[44] However, these options could be unappealing as they often necessitated moving from land held by families for generations, as Neldo had to do.[45] Moreover, only some types of settlements provided the definitive individual land title that many rural Amazonians sought—and even then with many conditions attached—while others only offered limited use rights.[46] Smallholders were thus often unable to access formal rights to their own land.

Scholarship on and discussions about REDD+ have often assumed that land tenure can be easily clarified in areas where it is unclear. This is part of why consultants at McKinsey, among others, believed that tropical forest carbon offered a cheap and simple way to reduce GHG emissions in the short term.[47] Yet clear property arrangements do not necessarily exist. More common are culturally specific, overlapping, and/or conflicting claims that cannot be easily or apolitically reduced to land titles, even when land claimants want them to be. In this sense, private property can be exclusionary not only in more dramatic moments of enclosure but also in quotidian class- and identity-based limitations in access to tenure.[48] It can be both the source of inequality and a result of it.

To attach forest carbon's new monetary value to land title, then, threatened to reproduce this inequality. Only those with the means to obtain land title, like Diego, would have access to it. Once they had title, they too could theoretically create private forest carbon projects to sell offsets to access that value.[49] Many in Acre anticipated forest carbon's exclusionary potential. Early consultations with nongovernmental groups and stakeholders (e.g., rural producers, extractivists, and Indigenous communities) held to develop what would become SISA, for example, emphasized that confusion and conflicts over land could limit *posseiros*' access to forest carbon's benefits. Moreover, while SISA developers and administrators told me that they thought that land rights were important, they did not see them as a necessary prerequisite to valorize the forest. In fact, such a prerequisite threatened to create a Euro-American-centric quagmire, one of its developers told me as we sat outside the GCF Task Force meeting in Rio Branco:

> We are in Latin America. When discussing a European model or an American model, you have centuries and centuries of structuring and regulation of property rights. And you fortunately succeeded in

Europe to title property.... Here you do not have it. So if you follow a path in Latin America applying European models of the conception of title... you will never be able to do emissions reductions or carbon programs linked to the land, because you need to solve an earlier problem, which is the problem of titles. Nor in the next 100 years, unfortunately, will we... be able to do this.[50]

Instead, the Acrean state pursued what officials saw as a more inclusive approach to valuing forest carbon by sidestepping these issues of property.

From Land to Labor

Private property is only one way of arranging relationships within landscapes, only one way to create and distribute value, and only one way to facilitate environmental protection. In its attempt to value the living forest, the Acrean government adopted a different approach to carbon and land ownership, and in doing so, advanced different socioenvironmental relations. It created a statewide "system"—the System of Incentives for Environmental Services—rather than the geographically circumscribed projects it had originally planned to develop in areas with high deforestation. This type of statewide approach is sometimes called "jurisdictional REDD+" in policy discussions and scholarship because it promotes, measures, governs, and values reductions in deforestation and forest degradation at a jurisdictional level and across that jurisdiction's territory.[51] Through SISA, Acre developed an influential jurisdictional REDD+ program, which instructs the state government to create economic and financial instruments to "promot[e] the progressive, consistent, and long-term reduction in greenhouse gas emissions."[52] The law also directs the state to "promote the sharing of benefits with actors who contribute to reducing deforestation and forest degradation and who conserve, preserve, and recuperate forest assets."[53] SISA also enables private landowners to develop private forest carbon projects, as Diego did, deducting the carbon sequestered in those projects from the state's tally. But there were only a handful of these projects in the 2010s, in part because land title was so difficult to obtain.

The state's approach had an important effect: it made the Acrean state the de facto owner of most forest carbon. To determine the level of monetizable emissions reductions, the state's first step was to determine aggregate deforestation across the state, using INPE data.[54] Comparing this number to the reference level and baseline gave the amount of reduced deforestation, which could be converted to carbon emissions via standardized formulas (see chapter 1). Doing so establishes the amount of emissions reduced throughout the state for which officials could pursue compensation (via offsets or payments for emissions reductions, for example) without having to determine the private ownership of any of it. The state then distributed any compensation they accessed to rural residents.[55] This approach effectively made forest carbon into a kind of publicly held resource rather than a form of private property. In other words, the Acrean state did not seek to secure rural people's private property rights so that they could privately reduce emissions and sell carbon offsets. Rather, it made forest carbon into a source of public wealth whose value it could redistribute as benefits.

This approach decoupled forest carbon's value from the geographical location of forest protection or destruction. As Andrew Mathews found in a REDD+ program in Mexico, "it does not matter whether trees here or there have died; it is at the scale of the landscape that knowledge is made."[56] Neither is monetary value attached to particular parcels of land; SISA did not "channel[] incentives and attribut[e] contributions of the individual land owner or territorial unit," KfW explains in a report.[57] Funding therefore did not have to be allocated based on land ownership, past deforestation, planned or anticipated future deforestation, or forest protection occurring in particular locations. Since revenue was generally not allocated to or withheld from landholders conditionally, as reward or punishment for reduced deforestation, there was "no need to determine 'who owns the carbon,' or to resolve all pending land tenure issues," a report on SISA by the NGO Environmental Defense Fund points out.[58] My observations of the daily work of making that carbon valuable confirmed this; there were few discussions of land, forest, and carbon ownership.

Property rights are a way to distribute value, among other things. But there are other ways to do it too. Instead of attaching forest carbon's new value to private land titles, Acre's jurisdictional approach attached much of it to something else: labor.[59] The concept of ecosystem service providers was, I found, key to this distributional logic. Ecosystem services are often understood as natural—on the other side of the nature/culture divide from humans. They are seen as freely provided by the forest and other ecologies in ways that are easily taken for granted. In contrast, SISA conceptualizes ecosystem services as generated through the human labor of providers of ecosystem services (*provedores de serviços ambientais*). In its legislation, SISA defines "ecosystem services providers" based on their "legitimate actions for the preservation, conservation, restoration, or sustainable use of natural resources."[60] This approach makes labor essential to creating and distributing forest carbon's new value. One of SISA's developers described the logic to me: "It is not through the simple fact of owning land that I have the right to earn from [carbon] emissions reductions. I was not born with that right." Rather, he explained, carbon sequestration can be framed as the result of "human activity that leads to the reduction of emissions in the forest."

This was a logic I heard frequently from SISA administrators and developers, and one that facilitated their emphasis on inclusive productivism: being a beneficiary was a status linked to activity or action. Many of these actions centered on increasing production in ways understood to decrease deforestation, such as increasing açaí collecting as I described in the last chapter and the aquaculture I explore later on. Whereas private property rights attaching carbon to land might incentivize landowners not to deforest, attaching forest carbon's value to labor was meant to incentivize more people to produce more in ways deemed to protect the forest. Positioning labor as the gatekeeper to forest carbon's value also made SISA more inclusive than property-based REDD+ because it enabled more people to benefit from forest carbon's new value. The SISA developer continued: "You open up the scenario, because you can have the owners, you can have the extractivist, you can have the guy who just holds land title, you can have indigenous communities. You work on the horizon that has to do with the nature of activities and services. This makes a very big difference." Excluded from the institutions of land ownership, *posseiros* could access forest carbon's value through their actions, through their labor.

In framing ecosystem service provision in terms of labor, Acre's approach to valorizing the living forest and the carbon in it usefully hewed to widespread Euro-American views of labor. In Brazil and elsewhere, people with divergent politics (including, famously, thinkers like John Locke and Karl Marx) have seen labor as value-creating.[61] As James Ferguson describes, it can be seen as "the wealth that is the true foundation of the economic system."[62] Relatedly, official workers in Brazil have historically been the primary beneficiaries of limited state welfare policies in ways that have also given work cultural status.[63] This status can be seen in characterizations of rural people as rural workers and rural producers, as well as in the characterization of the rubber tapper movement as a rural labor movement.[64] Conversely, such widely held views about labor and work may undermine people's ability to effectively make nonlabor-based claims for support within conservation efforts, including REDD+; that is, claims for compensation for forgoing land use and related forms of benefit-sharing—for not working their land.

Benefits from SISA, though, were not the usual sort of payments given in exchange for work. They were not wages or contract payments paid by employers to employees or contractors. Rather, the distribution of some of the forest's new value as labor-linked government benefits engendered different sorts of relations: that between the state and what officials called *beneficiaries*. Beneficiaries reported valuing many of the SISA-funded benefits they received, even when they refrained from using them to increase production or decrease deforestation, as state officials wanted them to. Those benefits included government-subsidized or constructed fishponds.

Green Redistribution

Water pooled in new fishponds (*açudes*) along the BR-364 and many of the dirt roads off it. Streams were blocked, holes dug, and rain collected. Some rural Acreans had built the fishponds themselves, while others had paid state workers building the BR-364 to do so, using the large construction machines to carve fishponds into their land. But many of the fishponds had been constructed with state support. A state-employed fishing engineer told me that 4,500 new ponds had been built for free or at a greatly reduced cost on cleared smallholder land from 2011 to 2014. That state support included money from SISA. The ponds were part of a larger effort to create a socially inclusive aquaculture industry in Acre that would create monetary value from pastureland that was characterized as degraded. The logic was that by intensively farming fish on already-cleared land, rather than raising cattle extensively on it, smallholders and ranchers would refrain from deforesting to create more pasture.

Through a public-private partnership, the state also supported a new industrial aquaculture complex on the BR-364 near the capital Rio Branco. Peixes da Amazônia (Fish of Amazonia) boasted on its entrance sign of being the most modern aquaculture complex in Brazil (figure 3.2). On the day I stopped by, the manager led me enthusiastically by large rectangular ponds (120 in total) where Amazonian river fish—*pirarucu, tambaqui*, and *pintado*—were bred. A processing facility, then under construction, would slaughter and freeze the fish for export. Their heads, tails, and other discarded parts would be turned into fish feed in another building in the complex. It was a cradle and grave operation, for the fish were meant to spend

Figure 3.2 "The most modern industrial aquaculture complex in Brazil." Peixes da Amazônia. Photo by Maron E. Greenleaf. Senador Guiomard, Acre.

the middle of their lives growing in private fishponds elsewhere, including those constructed by the state along the BR-364.

This approach to aquaculture aimed to be inclusive. Some large and medium-sized cattle ranching landholders had already embraced aquaculture, building hundreds of hectares worth of fishponds on former pasture land—a significant change for those responsible for most deforestation in Acre, given the cultural value of cattle there.[65] But to avoid wealthier people coming to dominate the new industry, the aim was for smallholders to participate in fish farming as well: to buy fry and fish feed from Peixes da Amazônia, grow the fish in their new government-subsidized ponds, and then sell them back for slaughter, packaging, and export. With this intensively productive and lucrative use of already-cleared land, they would refrain from deforesting. They would become not only fish farmers but also working ecosystem service providers—garnering access to more forest benefits.

However, it was not at all clear that smallholders would adopt fish farming as state officials planned. Lack of familiarity with fish farming, the cost of inputs, and the logistics of transportation made many wary of joining in. Among them was Flávia, with whom I stayed on the BR-364 in Feijó.

Figure 3.3 Fishing on water. Photo by Maron E. Greenleaf. Feijó, Acre.

A busy teacher and grandmother, she often still made time to fish from a small, partially submerged dock in one of the two ponds next to her house. I knew her as someone who always seemed to be moving, but while she fished, she was still and quiet, as if suspended on water (see figure 3.3). She was interrupted only by the sounds of her grandchildren, splashing in the welcomed cool of another pond nearby. The fish Flávia caught were not to sell but for the family to eat, and she scoffed when I asked if she fed them commercial feed. She told me that fish did not need expensive, purchased feed. They just needed some manioc and corn. A state fishing engineer in Rio Branco overseeing fishpond construction had a different view. He told me that, while that diet was fine for growing fish for personal consumption, it was insufficient to grow fish of the size or taste necessary to be of commercial quality. This standard did not matter to Flávia, however. She expressed no enthusiasm to me about the prospect of raising fish to sell. She had received a fishpond but did not want to do the forest-protective production envisioned to happen within it. And she and her husband Luis continued to keep hundreds of head of cattle.

Yet Flávia and many other smallholders valued their fishponds a lot, and not just for personal fishing. They were also a way to store water. In an area with insufficient water infrastructure, and where water could be in short supply outside of the rainy season, that mattered. Some smallholders collected rainwater in large blue plastic tubs. Those with the means and access might dig a well or rig an electric switch-operated pump to bring up water from a downhill stream or river. But when rain was scarce, as it often was in the summer, water could be scarce too. A fishpond could help. It filled up during the rainy season, and, if properly constructed, retained water through the dry summer. It offered water for bathing and washing clothes and dishes, allowing households to retain clean rainwater for drinking and cooking.

When the CIFOR research team I worked with stayed with rural families during the end of the dry summer months, I learned how important water was. One of the biggest strains we put on our hosts was our use of it. We tried to minimize our impact, limiting our consumption of their dwindling rainwater collections not only by bringing our own drinking water but also by doing our bathing and laundry in fishponds. The ponds also helped us, and the families we stayed with, cool down from the summer's baking heat. And they offered water to the animals that Flávia and many others were interested in raising: cows. Like us humans, they often congregated at the water's edge on hot days. It seemed to present a quandary for fish farming–based forest protection efforts. Watching a small group of them from the other side of a pond as I squatted on a dock, washing myself and my clothes, I wondered if the fishponds might support more cows, and with them, more deforestation, instead of less.

The fishponds, then, did not necessarily support fish farming, forest protection, or forest carbon sequestration. But they did provide water. In building fishponds, the government ended up building water infrastructure that many smallholders valued. In this way, forest carbon's new value funded and justified redistribution, social welfare provisioning, and infrastructural development in a region shaped by entrenched and unwelcomed governmental neglect.[66]

In rural Acre, these forest carbon–funded benefits appeared less as neoliberal conditional incentives and more as an environmental component of a larger suite of new state-sponsored benefits. The governing PT had created and expanded a host of social welfare programs in its redistributive approach to improving the lives of poor people. Many in rural Acre, like Neldo, had lived in unwelcome isolation through the rubber economy's collapse and governmental neglect in the twentieth century.⁶⁷ But especially since the PT came to power, in 1999 at the state level and 2003 at the federal level, government benefits had improved their lives in some ways. These benefits were financially underwritten by expanding agribusiness and extractive industries—an approach also embraced by other left-of-center Latin American governments at the time. Eduardo Gudynas, Maristella Svampa, and other Latin American thinkers have analyzed this approach as *neoextractivism* or *neodevelopmentalism*, showing how it repeats some of the entrenched patterns of resource-based economies, funds some measure of redistributive social support, and is meant to reinforce state sovereignty, in part through the populism that benefits can facilitate.⁶⁸ In developing environmental export commodities, like carbon offsets, that funded socially progressive benefits, Acre's forest carbon program offered an environmentalized version of this approach, one based on keeping carbon in place rather than extracting it.

For much of the 2000s and early 2010s, this approach was relatively effective. Inequality decreased, poverty declined, Amazonian deforestation plummeted, and the economy in Acre and elsewhere in the country grew.⁶⁹ While SISA benefits did not generally involve the transfer of large sums of money, and rural people were often critical of their insufficiency (as I discuss in chapter 4), cumulatively with other benefits, they still became important to the livelihoods of some rural Acreans, for whom the government had been largely absent before. The government could take forest carbon's international value from cattle ranchers with land titles, like Diego—who would have received it in a land rights-based approach to REDD+—and give it to, among others, *posseiros* who lacked land rights and other smallholders, like Neldo.

Neldo and other recipients of these forest and non-forest–linked benefits spoke about many of them approvingly. They told me about how, after difficult childhoods in which they lacked regular access to medical care, education, food, and material goods, their life had improved. Through school construction, job creation, and benefits, Neldo said, the government "gives" (*dá*) now. Others also used that verb *to give* to describe the government working well: *o governo deu certo*—the government worked. One woman told me that everything changed when the PT came to power. "There was food on the table," she said, and stocked medical facilities in town. For them, a good government was a giving government, and forest carbon's new value facilitated this giving.

Forest Statecraft

Benefit distribution is relational. It shapes not just the recipient but also the giver, in this case the Acrean state. Jurisdictional REDD+, in other words, was a form of statecraft. It was one that used some of the tools and concepts of neoliberal environmentalism to create its supposed opposite—a welfare state. The explicit goal of SISA was economic and environmental: the monetary valorization of forest ecosystem services with the aim of protecting the forest. But SISA simultaneously could create the "effect," though not always the full reality, of a more powerful and beneficent state in a place that only gained statehood in 1962, after decades of federal neglect as a territory.[70]

Even before its effort to sell carbon offsets and get paid for emissions reductions, Acre's Government of the Forest had undertaken forest-focused state-building. It created and expanded environmental and forest-linked bureaucracies, building or refurbishing buildings for them in downtown Rio Branco. The change was dramatic. One staff member told me that before, government entities like the state Secretariat for the Environment did not have sufficient desks and chairs, or even a single computer. Now the Secretariat occupied a well-appointed building in the center of town. The state also expanded and "professionalized" its environmental institutions, hiring staff away from NGOs, attracting outsiders to move to Acre, and training Acreans to work at them through new educational programs.[71]

The Government of the Forest also created environmental policies that built the state. It pioneered "ecological-economic zoning" (ZEE) in the 2000s, for example, a process that officials presented to me as a precursor to SISA.[72] In studying, mapping, and planning appropriate uses for Acrean land, ZEE not only envisioned economic growth and environmental protection—its explicit goals—but also conjured and to some extent enacted state territorial knowledge and control. Also in the 2000s, stricter enforcement of the Federal Forest Code's requirement that Amazonian landholders retain forest on at least 80 percent of their land often entailed officials issuing hefty fines, which were sometimes higher than the value of the land itself.[73] These fines not only reduced deforestation. They also asserted and expanded the state's authority to do so. One smallholder described to me his fear of the government's "computers in the sky." Moreover, this forest statecraft attracted domestic and international funding in the form of grants, loans, and partnerships through which the environmentalized state expanded further, well before SISA sought to monetarily value forest carbon. Previous administrations (at the federal and state level) had sought to control and govern Acre through deforesting it. The Government of the Forest largely took the opposite approach. They built the state to protect the forest and protected the forest to build the state.

The System of Incentives for Environmental Services extended this forest-based statecraft and territorialization in multiple ways.[74] On a literal level, SISA created new institutions. These included the IMC (SISA's main regulatory body), the State Commission for Validation and Accompaniment (for governmental and NGO input), and the Company for the Development of Environmental Services (a public/private company owned by the state and investors and charged with monetizing forest carbon emissions reductions). In addition, SISA aimed to shape government policy and practice more generally by working through and funding existing programs. My questions about whether a specific state effort to reduce deforestation was part of SISA were often met with answers like, "It isn't part of it in a strict sense." I came to understand this vagueness as indicative of what it meant to make REDD+ jurisdictional, what it meant to create a state system rather than a project. One high-level government official, endeavoring to explain to me what SISA was and how it was entangled with so many government programs, put it this way: "It would be like if you were to make a home, SISA is the cement. Would you say the brick is part of the cement? If you consider the wall, both the cement and the brick are part of the wall. And the cement ensures that the bricks are integrated." Another administrator described SISA to me as the electricity powering the existing environmental "system." It was supposed to hold the component "bricks" of government policy, programs, and bureaucracy together, making them into a structurally sound and apparently coherent whole, rather than an unstable and haphazard pile of bricks. And it was supposed to power them, turning inert separate parts into an integrated system.

Making forest carbon into a form of public wealth also empowered the state as its administrator and protector. Through SISA, the state seemed like the entity best able to access Acrean forest carbon's new international value. Simultaneously, it positioned the state as the regulator of private forest carbon projects like Diego's. When I asked officials about these private projects, they spoke carefully. Private projects were absolutely allowed, they emphasized, with any associated emissions reductions simply deducted from the state's total. Through SISA, the state could certify that these projects were in compliance with the law, including by protecting *posseiro* rights.[75] Yet some of those working to start private forest carbon projects in Acre grumbled to me that state officials had started to see them as competitors for forest carbon money, publicity, and outside buyers. The owner of the US company creating private REDD+ projects with landowners like Diego confided in me and a man from an Acrean NGO: state officials would not meet with him anymore. He thought that they were upset that one of the company's REDD+ projects had made the high-profile offset sale to FIFA, while the state had still sold none. The man from the NGO replied with sympathy: "That's how it is here. The state likes to be at the center of things."[76]

Some of those who had worked to create SISA agreed. One afternoon in Rio Branco over bowls of açaí, one of its original developers lamented to me that SISA had become a state system. He had been involved before SISA was SISA—when it was a nongovernmental program in the making. Then it was absorbed into the state. In Acre, the government was better than in other states, but it was also everything, he said: the visionary, the recipient of payments, the taker of loans, the executor of policies. There was nothing outside of it. He explained disapprovingly that when the PT was elected, it had hired everyone into the state government. Through SISA, forest carbon money became part of the government budget. Yet the structures of power were all the same; forest carbon had become a way to fund them.

Redistribution was also part of this forest statecraft. The imperative to reduce deforestation and the valorization of carbon created a means and narrative to constitute the Acrean state in rural areas. As Andrew Mathews writes, through measuring, modeling, subsidizing, and ensuring deforestation emissions reductions, "REDD policies make political authority emerge in relation to the production of value."[77] The Acrean state not only calculated and collected money from the emissions reductions achieved statewide, but it also had the discretion to decide who would benefit from that value. To receive benefits, smallholders had to be recognized by the state as ecosystem service providers. They had to undertake forms of labor the state deemed to be forest protective. And they also had to be "integrated" into state-run or recognized programs.[78]

Moreover, the state distributed forest carbon's new value as its own largess.[79] A state that redistributes its wealth differs from one that gives citizens back that to which they claim ownership. The latter is an intermediary carrying out its duty without discretion to mediate access. It gives to citizens that to which they have a right, that which is already theirs. In contrast, the former type of state appears benevolent, bestowing its wealth as it sees fit. In Acre, this environmentally premised welfare state did not replace an active or rights-protecting government, as the story of Neldo illustrates. Rather, giving benefits was a form of rural state expansion. The giving of benefits made the state appear more powerful than when it was absent or if it were to pass on forest carbon revenue to landowners as their rightful property. Administrators redistributed SISA revenue as state wealth, transforming it into contingent benefits through the act of giving.[80]

Conclusion: Land, Rights, and Green Capitalist Statecraft

Had the Acrean state focused on attaching forest carbon's value to private land rights or privatizing rights to forest carbon, its new value would have accrued to those with both land titles and access to carbon markets. That is, it would have accrued to those who were already relatively well-off, like Diego. In this, the process would have aligned with the predictions of a diverse literature about environmental markets and neoliberalism. Often implicit in that literature is an assumption that a property-making state is easier to create than a distributional one. But in the context of confusion and conflict over land rights, as well as the dominance of neoextractive PT politics in Acre, redistribution could seem more plausible. Many scholars focused on the Global South have argued that, as Verónica Gago writes about Argentina, "neoliberalism's opposite" is not the "return of the state."[81] Not only have states always been central to neoliberal capitalism through making and enforcing property rights and other market rules, but they can also implement redistributive policies that may be linked to neoliberal discourse.[82] And this redistribution can facilitate green capitalism by increasing environmental stability and social inclusion. In Acre, forest carbon's new monetary value articulated with PT politics. Forest carbon was made into a form of public wealth redistributed to smallholders like Neldo, including them in a statist vision of green capitalism. Relations between the state and beneficiaries, not the state and landowners, were enlisted to keep the forest standing.

In the process, the Acrean state became increasingly involved in providing for rural welfare—a contrast from its minimal role during the period of rubber's dominance and decline. Forest carbon–funded benefits were among those the state offered to improve rural welfare as part of the PT's approach to social inclusion via distributing material benefits. Whereas the state had expected smallholders to deforest in order to improve their lives, colonizing Brazilian territory in the process, now their well-being and status was to be linked to forest protection and the labor of providing ecosystem services. In this, the living forest and its sequestered carbon's new monetary value were the latest state scheme to govern and improve the rural people and territory it claimed. And it was one that simultaneously expanded the power of the state in rural areas as the provider and arbiter of that value.

Yet this inclusive approach could simultaneously perpetuate ongoing forms of dispossession. Scholars of neoliberal environmentalism are not the only ones focused on property, after all. For rural smallholders, property rights, for all their exclusivity, could be a sought-after means of inclusion.[83] Angelo and Lucia wanted title to their land. So did Flávia and Luis on the BR-364. They had plans, Luis told me, to hire a private assessor to take GPS points on the land they claimed, just as Diego had. They were sick of waiting for the government to do it.[84]

There were many good reasons for them to do so. In Brazil, land title has been a basis for securing access to credit, for example. Moreover, formal land rights dangle the promise of social and political status and belonging—of inclusion.[85] "In Amazonia," Jeremy Campbell writes, "the allure of property lies not only in controlling and potentially profiting from the land but also in land's status as a historical threshold that promises participation in a broader political economy."[86] In avoiding property, then, forest carbon's valorization in Acre aligned with a long history of Brazilian governmental avoidance of something else: redistributive land reform, the challenges to entrenched and powerful landed elites (like Diego) it would require, and the forms of political and economic inclusion it promised.[87]

Moreover, the redistribution of forest carbon's new value did not provide a ready basis to make other sorts of rights-based claims—for example, claims to that new value. Instead, that redistribution meant government discretion over who had access to forest carbon's value. Those more willing and able to adopt state-promoted land use practices could benefit more than those who did not. Smallholders like Flávia might start farming fish to access more of forest carbon's value through state-favored intensive land use, for example. Or, given her busy schedule and disinterest, she might not—but then risk being excluded from state benefits. Attaching forest carbon's value to labor was more inclusive of those without land rights. Yet it also created other kinds of differential access to that value based on the governmental valorization of certain kinds of labor and land use, and certain kinds of socioenvironmental relations along with them.

INTERLUDE III

The Rural Road, Part 1

When I asked about which government benefits had most positively impacted their lives in recent years, smallholders I spoke with often offered a non-forest focused example: the paving of the BR-364 and the building of more side roads off it. These roads offered them access to the education, health care, consumer goods, and connectivity that many rural residents wanted.

For one, the BR-364 promised an end to mud. "Mud," Laura Ogden writes, "resists the lure of endless forward momentum."[1] For those yearning for momentum, this resistance can be unwelcome. "Thank god for the road," Luis once told me as Carlos, the itinerant salesman I once hired to drive me on the BR-364, drove us to the town of Feijó from Luis and Flávia's home. By this, Luis meant the pavement. The road had been there for decades, but it had often been impassably muddy. What was now a forty-minute drive on the paved road, he told me, used to take him two days walking in sometimes waist-deep mud during the wet season. Some years, the road never dried out enough for vehicles to drive on, even after the rains stopped. People walked to town, if they really needed to. Or they might take a horse, if the mud was not so deep as to break its legs. Luis pointed to the home of an old friend with whom he would stay the night before heading out again early the next morning, when he would finally reach the river on which he could take a boat to town. He also pointed to a spot on a stream where a sick neighbor had died. He had been helping to carry her in a hammock to a health clinic in town, but they did not make it. I could not quite make out the spot; we sped past it too quickly.

In an incomparably more limited way, I too was familiar with some of the benefits of pavement. When I started traveling from Rio Branco to Feijó in the summer of 2013, a twelve-kilometer stretch of the BR-364 was still being paved. Whatever truck I was riding in would stop in a long line of other vehicles and wait for the crew paving the road to wave us through. Then we would slowly caravan, fishtailing on the slippery red mud if the road was wet. Conversation would quiet and was sometimes replaced by prayers mumbled to shepherd us through safely. To the side were the skeletal remains of abandoned vehicles. When I returned the following spring, in contrast, that last stretch of road had been paved, and we could drive straight through without stopping. The ease felt like evidence that the BR-364 fulfilled its promise of state integration (figure InterIII.1).

When I stayed with him, Luis liked to look out at the road from his front porch. The stretch in front of his house had only been paved three years before. Sometimes people passed by on foot, bicycle, or car. Sometimes they waved and sometimes Luis waved back. But often nothing was passing. There was just the empty road. It was enough to hold his attention. Perhaps he was making plans for the future: with the road now paved and his savings sufficient, he told me he was finally going to buy his own car—a new Fiat due to arrive soon in Rio Branco. Its purchase necessitated further paving, this time of his treacherously steep and mud-prone driveway. He described the plans that his son-in-law, an engineer, had drawn up to line it with R$12,000 worth of bricks. With a car and driveway, he would soon be driving the BR-364 himself. There would be no more walking through the mud and no more hitching a ride.

There were also other ways to ride on the BR-364, and doing so could be a way to get a formal education or get to work. Every weekday, a school bus passed in front of Flávia and Luis's home. It was often driven by one

Figure InterIII.1 Sign in Rio Branco touting the BR-364 as a way to integrate Acre. Photo by Maron E. Greenleaf. Rio Branco, Acre.

of their sons. From five in the morning until six in the evening, he would drive it back and forth. Flávia and one of her granddaughters rode it to the school that had been built in recent years down the road, where Flávia was a teacher. In the mornings, she taught children like her granddaughter. In the afternoons, she taught some of their parents and other adults. The school had not been there when they were young, and many had received little formal education. Now as adults, they could take classes from Flávia at this new school through a government program called Wings of Florestania (Asas de Florestania). For Flávia, Luis, and many others, the BR-364 was evidence that the government was doing its job, that it cared for them, that it gave, and that it worked. In this, the road seemed to facilitate their inclusion in Acrean society and the lives they wanted in it.

4

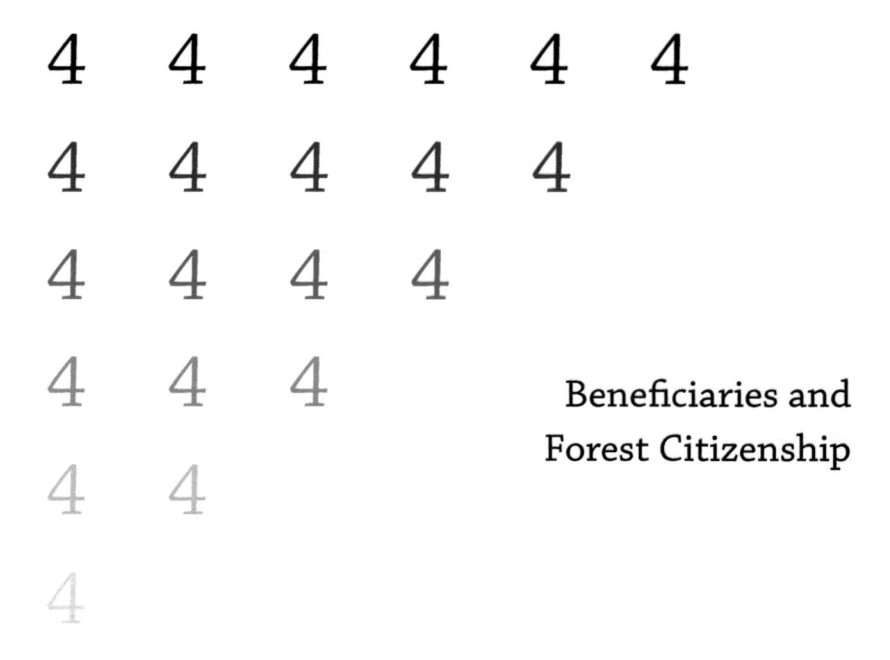

Beneficiaries and Forest Citizenship

A state agricultural extension technician named Edgar spoke to a group of about two dozen smallholders, members of the Feliz Rural Producers' Association. The meeting took place in the only public space in this part of rural Acre: a two-classroom elementary school located at the end of a long and often impassably muddy, unpaved road. The road was bordered by sparsely grazed cattle pastures and a few small manioc fields. In the distance, forest appeared to rim these cleared spaces.

Edgar wanted to talk to the smallholders about handheld trimmers they had recently received from the government. The trimmers were intended to boost the smallholders' agricultural productivity while also protecting that forest.[1] If they sought to control the regrowth on their land with fire, they might burn the forest too—on purpose to clear more land or by accident, as fire spreads more easily as the rainforest dries in the changing climate.[2] The idea was that they could use the trimmers instead.

Edgar asked the group what they would do if the trimmer broke. Would they buy another one? A man volunteered a response: "It's better if the government gives it!" Everyone laughed in agreement, including Edgar. Sure, he said, but would they wait for the government? The government might not give them another one, he warned. He described how some rural producers' associations had pooled their money so that they could purchase trimmers without government assistance. Those assembled quieted. They did not express enthusiasm for Edgar's vision of collective organizing and independence from the government. Instead, the vision loomed like a threat.

The government had a name for smallholders who received some form of assistance from the government, like those at the meeting with Edgar: beneficiaries. Through the Acrean state's sidestepping of property rights, beneficiaries had replaced landowners as the central rural figures in the governmental effort to make the living forest valuable. State-beneficiary relations were important to the effort to keep the forest standing and carbon sequestered in it. Administrators distributed benefits to beneficiaries, visited beneficiaries in rural areas to train them to use the benefits to increase production and forgo deforestation, and bussed current and potential beneficiaries into Rio Branco for consultations on SISA's forest carbon program, of which they were beneficiaries. At one such consultation, I watched an administrator explain how rural producers might be transformed into and become beneficiaries. Another administrator traced the trajectory for me with a flowchart on his computer. Arrows led from "rural producer" to "ecosystem service provider" and then culminated in "beneficiary."

Various types of beneficiaries appear frequently in late capitalist and Anthropocene governance worldwide.[3] The meaning of being a beneficiary in these contexts is always specific, of course. It is constituted through the particular assemblages of socioenvironmental relations of particular places and times. Yet the figure of the beneficiary is always a relational one—you are a beneficiary of another person or institution. Moreover, being a beneficiary also commonly indicates relationships with state or NGO institutions. In rural Acre, after the relative state neglect of much of the nineteenth and twentieth centuries, being a state beneficiary could be a relatively new status. Administrations led by the PT at the state (1999–2018) and federal (2003–2016) levels had expanded benefits or created new ones as part of their strategy of facilitating social inclusion via material redistribution. Some of SISA's administrators spent a good deal of time cultivating and governing smallholders as beneficiaries. Beneficiaries, in turn, frequently sought to make claims on and cultivate their relationships with the state. Such giving and receiving can enable and sustain relationships, and, the state was betting, it could help to sustain those socioenvironmental relations of the living forest as well.[4] In this case, in other words, the beneficiary's relationships involved not only state officials and those they governed but also the forested landscape in which they lived and the carbon that it stored.

Forest beneficiaries' prominence in the Acrean state effort to valorize the forest also suggested that a kind of nascent environmental citizenship was at work. During the period of state neglect and rubber's decline, Acrean smallholders had been regarded—to borrow Kregg Hetherington's description of *campesino* farmers in rural Paraguay—as "pioneer citizens": they were expected to deforest the countryside to live in it.[5] In the process, they would help establish Brazilian sovereignty and deforestation-linked forms of production and culture on the frontier. But, as part of its effort to create a forest-protective form of development in the early 2000s, Acre's PT-led government adopted the concept of *florestania*—a neologism combining the words for citizenship (*cidadania*) and forest (*floresta*).[6] Both types of citizenship entailed socioenvironmental relations, but oppositional ones. The former worked through forest destruction; the latter enticed forest protection.

A decade later, the forest beneficiary had become, I found, one important iteration of forest citizenship. The term *florestania* was used in various settings in which government services and benefits were distributed, as in, for example, the adult education program in which Flávia taught and adults on the BR-364 in Feijó were educated. It was called Wings of Florestania. Moreover, as the state sought to make forest carbon monetarily valuable as part of an inclusive green capitalist economy, the contours of rural Acrean citizenship became increasingly delineated through forest benefits. The forest—its protection, destruction, new value, and the benefits attached to it—had become essential to burgeoning state-citizen relations, just as those relations were essential to the forest and the carbon it stored.[7] Citizenship in rural Acre was a socioenvironmental relationship.

The meaning of forest citizenship, however, was unsettled. Beneficiaries in Acre often deemed the benefits they received to be insufficient, and the benefits often seemed on the verge of being cut. As Edgar warned, the government might not give another trimmer. Yet these benefits' precarity was different from that linked to declining state support and increasing job insecurity experienced by those elsewhere, those who had benefited from twentieth-century welfare state policies and relatively secure formal sector employment.[8] In contrast, in Acre the forest beneficiary was an emergent status engendered through new benefits rather than dissolved through the elimination of existing ones. This directionality mattered; it enabled a fragile sense of optimism that things were getting better and that the government would give more, since it was already giving more than it did in the past.

While beneficiaries and state officials often had conflicting understandings of what it meant, and should mean, to be a forest beneficiary, they did overlap in one regard: neither centered on the type of individual rights often imagined to characterize democratic citizenship. Yet the giving and receiving of benefits enabled other sorts of claim-making, much of which took place in the everyday negotiations surrounding benefit distribution. It was in these negotiations that the meaning of being a forest beneficiary was contested, and with it, the contours of forest citizenship. In this, forest citizenship resembled what Nikhil Anand describes as the "hydraulic citizenship" of Mumbai residents seeking access to water; it was "an intermittent, partial, and multiply constituted social and material process."[9] In this case, forest carbon's new international value and its distribution as forest benefits enabled political and material claims, in a kind of environmental "politics of distribution," to use James Ferguson's term.[10]

These green distributional politics tended to play out in contexts that became familiar to me, including those surrounding rural producers' associations. Benefit distribution and its negotiations often occurred through these associations, which organized rural Acre into usefully governable units. On visits to associations like the Feliz Association, state officials like Edgar promoted the kind of inclusive productivism I discussed in chapter 2. They positioned forest benefits as temporary allocations meant to enable beneficiaries to become productive independent farmers—the sorts of people who would organize to buy trimmers themselves. Contesting this inclusive productivism, in contrast, association members advocated for continuing and increased benefits as part of an ongoing exchange—reductions in deforestation in exchange for state benefits. "It's better if the government gives" things like trimmers, the man at the Feliz Association meeting said to Edgar. "More would be better," Neldo told me after approvingly narrating the state benefits he received. Getting some benefits enabled beneficiaries to envision, hope, and negotiate for more of them. Through this work, they cultivated inclusion of a different sort than that promoted by state representatives like Edgar.

These conflicting understandings of forest benefits and beneficiaries were also about the socioenvironmental relations and citizenship that they entailed. State officials like Edgar critiqued beneficiaries' dependence and saw benefits as forging a temporary state-citizen relationship that would dissolve into independent and forest-protective production. Beneficiaries, in contrast, asserted that the relationship should continue if the forest was to be protected and they were to continue to live in relationship with it. These claims not only embraced their own dependence but also implied that it was mutual. Smallholders' key role in forest protection, and forest carbon sequestration, meant that the Acrean state and the climatically changing world beyond were also dependent on them. The term *beneficiary* implies that benefits flow one direction: to the beneficiary. Yet here, the benefits move in both directions. The imperative to reduce climate-changing deforestation meant that the socioenvironmental relations of forest citizenship were about mutual dependence, as well as mutual benefit.

Benefit Brokers

For numerous reasons, many forest benefits were given through rural producers' associations (hereafter referred to as *associations*). These were often composed of a few dozen members (though they could be smaller or larger) and were important, if unevenly active, institutions in some areas along the BR-364 in Feijó—an area with few others. One reason for their importance was that they increased the reach of an understaffed and under-resourced state. The state could save money by giving some benefits (like trimmers) to groups rather than individuals. Having association presidents organize gatherings also reduced the number of meetings and field visits needed. Moreover, the approach aligned with the PT's roots in social movements, including the Acrean rubber tapper movement, and the associated belief in community organizing as a good way to arrange rural society and combat dyadic power relations.[11]

Yet associations were not grassroots community groups, and personalized and informal power relations were often part of the forest benefit distribution that occurred through them. Association presidents and agricultural extension technicians (hereafter referred to as *technicians*) were frequently key brokers, mediating smallholder access to forest benefits.[12] Groups of technicians employed by SEAPROF or similar institutions and other state officials went on daylong or extended trips from urban headquarters to what they called the *field*. In the dry season, they often drove white pickup trucks and other vehicles emblazoned with state logos up dirt side roads; in the wet season, when the roads were impassably muddy but the rivers were high, they traveled by boat. Associations' elected presidents and some members also sought out technicians in urban offices. The most effective president I knew at accessing state benefits—Manuel, president of the association to which Neldo belonged—seemed always to be traveling between his rural home off the BR-364 and "*a rua*"—the town of Feijó, Rio Branco, and even Brasília, where the government bureaucrats and officials who controlled access to benefits were based.[13] One friend who worked in agricultural extension explained to me that every association had a president who worked with state officials. Through these relationships, the state gave benefits to the association, and presidents might then encourage members to vote for particular candidates and parties at election time.

This approach led to uneven access to some benefits. For example, several technicians told me that they were more attentive to associations with tenacious presidents, and some smallholders described how association presidents channeled benefits to their relatives. Membership in associations that were successful at accessing benefits could swell, while membership could atrophy in those that were not. In addition, some of the association presidents I knew were the descendants of rubber barons, so that the new rural power structure could in some ways come to resemble the old one. In social science literature, these personalized political relationships might be termed, critically, "clientelism" or "patronage."[14] In the context of forest benefit distribution, these relationships were often entwined with, rather than replaced by, democratic politics.[15] They were neither purely exploitative nor ideally democratic, and they could include assertions of dignity, solidarity, vulnerability, and community power, along with invocations of hierarchy.

Moreover, the state's goal of reducing deforestation also shaped officials' relationships with associations and their presidents. In other words, beneficiaries were not just voters. They were also consequential rural actors who could either deforest or refrain from doing so. In association meetings, I often heard technicians explain to beneficiaries that even though each of them was not deforesting much, together they would destroy the forest. By working with associations, the state sought to more effectively counter the smallholder deforestation that was responsible for an increasing portion of Acrean deforestation. The only way to keep forests standing inclusively was to get them to cooperate. Recognizing their importance, beneficiaries often saw their role in forest protection as a way to maintain and strengthen their relationships with the state.

Some of the negotiations around benefits occurred in performative supplications, claims, and silences in meetings between associations and technicians. Below I examine two examples to explicate how the figure of the beneficiary was negotiated, and with it, forest citizenship. One of the meetings was with an association called Boa Vista and the other, which began this chapter, was with the Feliz Association.[16]

Benefits as a Path to Independence

On a long car ride to a meeting with a rural producers' association, Edgar described his perspective on benefits. Life had improved in recent years for many rural Acreans, he told me. When he began working as a technician for the state in the 1980s, the smallholders with whom he worked often lacked access to sufficient water and food. He feared that bringing his own lunch would be seen as a judgment on their poverty, on their inability to feed a guest. Sometimes they hid from him out of shame about their tattered clothes. Edgar insisted that this was no longer a problem. Now, rural people had plenty of food to share and clothes to wear. Many had lots of money, even if they appeared poor. He knew one man who bought a fancy pickup truck and another who had almost eighty head of cattle. He claimed that their phones were nicer than his, and said that they always wanted to take selfies with him. He was describing a contrast between past isolation and poverty and present-day material inclusion akin to what Neldo had shared with me.

Why were they better off now? I asked. Edgar's answer was the same as Neldo's, and he gave it to me quickly, as though it were obvious: government benefits. Then Edgar offered a critique. He said that benefits were important to help smallholders who had "nothing." But they should not become "dependent" on the state. In fact, it was his job to get them to no longer need benefits because they would be producing more crops and forest products, while deforesting less. This was the primary challenge of his work: nothing was produced in Acre, he said, echoing the common refrain. His charge was to change that. In other words, Edgar aimed to include rural people in green capitalism as productive independent farmers and forest collectors. In his view, a good beneficiary did not stay a beneficiary.

It was work he had been trained to undertake. A workshop at SEAPROF's headquarters that I attended, for example, instructed Edgar and the other technicians that their goal was not only to distribute benefits but also to teach rural producers to use those benefits in the right way, defined as producing crops for "the market" without deforesting. Forest benefits enabled an environmentally premised welfare state, but those benefits were still meant to facilitate entrepreneurial and independent farmers. Neoliberalism and environmental welfare state policies were not antithetical, in other words. This was how green capitalism was meant to work.

One strategy Edgar and other technicians used entailed promoting the cultivation and the use of mucuna as part of what was called *roçada sustentável*, or sustainable production.[17] Mucuna fixes nitrogen and is meant to enable smallholders to grow crops for many years in the same field, rather than clearing new forest. The government had distributed mucuna seeds to some rural producers' associations, including the Boa Vista and Feliz Associations. Edgar and Felipe, another technician I shadowed, pushed the seeds' use with two rationales. First, they sought to make it appear cost-effective via a collective budgeting exercise they had learned at the SEAPROF training. On large chart paper, they compared the costs of growing crops on one hectare of land using "traditional" versus *roçada sustentável* methods, tabulating the costs and the resulting yields of each in a kind of Weberian bookkeeping.[18] Edgar would field real-time cost estimations from each association with which he met. He also converted days of labor into money (e.g., R$40 per day for deforesting in traditional agriculture), even if—as association members seemed to inevitably maintain—they always did the labor themselves and so never had to pay. Including labor made the costs of the two types of agriculture appear closer than they would have otherwise. *Roçada sustentável* was still more expensive, but it would have appeared to be more so without the monetized value of deforestation labor. And, crucially, when the government provided mechanized tilling or mucuna seeds as forest benefits, the cost of *roçada sustentável* fell below that of traditional agriculture. Having seen that using mucuna would produce better yields and therefore higher profit, smallholders were supposed to see the rationality of *roçada sustentável*, and therefore want to adopt it. Whether *roçada sustentável* made economic sense depended on what was counted and how. The numbers at the end were a product of a long process of negotiation.

There was another part of the theory too. As rural producers earned more from selling their increased bounty to private buyers, state benefits would become unnecessary. Technicians promoted this vision of the future as a collective victory earned by smallholders working together with their associations. In other words, technicians were not promoting a top-down clientelism so much as they were advocating for rural organizing and collaboration to encourage independence from the state.

For example, at the Feliz Association meeting, Felipe asserted that the association did not need the government anymore. A trimmer cost R$2100, which, he admitted, sounded unaffordable. But if the association started to save now, they would be prepared to get a new one when they needed it in what he estimated to be three years. Felipe quickly did the math aloud: the cost amounted to R$58.30 per month, or R$2.65 per day, assuming twenty days per month of use. "That was affordable!" he exclaimed triumphantly. He maintained that it was not a hard calculation to do, and it could (and by implication, should) be done for other benefits they received as well. Edgar chimed in, chidingly, with the reminder that the association had to become self-sufficient. They could not rely on the government for everything. If the producers would only do what the technicians advised, he said, they would soon be on "their own feet." The technicians' goal, he explained, was for the association to have "sovereignty." It was to be achieved both through collective savings and by collectively selling crops to private buyers in ways that would cut out middlemen, who beneficiaries and technicians agreed bought at exploitatively low prices.

At a similar meeting with the Boa Vista Association, Felipe asserted that the real problem was that the association was not well organized. But he could help: he was working with a nearby association to produce high-quality manioc flour to sell commercially at a good price, and he could help the Boa Vista Association do the same. They did not have to wait for the government. For the price of one calf, the association could buy fertilizer and calcium for the soil. They could earn more money and become independent of state support. In this vision, Felipe advocated for an inclusive green productivism and citizenship enacted via rural organizing and market engagement.

Forest Benefits, Deforestation, and Smallholder-State Relations

However, neither association assented to this vision of productive independence. Instead, they used meetings with technicians to advocate for a contrasting understanding of what it meant to be a beneficiary. For them, it entailed a long-term relationship with the state based on the exchange of benefits for forest protection. This was the inclusion they sought and the vision of the beneficiary and forest citizenship they advanced.

The meeting with the Boa Vista Association, which took place under the shade of a thatched-roof community space and the mango tree that stood next to it, began with an association member thanking each technician by name and, in a telling assertion of kinship, calling SEAPROF the "mother of the community." Through the association and its relationship with SEAPROF, she went on, the community finally had "good luck," after years of having "nothing." This luck was not due to collective organizing or engagement with markets, I realized. Rather, as she said, it was because the association enabled its members to meet people in the government and to receive benefits from them. Vicente, the association president, then spoke up to thank the government for investing in his community.

As the meeting concluded, he thanked SEAPROF again and undertook a second common feature of such meetings: he recounted his effectiveness at accessing government benefits. He was giving a reelection speech (he was running unopposed) and articulated his role to date: he was not the "owner" of the benefits the association had received, but was instead the community's "representative" to the state. He again thanked SEAPROF, listing some of the benefits that the community had received. Then he articulated a vision for the future: accessing new benefits. With his reelection, the association would have another three years to "care" for what they had already been given and to "look for" benefits they had not yet been provided (they had heard of the fishpond program, for example, but had not received fishponds yet)—that is, as long as the PT government remained in power. Vicente warned that if the PT was not reelected in the upcoming gubernatorial election, his association would be left "outside" the "political process," as it had been before the PT came into office.

All in all, Vicente presented the association as both a capable beneficiary, able to care for the benefits it received, and a needy one. In these performances of gratitude—a common feature of such meetings—community leaders positioned the association as a willing forest beneficiary, with the government as its benefactor. Being a beneficiary could constitute a welcome means to political and material inclusion for communities who had long been marginalized.

In between these two benefit-focused recitations of kinship and gratitude, though, Vicente pushed for a different version of what it meant to be a forest beneficiary than Edgar and Felipe. For Vicente, it entailed a long-term relationship with the state based on the exchange of benefits for forest protection. This was the inclusion he and many other smallholders sought and the vision of the beneficiary they advanced.

Mucuna was too risky, many believed.[19] Vicente, for example, described mucuna as something for those who could afford it. When Edgar and Felipe praised a nearby smallholder for his use of the legume, Vicente countered that the man was single, whereas he and the others in his association had families. Using mucuna would leave them vulnerable. In contrast, their traditional deforestation-based production was an inheritance from their parents. It was sure to produce something, even if the yield was less. At the meeting with the Feliz Association, its president—a fast-speaking former rubber tapper and social movement organizer named Pedro—offered to show Edgar and the rest of us five hundred kilos of state-given mucuna. It was sitting in a nearby shed because no one wanted it. He himself had tried mucuna with disastrous results, he exclaimed with dramatic flair to appreciative laughter. Echoing what I had heard from other smallholders, he called it a "plague" that "invaded" all his fields and destroyed his crops.[20]

In interviews, some rural producers told me that mucuna was more laborious than deforestation, requiring significant weeding to keep it from overtaking other fields. They also asserted that they did not receive sufficient training on how to use it effectively. Such critiques reflected both the inadequacy of government training and how busy many smallholders were with livelihoods that were often not exclusively agricultural. Labor-intensive interventions, such as mucuna, could disrupt their ability to engage in other livelihood practices, including local wage labor, hunting, and collection of forest products.[21] Moreover, smallholders knew that it could be difficult to sell any surplus yield that they might obtain by using mucuna. They sold crops to the middlemen who visited their land in part because they lacked other transportation options. Many described letting some crops (especially unprofitable fruit, like papaya) rot or planting less because they had no way to transport their harvest. Why risk growing nothing by using invasive mucuna when it would be difficult to sell any bounty?

What was needed, Pedro asserted in the Feliz Association meeting, was not increased independent production but rather a deeper relationship with the state. The government should buy more of their crops at a "guaranteed" price, he argued repeatedly.[22] This stability was more important than the price itself.[23] So was the relationship with the state that it would forge.

Pedro's vision of government crop purchases was one iteration of a broader understanding of government benefits as given in exchange for reduced deforestation.[24] Pedro and Vicente understood that the government wanted them to forgo deforestation. This knowledge created the possibility for exchange. Vicente did not assert a right to deforest. Rather, he sought more benefits for not doing so. He decried the government's enforcement of deforestation restrictions against members of his association, expressing indignation over the expectation that they would stop deforesting without sufficient government support. He said that they were paying the price for the large-scale deforestation of "*os grandes*"—large landholders and cattle ranchers—that began in the 1970s.[25] He then recounted angrily, his voice rising, how the "environmental police" had come to the area two times in the past week, wielding accusations of illegal logging and threatening fines. One woman was so distressed by their visits that she had to go to the hospital in Rio Branco, about two hours away. Vicente's voice rang out defiantly, "We'll go to court! We are old residents of this area who need to be respected!" He then lowered his voice and offered an alternative: their willingness not to deforest.

This was something I heard frequently from rural producers, and not just because they saw it as a means of receiving benefits. It also reflected a cultural valorization of the living forest based on rubber tapping, the experience of bad health effects from forest fire smoke, and an interest in forgoing the arduous labor of deforestation. Some also expressed concern about climate change (*mudanças climáticas*), which they connected to their experience of more extreme and unpredictable weather, such as increasing midday heat that made their outdoor work untenable. Forest protection, then, was not simply an imported priority but one entwined with local history, culture, and the changing climate.

Yet forgoing deforestation was not something smallholders could afford to do on their own. Vicente positioned it as part of an exchange. His association members were willing, even eager, to stop deforesting, but in exchange, they needed more from the government—more money, more agricultural trainings, more supplies, more assistance transporting crops, and more government purchases of them. Vicente could have claimed that association members had the right to deforest or otherwise determine how they use their land. Yet he did not utilize a language of individual citizens' rights or their "right to have rights."[26] After all, the state did not attach forest benefits to formal land rights, and rural people did not have guaranteed rights to forest benefits, even if they stopped deforesting. Even Vicente's threat of going to court focused not on rights, but on a demand for respect, both in the present and as part of a long-term relationship with the state. That relationship was predicated on and facilitated by the exchange of foregone deforestation for forest benefits.[27]

Vicente couched this relationship in terms of need, imploring the technicians to give them more. Their children, he said, had no pants to attend school. Without more government help, rural producers like them would be done for ("*vai acabar com o produtor rural*"). They would not "survive." Vicente thus paired his assertion of dignity with an articulation of need. Association members could either leave their land, he implied, or continue to deforest. If Edgar and other state officials wanted them to grow more crops and not deforest, they would need to give more forest benefits.

For those for whom the state had been largely absent, getting some benefits did not generate unquestioning loyalty or complacency. Rather, it brought not only performances of appreciation but also the desire for and ability to envision receiving further benefits in exchange for something smallholders knew had value: forest protection. Even insufficient state support provided a basis for claiming more—a basis to cultivate relationship. While state technicians pushed beneficiaries to work together for their independence, beneficiaries themselves foregrounded exchange and the relationships formed through it. As long as the state wanted them to forgo deforestation, that relationship would continue.

Conclusion: Mutual Dependence

The socioenvironmental relations entailed in being a forest beneficiary were not predetermined. Smallholders, such as Pedro and Vicente, sought to define them differently than state representatives like Edgar and Felipe. In the face of inconsistent, inadequate, and temporary benefits—and the risks some of them (like mucuna) posed—beneficiaries negotiated for more benefits and for continuing relationships with the state. Association meetings with technicians, like those analyzed in this chapter, provided the space for some of these negotiations. In other words, forest benefits were not a depoliticizing, complacency-generating salve. Rather, getting some government support enabled beneficiaries to claim that they needed and deserved more of it—and to envision a state that would give it to them.[28] Some smallholders positioned benefits as part of an exchange that forged and enabled the cultivation of reciprocal, lasting, and sometimes personal relationships. As part of this, forest benefits could enable a sense of being included in the political process, perhaps for the first time, and some measure of material inclusion as well, as precarious as it may be. Being a beneficiary could thereby provide smallholders with a way to engage with the state, cultivate relationships with state officials, and thereby negotiate their own version of forest citizenship in everyday life.

The particularistic, relational work of claiming benefits is different than that of claiming universal rights, often assumed to be the proper politics of democratic citizenship. Benefit distribution can work through personalized political relations, sometimes condemned as patronage or clientelism.

Yet, as James Ferguson argues in the context of southern Africa, "where social assistance looms so large in the state-citizen relationship, the state may well appear to the citizen not principally as a protector of equal rights, but as a material benefactor or even patron."[29] Neoliberal capitalism has undermined state social welfare programs in many places. But it is also linked to emergent forms of state support—such as Acre's forest benefits—in places and for people that never benefited significantly from state welfare policies. In such places, the state may no longer be primarily an absent or tax-extracting authority, but rather an increasingly important source of social support that is sometimes mediated through personalized politics. Such support could make marginalized people into state beneficiaries—sometimes for the first time. Receiving benefits could, then, form a basis of their relationship with the state, and with it, their experience and practice of citizenship. The citizenship of forest beneficiaries was therefore forged not only through receiving and claiming more traditional forms of social assistance and services.[30] It was also created through the distribution of new forms of value conjured in the green capitalism of the Anthropocene, such as that ascribed to the carbon sequestered in living forests.

What was at stake in forest beneficiaries' negotiations with technicians like Edgar were the contours and socioenvironmental relations of this forest citizenship. Ferguson argues that while liberal thought tends to condemn it, dependence can be central to democracy: "The positive content of citizenship itself may increasingly come to rest precisely on being a rightful and deserving dependent of the state."[31] Being a beneficiary could put smallholders in such a position of dependence. It was a position some of them embraced, and they countered technicians' promotions of independent production. Invoking kinship, gratitude, responsibility, and need, Vicente, Pedro, and others, for example, performed dependence in encounters like those I analyzed above.

Yet the fact that beneficiaries' socioenvironmental relations were meant to hold forest carbon in place meant that dependence was also mutual. Beneficiaries often saw the forest benefits they received as given in exchange for the forest protection they knew the government, and indeed the larger world, wanted in the context of the climate crisis. The state's pursuit of inclusive green capitalism meant that it was reliant on beneficiaries to forgo deforestation. In this "carbon democracy" of the Anthropocene—centered on carbon's sequestration rather than extraction—citizenship entailed mutual dependence.[32]

INTERLUDE IV

The Rural Road, Part 2

The BR-364 connected rural and urban areas in ways valued by many residents of both. Yet, in rural areas at least, it also indexed and aggravated inequality. Most smallholders who lived on the BR-364 near Feijó, and on the unpaved side roads and small paths that branched off it, did not have private vehicles. While some, like Luis, were planning and saving to buy one; for most, such a purchase was far out of reach. Nor could they rely on school buses to take them to work, as Flávia did, or all the way to the city of Feijó, where they needed to go to collect some government benefits, visit the doctor, or buy what could not be made, grown, or collected. Sometimes, agricultural extension technicians or other state workers, NGO employees, or researchers like myself could offer a lift. If not, to ride on the BR-364 might require them to pay for a seat in a *freteiro* (a truck with benches in the back) or another private vehicle. The price could be burdensomely high: R$20 from near Flávia and Luis's house to Feijó, for example. Paving the BR-364 did not seem to reduce this exorbitant cost.

Other differentiating effects were not so apparent. They were slower and harder to see. I could hear them within Luis's narration of the highway landscape as we drove to Feijó on the recently paved stretch of the BR-364. As we neared the town, his narration changed. No longer was he pointing out the homes of family and friends known for many years. Instead, he was pointing to the land of *fazendeiros*, many of whom had moved in quite recently or, more frequently, had bought land along the newly paved road, even as they continued to live elsewhere. In his narration of the landscape outside the truck windows, Luis often included the year a particular *fazendeiro* had bought in and sometimes the often distant place where he had heard they actually lived, but nothing more. He did not know them. Their presence registered as absence.

Their purchases also offered a preview. Land speculation grew with the prospects of the BR-364's paving. I wondered whether Luis too would soon be offered cash for his land now that the road was paved past it and all the way through to Rio Branco, and the rest of Brazil beyond. And I wondered whether deforestation would grow accordingly, with the forest pushed back farther and farther from the road. I wondered too about how the road's paving would impact those who lived farther afield and how speculation and deforestation might ripple outward from the road, reshaping socioenvironmental relations in the process.[1] Inclusion and dispossession were entwined in the BR-364's paving.

Yet the newly paved road was already crumbling. Potholes formed in the middle, and pavement washed out on the edges (figure InterIV.1). One of the workers that paved its last section in Feijó joked to me that he would always have a job, since the road was always falling apart. There were different theories as to what, or who, was to blame for its rapid deterioration. Surely it was the incompetency and corruption of the government, many thought. Aggravating this was the weight of frequent trips by heavy trucks. Some of these were laden with tree trunks (legally and illegally logged), goods and food being brought north to Feijó and the towns beyond, and, when the 2014 flood cut off the BR-364 in Rondônia, goods and food being brought south after being shipped in by river. And then there was the Acrean earth itself. I was told that it contained few rocks. Without them, the ground under the road was easily compacted. It sank quickly, creating subsidences for which pavement was no match. It seemed to simply fall into the earth. Proponents

Figure InterIV.1 The edge of the BR-364. Photo by Maron E. Greenleaf. Feijó, Acre.

of this explanation included Jorge Viana, who wrote in a July 2014 Facebook post about repairs on the BR-364 that "roads in Acre are a tremendous challenge, but maintaining them is no less of a challenge: we have no rocks, our soil is bad for roads and the best thing we can do is take good care of the ones we have."[2] The BR-364, then, was at once a symbol of the promise of Acrean inclusive integration and development—green or otherwise—and the evidence of its uneven impacts and broad failures.

5

The Urban Forest

Capitalism is cultural, and so the effort to make it green is necessarily a cultural project as well.[1] Part of what distinguished Acre's forest valorization effort—and for a time, at least, made it quite compelling to some—was its embrace of this cultural dimension. Building on previous chapters' analysis of Acrean efforts to make the living forest valuable through performance, productivity, and politics—and the socioenvironmental relations surrounding them—in this chapter, I trace connected efforts to make the forest culturally valuable within Acre. These efforts aimed to shift what the forest meant to those who lived in the state.

Cities were a particular target of this cultural valorization, particularly the capital city, Rio Branco, where close to half of Acreans lived. I lived there too a lot of the time I was in the state, and I came to understand that much of what was going on around me there was inextricable from the more explicit efforts to protect and valorize the forest in rural areas. In other words, I could not understand the Acrean effort to valorize the forest without also understanding what was happening in the city. This was because increasing the forest's cultural value in Rio Branco was meant to facilitate its protection in rural areas, and thereby increase and enable access to its new international monetary value.

Rio Branco did not have an urban forest of the usual kind. With a few exceptions, living urban trees were not central to it. A Brazilian forest researcher told me over dinner on his first night of a trip to Acre that the city looked less like the Amazon than the Cerrado—the grassy savanna ecosystem to Acre's south. He compared it to Manaus, a major Amazonian city in the state neighboring Acre. There, the city's trees had broad branches that created their own canopies, he told me. They offered merciful protection from the hot sun and pouring rain. His comparison was not meant to be complimentary to Rio Branco. The city also was not an urban forest in the sense that Bertha Becker used the term.[2] For Becker, the "urban forest" captured both the expansion of Amazonian cities (as some 70 percent of Brazilian Amazonians came to live in urban areas in recent decades, including in Acre) and also the urban-driven transformation of rural areas and attendant spread of urban "values."[3]

I came to understand Rio Branco's urban forest also in a different sense, and once I knew what to look for, I began to see it all over the city. It was present in the city's statues that celebrated rubber tappers and in buildings like the Forest People's House (Casa dos Povos da Floresta). It was in well-appointed and well-staffed state environmental agencies. It was in some urban Acreans' environmental jobs. And it was in their newly frequent trips between the country and the city on the BR-364. All of these were features of an urban forest of a different sort than I imagined encountering. It was one that invoked the living rainforest to increase its cultural value in the city. In Rio Branco, this urban forest improved lives in broad, though unequal, ways.[4] It worked through facilitating gainful and meaningful employment and underwriting material improvements to city life. It also featured in state-sponsored art, celebrations, and public spaces to generate pride in being Acrean. This was meant to counter the state's reputation as backward, poor, and perhaps even nonexistent. These forest-themed urban spaces, jobs, and identity made city residents into forest beneficiaries and forest citizens, but of a different sort than Luis, Neldo, and other rural benefit recipients.

Through this cultural valorization, Acrean officials and nonstate allies sought to reposition the forest as central to what it meant to be Acrean, and to reframe "forest people" as the "true" Acreans, as Maria de Jesus Morais explores.[5] It was an approach that aimed to address a particular version of what Evelina Dagnino describes as Brazil's "culture of social authoritarianism," one that excluded and devalued the forest and those associated with it.[6] Both historically and contemporarily, parts of dominant Brazilian culture and politics have framed the Amazon as a kind of "green hell" to be replaced by fields and cities.[7] Forest cultural valorization aimed to instead center the forest and its residents within Acrean culture—a logic of social inclusion via cultural change. Forest citizenship, then, was not just about the direct citizen-state political relationships and material redistribution explored in previous chapters. It was also cultural, aiming to foster both belonging and forest protection by changing the culture of which these citizens were a part.[8]

For a time, this approach seemed to be working. It seemed to be enabling the socioenvironmental relations necessary to hold forest carbon in place in the living and threatened Acrean forest. For two decades (1998–2017), many of Acre's urban forest beneficiaries at least nominally supported forest protection by repeatedly reelecting PT politicians to state office. This support enabled some officials to pursue forest conservation. Deforestation declined and stayed relatively low between the mid-2000s and mid-2010s, allowing them to then access some of forest carbon's new monetary value via policies like SISA. They then channeled some of the resulting money into the forest, both urban and rural. In this version of green capitalism, cultural value begot monetary value, which then underwrote cultural value once again.

Yet Acre's urban forest also indicates how cultural valorization can fall short within green capitalism. Recalling Raymond Williams's analysis of the country and the city, Acre's urban forest was mostly an urban projection.[9] In the city, the forest could seem, as Michael Bunce describes the countryside, like a "cultural construct and a social ideal, forged by the historical processes of a metropolitan-dominated society."[10] In Acre, the state-sponsored urban forest diverged consequentially from the continued difficulties of rural life, as well as intransigent and culturally dominant views of the forest as a cause of privation and something to be avoided. Acre's urban forest was an effort to change the dominant culture by including the forest and forest people within it. Its failure to fully do so points to the limitations of green capitalism in shifting not just economic dynamics but also the cultures of which they are a part. Ultimately, this failure undermined the project of green capitalism in Acre.

Forest Urbanism

The Forest People's House was located in an urban park in Rio Branco. Almost circular in shape and constructed of dark patterned wood and a thick peaked thatched roof, it looked nothing like the homes of rural smallholders I knew. Those tended to be constructed of brightly painted, roughly cut boards or coveted government-financed cinder blocks. Neither did the building resemble those in the concrete low-rise cityscape that surrounded it. It was a forest people's house for the city and its residents.

I walked past the building frequently since it was near my apartment. Its doors were always closed, and it seemed empty. But on one evening in April 2014, people spilled out from a standing room–only interior. Live music set a celebratory mood. Attendees hugged in warm greeting and milled about among statues, photographs, and large printed texts about Acrean Povos da Floresta—Indigenous people, rubber tappers, and others. A giant sculpture of a snake-like animal was suspended, slithering, above the crowd and another sculpture of a *mapinguari* forest being loomed over us.[11] The local press was there too. As I entered, a tall, blonde television reporter was interviewing the event's central figure: Jorge Viana, former governor in the Government of the Forest and now federal senator. Jorge then worked the crowd with joyful and seemingly inexhaustible energy. He imparted rapid-fire hugs that were documented by a trailing photographer. I got a hug too. It felt genuine. I understood what a friend meant when she praised him as being "human"—a description she offered as a contrast to other politicians she had met.

The event was a celebration of a new publication called *Acreanidade* (figure 5.1), named for a concept adopted by the Government of the Forest when Jorge was governor (1999–2006). *Acreanidade* named an essential Acrean-ness related to the forest and struggle: the struggle to tap rubber starting in the nineteenth century, fight Bolivian claims in the late nineteenth century, become part of Brazil at the beginning of the twentieth century and a state in the mid-twentieth century, organize against deforesting ranchers in the 1970s and 80s, and now redefine development and make the living forest valuable.[12]

An MC quieted the crowd and began describing Acreanidade and the related concept *florestania* as part of a "profound change" in the state. He sang of migrants who fled the 1877 drought in Brazil's northeast to tap rubber in Acre. This was followed by a poetry recitation featuring Acre as a good place to live—one that offered a "new way to be human" through Acreanidade

Figure 5.1 Photo of the cover of *Acreanidade*, with Lula (*left*) and Jorge (*right*).

and *florestania*. He recounted stories of Chico Mendes, the rubber tapper movement, and the Acrean Revolution. Later, journalist Elson Martins, introduced as the "master of Acreanidade," described how Acreans were being kicked out of the state in the 1980s because of incoming ranchers, that is until Jorge and others rescued the state through *florestania* and Acreanidade. Perpétua Almeida, a politician running for a Senate seat, teased Jorge about how much younger he looked in the photo on the *Acreanidade* cover, which was of him and union leader-turned-president Lula back when both were organizing against the dictatorship (figure 5.1). Then she told a story about a poor woman she met who said that Jorge made her proud—for the first time—to be from Acre. When Rio Branco's mayor Marcus Alexandre spoke, he described himself as a son of Acreanidade. He said he had moved to Acre after Jorge told him the state's "beautiful" story.

Finally, Jorge spoke. Though still performative, what he said felt more personal than his articulation of the rubber narrative a few months later at the GCF Task Force meeting. He too retold the story of Acre, interweaving it with that of his family: Northeasterners like his grandparents came for the rubber and to escape the drought. Bolivians, also among his relatives, claimed the territory as well. Later, they struggled and triumphed against the deforesting ranchers. His family was Acrean by "genetics." But he offered that Acreanidade was also a feeling. It could become part of any of us through listening to music and poetry. By experiencing those forms and being together, he said, perhaps we could all feel it now. Jorge ended by memorializing three people who had recently died. Family members came to the front of the room to receive a hug and a copy of the magazine, which ended with Jorge's personal remembrances of each of the men. The family members were crying. Jorge was crying. The musicians and MC were crying. The reporter began to cry too.

This event was not just a celebration of Jorge but also of the forest-themed political institutions, community, and sense of belonging that he and those around him had cultivated in Rio Branco. These things had benefited many of the city's residents, including those gathered at the Forest People's House that night. Even those who were critical of Jorge's administration and subsequent allied ones tended to agree that they had made Rio Branco a better place to live. In the late twentieth century, the city had grown rapidly, largely due to an influx of former rubber tappers from rural areas.[13] By all accounts, Rio Branco did not accommodate the growth well. Longtime residents described to me how, by the late 1990s, it was strewn with trash and lacked working traffic lights, decent health services, restaurants, or stores with fresh fruits or vegetables. They described hospital halls infested with rats and the gruesome vigilante "justice" of a powerful and politically connected paramilitary and drug-trafficking "chainsaw" gang.[14] The Government of the Forest and related city administrations addressed many of these issues. They improved city services, administration, and infrastructure, and they extended some social rights to city residents, though they benefited unequally.[15] When I lived there, the city's giant supermarkets were well stocked (except when the BR-364 flooded). Trash collection, street cleaning, electricity, and internet were reliable in much, though not all, of the city. Public spaces in the city center had been refurbished into pleasant gathering places. The chainsaw gang had been prosecuted and dismantled.[16]

The forest helped to underwrite this urban improvement, financially and culturally. For one, the government used the forest to attract outside funding. Some of this money came through SISA and forest carbon's new value, but some was less direct. The state's growing reputation as a forest conservation leader—along with improved public administration that rehabilitated Acre's credit—allowed the state to access both loans and aid from multilateral, bilateral, and nongovernmental sources.[17] This outside funding, coupled with economic growth and federal cash transfers, meant that the state government had some money to spend during the early twentieth century. And it spent some of it on a kind of forest-themed urbanism that centered the forest and forest people within Acrean identity.[18] As Maria de Jesus Morais writes, these urban spaces helped to construct the city's memory and understanding of Acre as a place, making the forest into a kind of state-sponsored heritage.[19]

When I lived in Acre, a significant amount of public urban life in Rio Branco took place within forest-themed spaces of the state's creation—a kind of simulacrum of the forest in the city. There were other government-constructed Amazonian-themed buildings near the Forest People's House in the Rio Branco park (itself an emblem of urban improvement, created by the government in the 2000s from what had been, residents told me, a swampy, crime-ridden, and mosquito-infested expanse). Replete with decorative thatched roofs, one sold traditional Amazonian fare (like *tacacá* soup), while another offered Indigenous and rubber handicrafts. The Forest Library was located in the park too, with exhibits recounting the history of the rubber tapper movement and Indigenous Acrean cultures.[20] Nearby was the Rubber Museum and the Forest People's Plaza, with its statues of Chico Mendes and other rubber tappers. Across the street was the state assembly, where legislators met in a room lined by large wooden carvings of the state's forest history. The Forest People's Plaza also hosted the Rio Branco Palace, an ornate building from the rubber boom that had been largely abandoned until the Government of the Forest renovated it into a museum celebrating Acre's history of forest use and struggle. It included a two-story high mural commissioned by Jorge's administration featuring rubber tappers-turned-soldiers fighting for Acre, a room focused on the Acrean Revolution, rooms with photos and descriptions of different Acrean Indigenous peoples and immigrants who came to tap rubber, and a room focused on Chico Mendes titled "The Defense of the Forest." A historian who had worked on the renovation described its transformation to me as a "retaking" of Acrean history and the forest itself.

Downhill from the Rio Branco Palace, on the riverbank was the refurbished open-air Old Market—a pleasant place to peruse rubber and

Indigenous crafts, forest herbs, and souvenirs, as well as to eat lunch and drink a beer after work. Across the river was the renovated Gameleira—the city's oldest neighborhood, where Acrean rubber used to be amassed to be shipped downriver to Manaus. It was named for its old *gameleira* tree—one of the few left in the city from that era and a "symbol of the perseverance and resistance that characterize[s] the people of Acre."[21] The forest was also dispersed through public spaces on the outskirts of the city in places like the Arena of the Forest, a sports stadium.

State officials also foregrounded the forest in other cultural work.[22] For a while, a government historian published a newspaper column about Acre's past called "History from the Margins" (História das margins) and performed historically focused radio skits called "history chats" (*papos histórias*), both of which often featured the forest.[23] Local television stations played a new forest-focused state anthem, and traditional songs and dances were performed at public events. The state facilitated a nationally aired television miniseries called *Amazônia: de Gálvez a Chico Mendes*—an acclaimed celebration of Acrean history beginning with the rubber boom. The film set was then made into a pleasant park on the outskirts of Rio Branco, with reconstructions of rubber boom–era buildings and dioramas of daily life. Chico Mendes was memorialized throughout Rio Branco in statues, in banners, and in the name of a central Rio Branco avenue. Each year, the government awarded the Chico Mendes Prize in recognition of efforts to sustainably develop the state. Just outside town, in the Chico Mendes Environmental Park, dense foliage, caged jaguars and other forest animals, and placards about Mendes and the rubber tapper movement promised the mostly urban visitors a curated glimpse of the time before the BR-364 was paved and the land around the park was cleared, when forest still dominated the landscape.

The forest also featured in the citywide Floresta Digital (Digital Forest) public Wi-Fi network, whose 2009 creation made Rio Branco the first Brazilian capital to provide free wireless internet. Small pillars announcing its availability punctuated Rio Branco's downtown (figure 5.2). Elsewhere in the city was a Digital Forest Center with computers available for free use. These were part of the state's larger 2010 Digital Forest Project, which promised to bring free internet access throughout the state—a first in Brazil. It was framed as a project of "digital inclusion" (*inclusão digital*): "this program is dedicated to the excluded," Governor Binho Marques proclaimed of the Digital Forest at its launch.[24]

Figure 5.2
Advertising banner for the Digital Forest free public Wi-Fi network. Photo by Maron E. Greenleaf. Rio Branco, Acre.

Such public spaces, performances, and displays were part of the larger project of *florestania*. As Marianne Schmink writes, they were meant to "foster[] a pride of place and a sense of belonging."[25] They could produce a kind of forest-themed "imagined community" that could feel quite meaningful, at least for some, like those at the Acreanidade event at the Forest People's House.[26] In the city, in other words, the forest did not comprise messy and contested socioenvironmental relations among plants, animals, humans, roads and other infrastructures, and governing institutions. Rather, it was at the center of an identity conjured in song, publications, and public spaces in ways that could make urban life more meaningful and enjoyable.

The forest's urban beneficiaries were then meant to support the continued protection of the living forest in rural areas, enabling state access to its new international monetary value. Through this linked cultural and monetary valorization, the forest was to be a "passport to the future," as the Acrean

Secretary of Forests and Extractivism put it.[27] It was meant to improve and redefine Acrean life and landscapes in both rural and urban areas.

The Forest's Urban Command Center

The urban forest particularly benefited a specific group of Rio Branco residents: environmental professionals, who governed it from what became, as Marianne Schmink puts it, the Acrean forest's urban "command center."[28] For them, the urban forest not only supported an improved cityscape and state identity and pride, but it could also provide them with gainful, secure, and meaningful employment partially funded by the forest's international monetary value. State and private employees did not think of themselves as similar to the rural forest beneficiaries with whom they worked. Yet their employment was, in a sense, also a forest benefit.

This group was quite extensive. I seemed to meet people who worked on or studied issues connected to the forest everywhere I went in Rio Branco. Many had studied forest engineering or related fields at the Federal University of Acre and other local universities. Among them were members of the CIFOR research team with whom I worked, employees of state environmental agencies and environmental NGOs, and former forestry students who now worked in unrelated fields, such as policing. For a time in the 2000s, a number of them explained to me, forestry had been seen as the "course of the future." Many students pursued it, making it one of the university's largest programs. Its status was linked with the state government. Jorge Viana had trained as a forest engineer and brought new focus on the sector when he was governor, including through the development of the Federal University of Acre's forestry program. Some students there happened upon the subject and ended up falling in love with it; one friend told me that forestry "conquered" him. Many others described choosing to study it after being drawn to the state-promoted vision of the forest as central to Acrean history, economy, and identity.[29] And, at least for a time, studying forestry promised social prestige and good job prospects. Many of those who studied it hoped to work for the state's expansive environmental bureaucracy—the forest's command center.

Much of that command center was housed in well-appointed new or refurbished administrative buildings. These were part of the forest statecraft analyzed in chapter 3, but I also came to think of them as part of the urban forest. In them, environmental professionals and other staff worked to govern the rural forest. Those I spoke with offered me multiple reasons for their decision to work for the state government. These included their access to resources and staff, which some felt enabled them to more effectively protect the forest than if they worked outside of government. Moreover, working for the forest's command center was attractive because state jobs seemed to offer access to a comfortable, middle-class urban existence.[30] These jobs were secure, offered good benefits, and had the reputation of being easy (though many of those I knew worked very hard).[31] But even those who worked for environmental NGOs or private businesses were often connected to the state government via contract work and other forms of engagement.[32] They too were part of the forest's urban command center.

Many urban environmental professionals appreciated their jobs not only for the money and security, but also because they valued working with rural people and in rural landscapes. Some liked their work trips to "the field" (*o campo*) and the experiences they had there—trips that were enabled by the paving of the BR-364 and the construction of more side roads off it.[33] Many days, I saw white pickup trucks and other vehicles emblazoned with government and funder logos heading from the city to the country. These were trips that environmental professionals I knew often described as the best part of their work. Like other Acreans, most of them were the children or grandchildren of rubber tappers, and many recounted fond memories of childhood visits to family members' forested homes as motivating their work. They recalled those visits fondly: playing outside without shoes, for example, and a "freedom" not possible in small urban apartments and houses in a city many deemed to be dangerous.[34] When they visited rural areas, they told me, they sought out the "tranquility," "fresh air," "quiet," and "climate" of "nature," and life without the constant distraction of internet, cell phones, and television. The trips allowed them to leave behind "all that isn't necessary."

Environmental professionals often also valued their interactions with rural people. Being in "the field," one forest engineer told me, was the most important part of his work: "You learn from the small producers, from how they use land, how they live. It's an exchange," he explained. In these exchanges, he learned about the "real forest" and "unscientific" and "natural knowledge" about it—beyond the scientific names for trees and the GPS and remote sensing tools he had learned about in school. He and others shared photos of these exchanges on Facebook with pride. Some also felt they were helping rural people to "maintain their culture" and to not "have to go to the city to buy industrialized things" or join the ranks of the unemployed urban poor. Working to protect the forest could be a path to meaningful and fulfilling employment in part because of these urban-rural interactions.

At the same time, their employment in the forest's urban command center also seemed to benefit the forest: the policies they created and administered, the maps they made, the meetings and workshops they held with rural forest beneficiaries, their performances of the rubber narrative—all this work could feel important to the governmental project to valorize the living forest and hold its carbon in place. In this way, the urban and rural forests seemed to support each other, as did their cultural and economic value.

Robbery and Lies

Yet there was also a consequential gap between the urban and rural forests. With all the city's improvements and employment, the forest's new monetary value could seem to benefit urban areas more than rural ones. For environmental professionals, this gap was part of what made their work meaningful. They valued their trips to rural areas because they were so different from their air-conditioned urban offices and small city residences. The contrast between their classroom-based knowledge and rural people's forest-based knowledge made their interactions into valuable exchanges (rather than being an indication of a lack of access to formal education in rural areas). They enjoyed having no electricity, phones, and television because they would soon return to them and other urban comforts. To them, the value of the rural forest depended on it being different from the city and its urban one.

Yet where urban environmental professionals saw the "simplicity" of rural life and people, as some put it, others saw neglect, urban dominance, and corruption. One night at Luis and Flávia's house on the BR-364, for example, one of their grown sons who was there for dinner explained local politics to me in terms of the gap between the country and the city. He was a low-level local politician in Feijó and a member of a party allied with the PT. Local smallholders consistently told me that the BR-364's paving had improved their lives significantly. But he noted that, here in this rural part of Feijó, the new asphalt had already started to crumble. People were always complaining about its potholes. Yet, he noted, they would still vote for the PT because the party had been good, even if PT politicians "stole" some. When I asked him for clarification, he offered Rio Branco as his evidence, rather than any governmental improvements in rural areas like Feijó's where his parents lived. The city was entirely different than it used to be, he said. It was so much nicer. He told me that this was because of the PT.

At dinner with Luis and Flávia a different night with a different son, my research assistant, Fernanda, asked the group what they thought about Chico Mendes, Acre's famous slain organizer. The son, who was studying history at a university in Feijó, recounted the story of Mendes's murder with respectful solemnity. But Luis offered a different opinion, one that I had heard before: he said that Chico was the laziest person in Acre. He just did not want to do the hard work of tapping rubber, so he became an organizer. The dozen or so family members in the room burst out laughing. But the history student seemed taken with Chico and gently pushed back at his father. He quietly inserted that without Chico and his struggle, Acre would not have any forest left. Luis continued to tease him: he just liked Chico because he was lazy as he was—a jab met with more hearty laughter. Flávia interjected some diverting nuance: the problem was not Chico but the way that he had been used by the PT. In the past, money from the United States had funded the *fazendeiros*' deforestation. After Chico was killed, the foreign money kept coming, but this time it went to the government to protect the forest. But what was there to show for it around them? Not much, was her implication. An opposition party had recently put a dead jaguar in a nearby pothole on the BR-364 to draw attention to just how bad the government was. The road had just been finished, and yet it was already falling apart! Flávia too was rewarded with laughter. Luis followed her lead: there was so much corruption, crime, and robbery. There had never been more robbery, he said, than under the PT. It was almost better when the chainsaw deputy was around. At least he did not kill "normal men," only criminals. Rural areas—like where they lived on the BR-364—suffered most of all, he maintained. The crumbling road was a testament to it.

In Rio Branco too, people pointed out the gap between the urban and rural forest. Almir, a motorbike taxi driver in Rio Branco I once caught a ride with, was one of them. Upon learning that I was a foreigner studying the Acrean effort to protect the forest, he asked if I wanted to know a secret: the government here was "no good." And he had proof. Later, he sought me out to loan me a DVD. It was a recording of a national newscast about the area near his land (what he called his *colônia*) outside the city on the BR-364. The video showed a haggard reporter from another part of Brazil tromping through the forest to interview residents about their difficult lives. To me, the clip seemed to be a routine account of Amazonian poverty and remoteness pitched to a national audience who already understood the Amazon in this way. But to Almir, it was something different: a locally relevant exposé that showed that the state government did not help forest people as it claimed. The newscast was filmed near a "sustainably" managed state forest near his property, whose proceeds were supposed to be shared with local families. But the families did not benefit from it, he told me, indignantly describing a child he knew who had to walk twenty-four kilometers to get to school. In Almir's estimation, the continued poverty exposed that the government "lied": it talked about valuing the forest, it talked about forest citizenship, but it did not help forest people.

Flávia, Luis, and Almir were expressing a widespread critique articulated in different ways by those with divergent political views. Ilzamar Mendes, Chico Mendes's widow, for example, said in a 2010 interview that the PT "took his [Chico Mendes's] fight for itself," leaving rubber tappers, Indigenous Acreans, and rural people "forgotten" and living "in miserable conditions."[35] Some Acrean academics, NGOs, and those connected to the rubber tapper movement articulated related critiques.[36] Morais, for example, writes about the "uses and abuses" of Mendes's image and assesses the governmental effort at constructing a "collective memory" as problematically legitimating Acrean state policies while insufficiently benefiting rural people.[37] In this view, culturally valorizing the forest in cities exacerbated the gap between rural and urban life in Acre in ways that rendered that cultural value suspect.

Avoiding the Forest

The forest's cultural valorization was also undermined by its entrenched devaluation. While some environmental professionals and others revered the forest or were amenable to its governmental valorization, many others in Acre had a very different relationship with it. Even as its cultural valorization seemed to have made some of them less antagonistic to it, they continued to try to avoid it. For some, the forest reminded them of a poverty they felt they or their family had just recently escaped. When they spent time in rural areas, they cleared the forest, avoided it, or saw it as a backdrop to a life best lived outside of it. Many still valued the country, but envisioned themselves not as forest citizens but as ranchers and rural Acreans whose life was distinct and distant from the forest that lived and loomed around their cleared land.[38] In other words, there was another persistent gap, here between the governmental effort to make the living forest valuable and many Acreans' relationships with the forest.[39]

These views resonated with persistent mainstream currents within Brazilian culture about the forest and nature more broadly. Negative views about the forest brought to Acre by those who migrated there—in the late nineteenth and early twentieth century to tap rubber or in the twentieth century to raise cattle—were reinforced by dominant Brazilian discourse, media, policy, and practice. For centuries, the forest had been devalued for reasons I have explored in previous chapters: timber and other natural resource extraction and the conversion of forests into fields have been crucial to Brazil's colonization and the pursuit of economic growth, as well as many people's quest for livelihoods, land, belonging, and wealth.[40] As Jeff Hoelle has shown in his analysis of cattle in Acre, this work is also cultural. "The growth of cattle raising had been accompanied," Hoelle writes, "by a 'cattle culture,' a cattle-centered vision of rural life that was increasingly celebrated, in both the countryside . . . and the city."[41] In both landscapes, Hoelle found people of diverse social classes looking to live the *contri* life in music (*sertanejo*), food (barbequed beef or *churrasco*), and fashion (*cauboi* hats, belts, and boots). "The rancher . . . became the paradigm of success in rural Acre," while rubber tapping and the forest continued to be associated with poverty.[42] The urban forest failed to displace the cultural dominance of cattle and associated cleared land, though it did encourage some people to be less antagonistic to the forest and try to minimize forest clearing.

Take how two men, Edivan and his son Marcelo, viewed the forest. The father and son both raised cattle in areas that they cleared. Edivan was a former rubber tapper in his eighties who—to the chagrin of his Rio Branco–based children—lived alone in rural Xapuri. Edivan felt that ranching had given him a leg up from rubber tapping's poverty and toil. Marcelo was one of Edivan's urban children, but he also loved spending time in the country. A retired police officer, he kept a property (*chácara*) about forty-five minutes from Rio Branco where he spent many weekends. There, his police pension let him try his hand at ranching. He could enjoy it and access some of its social status without incurring too much financial risk.

I met them both through Edivan's son-in-law, a friend of mine. He introduced us at a relative's house in Rio Branco one evening as Edivan and Marcelo were finishing a game of dominoes. Edivan was in town on one of his periodic visits to see a doctor about chronic eye problems and to make his monthly payment on a fine he had received for deforesting (he had a five-year payment plan). His daughter Ana later told me that Edivan was, of course, annoyed to have to make the monthly payment but was happy to have so much open land on which to raise cattle. The fine was worth it.[43]

Pausing their game, Edivan and Marcelo playfully needled me about Amazonian politics and the forest, based perhaps on what they imagined an American studying deforestation might believe. Edivan complained that restrictions on deforestation had made his life in rural Xapuri more difficult. "It is no longer possible to grow anything!" (I did not probe as to why his extensive pastureland was, apparently, unavailable for growing food.) Marcelo joked that this situation was my fault. Everyone laughed, and I offered a performatively innocent "Who me?" Yes, he laughed. Yes, as an American, it was my fault. We in the United States had already cut down all our trees so now the Amazon had to be "the lungs of earth." With exaggerated disapproval, he said that Americans seemed to think that the Amazon was part of the United States. The mood was playful, but the critique—a familiar one to me—was real.

Both Marcelo and Edivan then invited me to visit them so they could show me what the Amazon was "really" like. Both had homes that were adjacent to the forest, yet far from it. Edivan's property was located about three hours from Rio Branco. His home was small and wooden, resembling that of poor smallholders I saw in Feijó. One of his children told me that it had changed little in the decades since everyone else in the family had left for Rio Branco, save for the notable addition of electricity. The tin roof stretched out slightly from the house, creating a small shaded area with room for a table and two chairs, where he liked to play dominoes with any visitor who would accede. But Edivan was not a poor smallholder. Over many years, he had cleared a lot of forest and accumulated hundreds of head of cattle, as well as a number of fines for deforesting. Still, he rejected the *fazendeiro* designation and laughingly demurred when his daughter teased him about getting rich from cattle. His cattle not only made him money; clearing the land also created a place of retirement that was distant from the city and the forest, both of which he sought to avoid.

The way Marcelo kept his land and cattle spoke to a different relationship with the forest than his father's, but one that still kept it at a distance. His land was about forty-five minutes from Rio Branco. On it, he had built a large house out of expensive imported cement and tile with a spacious porch wrapping around three sides. There was also a well-maintained corral for a few horses and some four hundred cows. As he showed me around, he told me his plans to double his herd size to eight hundred—just enough to support what he considered to be a "good life." But these he would ranch intensively, he told me, boasting that he could raise up to thirty head of cattle per hectare (compared to the usual one or two) via new rotational grazing techniques he knew. This would obviate any need to deforest more. Despite our initial conversation, he told me that he believed environmental laws should be followed. He had been fined two times for deforesting—the land had been all "*mata, mata, mata*" when he bought it a dozen years earlier, he said, using the dismissive term for forest.[44] But he now had all the requisite permits and did not plan to clear much more. He told me about conserving the forest that lined the nearby river to protect the water and the people downstream and to give the birds a place to live. Perhaps because two of his children were environmental professionals in Rio Branco or because of the effort to culturally valorize the forest (or because he was telling me what he thought I wanted to hear), Marcelo was not openly hostile to the forest. But it primarily served as a backdrop to his planned future. He valued legality and propriety, but not the forest itself.

Both Edivan's and Marcelo's properties served as important places for their Rio Branco–based family, not because of the forest, but rather despite it. Sometimes they invited me to spend a weekend day with them. We would gather on their cleared land, carved from the forest, to escape hot and cramped urban confines and their persistent fears about crime in the city. Women cooked in the kitchen and gave each other pedicures and manicures. Men grilled outside and watched soccer on TV. Adults played on their phones while lounging on porch hammocks. Children played outside, but in the cleared area close to the house, not in the forest. They rode horses and visited cattle, donning cowboy hats and boots for the occasion. Days at Marcelo's and Edivan's enacted culturally dominant and gendered practices of rural tranquility and plenty.

Distance from the forest was integral to this rural tableau. No one wanted to go near it, though it was always visible at the field's edge. Edivan's house was located close to the renowned Chico Mendes Extractive Reserve and only a few kilometers from a famous ecotourism site, where guests stayed in a forest-themed inn and could accompany rubber tappers on their treks through the forest. But no one in the family had visited the extractive reserve. I asked Edivan's daughter Ana about this as we drove home to Rio Branco from Edivan's house one Sunday evening. She recalled an arduous childhood. There had been no electricity or running water. They had washed clothes by hand at a nearby spring and carried water from it in buckets on their heads to the house uphill. They cooked over a charcoal-fired stove and walked two hours to get to school. The drive to Rio Branco that now took two hours then involved an eight-hour bus ride. Her childhood took place in the same home as Edivan now lived, but it looked very different from the leisurely day we had just spent there. Her memories served as her explanation of why she and her family had never visited the extractive reserve.

Those from the "*mata*" did not "value" it, she summarized. She was happy to be driving back to the city.

I thought about Ana's childhood while sitting in a new, upscale full-service laundromat that opened near my apartment in Rio Branco the week after my drive with her. A small troupe of women kept the place spotless and all the shiny new washers and dryers spinning and humming in the cool of blasting air conditioning. Pressed shirts, dresses, and slacks hung crisp on hangers waiting for the laundromat's professional clientele to pick them up. I had been eying the place since it opened, and when the washing machine in my apartment broke, it offered an occasion to venture in. As I sat waiting for my laundry and enjoying the artificially cool air, I watched a young professionally dressed couple carry in some comforters to get cleaned. They paused for a long moment, arms around each other, to look at a large photograph that hung on the wall. I looked it at along with them. It was of two women crouching on a dock that was partially submerged in a river or pond lined by trees. The women were washing clothes, one of them with a shirt, perhaps, raised high above her head to bring down hard against the wood—the wash cycle. Either of the women in the photo could have been Ana, I realized, when she was younger and still living with her father; when she, her mother, and sisters had had to wash the family's clothes by hand. "Cool" (*legal*), the man in the laundromat said as he gazed at the photo with some awe. But his companion did not seem to agree. "I couldn't do it," she replied, perhaps seeing in it an alternate life in the forest or the lives of her mother or grandmother. Despite efforts to increase the forest's cultural value in the city, for many it recalled often gendered familial histories and presents of toil. It was to be avoided or captured in an image to be hung on the wall and looked at from the comfort of air-conditioned urban spaces.

Cinema in the Forest

Some rural people also avoided the forest, including some forest beneficiaries. Manuel, the president of a particularly active rural producers' association, spent a good deal of time in cities—Feijó, Rio Branco, Brasília. He went to them to lobby for benefits and to visit his mother and girlfriend. But his true home, he told me, was in the "country." Manuel described his understanding of the value of country life in a community meeting that the CIFOR research team held with his association, attended by about two dozen association members. The government, he said, was "giving more" every day because of the work he did to access benefits. Before, their rural community had no lights, no road, no school. Now, he said, they had almost everything. In the country, there was no hunger anymore, whereas people who moved to the city had nothing to eat. Moreover, heat, crime, and daily struggle made urban life uncomfortable, he said. In contrast, his association members could keep their windows open to breathe fresh air and "feel good." Association members I interviewed later often agreed with this characterization: with electricity, the paved BR-364, and government benefits of all kinds, their rural area had become a better place to live than the city.

These rural improvements were, however, separate from the forest. Manuel and the members of his association whom I spoke with made similar assertions. Life was good, better than that in the city, but not because of their relations with the forest. In fact, Manuel told me, thanks to electricity and other improvements, the country was "basically urban." His bedroom had an air conditioner and a satellite dish to get HBO. Manuel also dreamed of ushering in other improvements—a soccer field and a "cinema in the forest." That cinema would actually be in a field next to the forest, Manuel clarified. There, people could gather to watch movies, have barbecues, eat popcorn, and drink soda. The forest was sidelined in this cinematic vision. It served as a symbolic backdrop and catchy name.

This marginalization was consistent with the forest's comparatively small role in the lives of many association members I spoke with, even though they were all former rubber tappers. Thanks to Manuel's efforts at accessing government benefits, some reported no longer needing to cut the forest down or to collect products from it. Many did not have to go into it at all. One former tapper described how he was now "afraid of going to the forest." Several others explained that they used Bolsa Família and other monetary benefits to buy food they felt that they could no longer grow because of deforestation restrictions, as well as other necessities (e.g., clothes and school materials) for which they would have had to sell crops or cattle to pay. The country was becoming a place to live an "urban" lifestyle, as Manuel put it, not in relationship with the forest—or by producing more—but by receiving forest-funded and other government benefits. The benefits enabled some rural residents to avoid the forest, like their urban counterparts.

Forest as Waste and Wealth

Related views infiltrated the ranks of some of the environmental professionals implementing SISA. Bruno, for example, was a fishing engineer from southern Brazil lured to work for the Acrean government by the prospect of building something new in Acre's burgeoning aquaculture sector—a key part of the state's forest protection strategy. As we drove to visit some fishponds, Bruno started talking about where he came from: the south of Brazil. It was like a "first world country, like the United States." Unlike his part of Brazil, Acre was poor. And it was poor because of the forest, he explained, gesturing to the trees left between stretches of cleared pastureland as we drove. He clarified: Brazil made its money from "*agropecuária*"—industrial agriculture and cattle ranching. But now, people could no longer clear the forest to plant crops or raise cattle, as they had in the south of Brazil, he said regretfully, because the forest was protected. Restrictions on deforestation were the cause of Acre's poverty. The forest was evidence of wasted land.

Yet, just a little while later, as we drove through a particularly forested stretch of the road, Bruno's take on the forest seemed quite different: it had a lot of "biodiversity," he marveled. It was "incredible." The nearby Purus River, he went on enthusiastically, had the most fish species in the world. Then he mused: maybe someday they would figure out how to turn that "*riqueza*," that wealth, into money. The seeming contradiction within Bruno's views reflected the gap between the urban forest of which he was a part and dominant views of the rural one—a gap that could persist even for those government employees working to protect the forest.

Conclusion: The Country and the City

Making the Acrean forest valuable was a cultural project as much as it was an economic one. As conjured in urban public space, culture, and employment, the Acrean urban forest sought to reposition the forest, its history, and people at the center of Acrean identity. This effort was premised on the idea that culturally valuing the forest in the city would enable the protection of the living rural one outside of it. It was a value-making proposition, given the international rise of market-based efforts to address the climate crisis and the attendant interest in paying for emissions reductions and in new environmental commodities, including forest carbon offsets. The forest helped to make Rio Branco a nicer place to live—the outside money the forest garnered, the jobs and travel its governance entailed, the forest-themed heritage it inspired, and the sense of meaning that it could bring to urban life.

Yet the urban forest could also feel divorced from both the ongoing difficulties of rural life and many urban and rural people's divergent socioenvironmental relations with the living forest, which often entailed avoiding it. The urban forest seemed to conjure an idealized forest of the past, and associated forms of rural life and meaning, as if they were ongoing—unaffected by changing socioenvironmental relations in rural areas. These relations included those surrounding rubber's decline, cattle ranching's rise, new and expanded government benefits, new roads and other infrastructures, declining but continued poverty, and a deforestation rate that was, when I was in Acre, starting to creep back up.

Moreover, while many rural people benefited from forest protection programs and the PT's larger emphasis on social inclusion and redistribution, they generally did not seem to gain as much as many city residents did. This is not to say, though, that rural benefits were meaningless. For example, the *florestania*-named adult education program in which Flávia taught, the school building in which the program took place, and her own education to become a teacher were all linked to governmental efforts to benefit rural people—to create citizenship in the forest through extending social rights and education into rural regions. Yet the gap between rural and urban life increased through the years of governmental efforts to valorize the forest, including culturally.[45] The urban forest was for those who could attend sports events at Rio Branco's Arena of the Forest, surf its Digital Forest Wi-Fi on a smartphone, or lounge in its Forest People's Plaza. It was for those who could study and govern forests in urban classrooms and air-conditioned city offices. And it was for those who could travel on the BR-364 to rural areas for work or a relaxing weekend, and then return to the comforts of the city. In response to, among other things, this urban-rural inequity and entrenched rural poverty, rural Acreans continued to migrate to the city. In this, the urban forest exacerbated ongoing dynamics of urbanization and rural exodus as people sought out jobs, infrastructure, social services, and education in the city—as they sought out inclusion in the urban life that the forest helped to create.[46] The urban forest drew people away from the rural one. In the gap between them, Acreans like Luis, Flávia, and Almir saw robbery and lies.

Beyond elucidating contemporaneous Brazilian politics and demographic change, though, the urban forest also points to the difficulties of cultural valorization as a strategy of green capitalism. Economic change is necessarily cultural—a necessity embraced in the Acrean effort to culturally valorize the forest in its capital city. Yet that cultural valorization effort also shows how difficult cultural change can be, particularly when the dominant culture is tied to profit and power. In Acre, the urban forest failed to transform the culturally dominant valorization of the city and the cleared countryside. It failed in its goal to center the forest and forest people within Acrean culture. Urban and rural people alike still sought to avoid or escape the forest, even if they no longer cut it down as readily. With the cultural status quo insufficiently changed, social inclusion and forest citizenship failed to fully manifest. Moreover, the gap between rural and urban life also indicated how this dominant culture defined the Acrean effort to make the forest valuable. In other words, cultural valorization is cultural as well. It is shaped and constrained by the culture in which it occurs. These cultural constraints point to the limitations of green capitalism more generally to effect cultural and economic change and to reshape socioenvironmental relations.

In Acre, these constraints undermined both the urban and the rural forest. Rio Branco's Digital Forest was barely functional when I was there. The Arena of the Forest was run-down, and the Forest People's House was often shuttered. Newly paved one year, the BR-364 that connected rural and urban areas was crumbling into potholes by the next. The living forest was also in trouble. Even though the government worked to valorize it culturally and monetarily, deforestation was increasing. Smoke hazed and polluted the air all around during peak moments of the summer burning season, in both rural areas and urban ones.[47] In Rio Branco, friends worried about the air quality. Those who could afford it started talking about buying air filters for their childrens' bedrooms and maybe even their own. Ash fell softly, like snow, on the city's cars, streets, and forest monuments alike. The sun glowed smog-orange. Both the urban and rural forest could seem to disappear into the smoke.

Afterword

CARBON BUST

A few days before I left Acre in the fall of 2014, I had lunch with a friend. He drove me from the IMC's centrally located office in Rio Branco, where he worked and where I had spent the morning, to his house on the outskirts of the city. As we drove, I noted charred land on the side of the road. It was early September, and the air was hot and dry. The rains had not yet begun (they would, everyone hoped, arrive later that month or the next). But some people in the areas surrounding Rio Branco were not waiting to start burning, as was custom. Instead, they were already starting fires. The burning season had arrived early. At a government workshop I had attended the previous day, the State Secretary of the Environment, Carlos Edegard de Deus, had talked about it—how there had already been twenty straight days of fires, and how he was meeting with "rural producers" right after the workshop to ask them to refrain from burning until the first rain. They were legally allowed to burn up to a hectare, he emphasized, since it was part of their traditional cultivation practices. But he would ask them to wait to start burning in order to protect "human health"—a nod to fire's known and feared degradation of local air quality, more than the climate crisis. But it was an election year, and so people thought they could get away with burning early and more, my friend told me as we drove to his home. Government officials did not like to issue fines right before a vote.

Some of my friend's neighbors were among those who were burning early. Their goal was primarily to maintain their cattle pastures by holding the regrowing forest at bay. He worried that one of their fires might jump the road and reach his land. In heat like this, fires could be hard to contain. Even if they did not spread, the fires still impacted him. He said that even a brief fire made the smell of smoke stay in his nose. It lingered there, a reminder of what is sensed and felt but cannot always easily be seen—harm to land and lungs from past fires, as well as the anxious anticipation of future harms that each burning season increasingly seemed to foreshadow.

When we arrived at my friend's house, though, there were lots of trees. He told me that he had recently seen an old aerial shot of the land where his house now stood. Thirty years ago, when the image was taken, it had all been cleared as pasture. Now the forest grew thick around us. It quelled anxious fire-filled thoughts. Trees shaded and transpired, making his home and small yard notably cooler than the city we had just left and the road on which we drove from it.

This drive offered a premonition of what has happened in Acre since then. In the years since I left, deforestation increased significantly there, as it did in many other parts of the Brazilian Amazon.[1] As a state law, SISA continued and the Acrean government continued to distribute some benefits meant to incentivize reduced deforestation and increase production of all manner of crops and forest products. Officials continued to hold meetings and workshops that sought to expand smallholder production of forest and nonforest products, like açaí.[2] I have explored such efforts as part of an inclusive green capitalism. Yet in the late 2010s and early 2020s, many Acreans opted instead for another kind of inclusion: inclusion in the dominant deforestation- and fire-linked forms of development and land use championed by Bolsonaro and the Acrean politicians elected to run the state government in 2018. At the same time, there are many others—within and outside of the state government—who also continued to work for different kinds of futures that value living forests. Moreover, trees continue to grow and live in places like the land around my friend's house. In this, what has happened in Acre is not so different from what is happening all over the world, with continued governmental and corporate investment in and use of fossil fuels coupled with the continued expansion of green capitalism. And it is not so different from those of us who fly in planes, drive cars, or eat meat, while lamenting and suffering from climate change. Such contradictions are both relatably human and reflective of the entanglement of green and more traditional forms of capitalism. They mirror and deepen the climate crisis during this period of early climate change.

This book's exploration of the Acrean effort to valorize the living forest and its carbon seeks to elucidate this period. It asks what happens when we do not simply view carbon as something bad that is changing the climate, nor as just a standardized commodity that epitomizes green capitalism. It explores what happens when, instead, we view it relationally: how carbon's state of being—in trees, in the subsurface, or in the atmosphere—comes into view as an outcome of and a way into understanding socioenvironmental relations that vary from place to place, even as they may be broadly patterned. This approach, I suggest, illuminates how green capitalist efforts to produce less greenhouse gas emissions can entail trying to produce more—hewing to the familiar productivist focus on intensification and economic growth. It demonstrates how making the forest and its carbon monetarily valuable in Acre entailed creating an environmentally premised welfare state and associated forms of green redistribution and politics of citizenship, instead of anticipated forms of private property and markets. It shows how green capitalism expands through inclusive politics and linked forms of exclusion. It points to the promise and limits of culturally valorizing the forest. And it elucidates how the socioenvironmental relations enlisted in forest carbon's valorization are entangled with past and ongoing land use histories.

At times, those histories could seem to haunt, ghost-like.[3] Some of them were actively conjured, while others could not be avoided. The rubber narrative and urban forest invoked rubber for international and local audiences, but they downplayed many other things: the colonizing violence of rubber extraction and the impoverishment surrounding its collapse; the inaccessibility of secure land rights in a place where land and rights to it can be economically, culturally, and politically vital; cattle, their cultural and economic value and extensive pastures; and deforestation, seen as the prelude and prerequisite to existence itself.

Forest loss also seemed to haunt from the future. Deforestation seemed to loom ahead when I was in Acre. The forest was not gone. It was not a relic of the past. Rather, it was all around, covering more than 85 percent of the state. Yet it was hard not to anticipate it. Once, after walking out of the dense dark of forest canopy into a recently burned field off the BR-364, I was surrounded by the ghostly charred remnants of trees. They were not cut down to low stumps but rather stood at human height, smoldering black apparitions against the bright green of the sprouting pasture grass that surrounded them. Cattle grazed peacefully around them. The smallholders who did the burning were, I assumed, following their predecessors' example of how to use and care for the land. They were likely looking forward toward what they hoped would be a more prosperous future. But I found myself pondering a different future—one that seemed to also await the fragment of forest from which I had just emerged. To me, the charred trees appeared as an indication of the end of a powerful socioenvironmental movement, colonization that enticed clearing land to claim it, forest loss, and climate crisis. Here I was thinking in an elegiac register that anticipates future ecological devastation with alarm and dread. It is a common one in this time of early climate change. "Deep green elegy," Timothy Morton writes, "presupposes the very loss it wants to prevent."[4]

A mixture of nostalgia, commitment, and elegiac dread colored much of my research. As in Brazil more broadly and some other parts of South America, the 2000s had been a boom time for left-of-center political power and economic growth in Acre. It was the heyday of the Government of the Forest. It was also a time when carbon markets, forest carbon offsets, and other forms of environmental marketization seemed to promise tantalizing new possibilities for stabilizing the climate via green capitalism. It could all be enchanting.[5] Yet, by the time I went to Acre in 2012, all that was beginning to change.

I sometimes felt that I arrived late, at the end of a really good party, when the music has been turned down, the lights have been turned up, most people have left, and the floor is a little bit sticky. If I had lived in Acre during the rubber tapper movement in the 1980s or during the height of the Government of the Forest in the 2000s, I would have written a different book. I might have written about the rubber tapper movement as forging multispecies relations and lifeways among "capitalist ruins."[6] Instead, most of my research occurred in a time when achievements of the past decade, in terms of reducing deforestation and poverty, were starting to feel precarious, even as state efforts to make the forest and its carbon valuable intensified. While Acrean officials articulated the rubber narrative internationally and constructed the urban forest internally, privately some of them recognized that the anticipated carbon boom was not about to happen. California was not going to accept Acrean forest carbon offsets anytime soon, and deforestation was on the rise. Even as those in the Forest People's House in 2014 celebrated Acreanidade and *florestania*, others who worked in the state's environmental bureaucracy and allied NGOs worried and whispered about rising deforestation, and about the viability of linked efforts, like state-subsidized aquaculture. I was not the only one envisioning future forest loss.

In the years since, Amazonian deforestation increased. The climate crisis has deepened. And, at the same time, green capitalism has expanded. The early 2020s saw a second wave of interest in forest carbon offsets and credits, other forest-focused "nature-based solutions," and the development of a "bioeconomy" in Amazonia.[7] In particular, the voluntary offset market grew quite precipitously as concern about the changing climate deepened, with the prospect for green capitalist profits seeming to increase along with it.[8] Despite popular exposés of their problems, offsets appear with some frequency in daily capitalist life.[9] We are routinely prompted to spend a few extra dollars to offset our carbon pollution when buying airplane tickets or walking through airports, for example (figure A.1). With the click of a mouse, or the tap of a finger to a phone screen, we can say "yes" to the extra charge and perhaps feel uneasily, briefly absolved, before scrolling on. The work it takes to actually hold forest carbon in place remains obscured in these easy gestures.

The Acrean effort to hold forest carbon in place elucidates some of that work and its challenges. In this, it holds lessons for this time of green capitalist expansion in early climate change. As I have explored, some of that work entailed trying to support and create certain types of socioenvironmental relations. These relations, though, did not involve just the forms of standardization and privatization often associated with the development of markets and commodities. The Acrean state did not simply pay people to forgo deforestation, or assume—recalling Jens Stoltenberg's quotation with which I started this book—that they knew how to not cut down trees. Neither did it seek to maximize carbon sequestration in people-free protected areas or tree plantations. Such dispossessive forms were not its major key. Instead, to valorize the living forest and its carbon, state officials sought to construct a form of green capitalism that included smallholders, nonhuman species (like rubber trees, mucuna seeds, Amazonian river fish, and açaí berries), roads like the BR-364 and other infrastructures, an urban forest composed of forest-focused public spaces and employment, forest citizenship, and a nascent environmental welfare state. These were among

Figure A.1 Airport carbon offset advertisement. Photo by Maron E. Greenleaf.

the relations posited as capable of effectively holding forest carbon in place, and of doing so inclusively—in ways that would benefit smallholders and enable sustainable low-carbon rural development. The inadequacies and contradictions of this inclusive green capitalism may be taken less as an indication of failure than of how capitalism is both changing and remaining the same in the Anthropocene.

The Acrean approach to valorizing the forest and its carbon entailed making speculative promises, both externally and internally. Trying to sell carbon offsets or otherwise attract outside funding to keep forest carbon sequestered required making a conjecture: that the socioenvironmental relations enabling forests to live and carbon to remain sequestered in them would continue. Attempting to maintain these relations and create new ones involved promising smallholders and other rural forest beneficiaries that refraining from deforestation and using prescribed methods to increase production from forests and fields would pay off—that the forest would, finally, succeed at being the long-promised "passport to the future," as the Secretary of Forests and Extractivism under the Government of the Forest had put it, and the way to ensure economic security and belonging.[10] Each promise relied on the other: forest carbon's continued sequestration would fund forest benefits, which would then enable further sequestration. Both entailed positioning the Acrean state as forest carbon's best arbiter and steward—the entity most capable of increasing, accessing, and then effectively distributing its new international value, facilitating its continued sequestration in socially inclusive ways and further augmenting its value.

That neither promise panned out as planned should not be a surprise. Capitalist processes continue through inclusion—of new people, places, and valorizing practices. And states often fail to achieve their stated social and environmental goals, even with the best of intentions and the most competent of staff. They often do not have the power that they appear to, especially over landscapes like forests that are still more valuable dead than alive, and when reliant on international markets and public funding.[11] Forest carbon offers a case in point. As a strange new tropical forest resource, forest carbon went bust in Acre before it ever really boomed. Accessing international carbon financing was part of the left-of-center Government of the Forest's claim to legitimacy. But international institutions, offset buyers, and funders did not provide anticipated levels of funding, either through carbon markets or other means.[12] I set out to do a supply chain analysis of forest carbon offsets, but the chain never quite formed. Acre did not sell any offsets. California did not create a legal mechanism for allowing them into its carbon market until 2019, by which point deforestation was already surging in Acre and elsewhere in the Amazon.

The impact of carbon markets' failure to manifest on SISA and connected government officials could be palpable. They knew that they were at the forefront of global REDD+ efforts, and yet they were still just getting by, with an overworked staff and limited access to international funding. A former high-level SISA administrator described his experience to me this way: he believed the state government had completed all the "technical, scientific, institutional, political, and legal" components needed to sell offsets, but "no one [was] buying" them. It was frustrating:

> I'm perplexed. I go crazy seeing a state like Acre invest more than other states and countries in doing this process of transition [from deforestation] ... and no one wants to help us ... because no one wants to pay. And we are paying. 700 thousand people [the approximate population of Acre] are spending billions of dollars to reduce deforestation. And no one pays for this. Only us.[13]

Acre had only received funding for a little bit of the state's reductions in carbon emissions, most prominently through the KfW payments. But that had only paid for a "*carboninho*," as the former administrator put it. The failure to sell offsets or be otherwise compensated for reducing forest carbon emissions meant, to him, that Acreans had paid for the rest. Those of us outside of Acre were (and are) dependent on the Amazon for any semblance of climatic stability, and therefore on those who protect it or at least forgo felling it. Yet we were unwilling to pay.

Or at least unwilling to pay much. At a small meeting in Rio Branco in 2014, an American from the California Public Utilities Commission gave a presentation on California's cap-and-trade program to Acrean officials working on SISA. The challenge, he said, was that the price of carbon in the California market was so low, only US$12 per ton. Many of the bureaucrats smiled and chuckled. To address his confused look, one of them explained, "You say that US$12 per ton is low, but here we think, *no*! That is really high!" The voluntary carbon market was only paying US$5 per ton, he said, and Acre had only received US$5 per ton through the KfW payments (and, since the agreements required the Acrean government to retire an equivalent amount of carbon, the price was effectively US$2.50 per ton). Another former SISA administrator told me over lunch that without the admission of Acrean offsets into California's or another compliance carbon market, SISA would remain small and frail. The value of Acre's forest carbon would continue to be no match for that of cattle and soy. In its capacity to protect the living forest, forest carbon might not be an extractive commodity, then, but it echoed extractive commodities in its reliance on powerful, outside buyers and in the socioenvironmental impacts when those buyers failed to appear. First rubber, then forest carbon. Their booms, quasi-booms, and busts developed through and re-created global inequalities between places like Acre and California.[14]

In the wake of forest carbon's low monetary value, many Acreans I knew had grown wary of government discourse and promises about the value of the forest—even those who agreed that the forest should be protected. Vote them all out, as one friend, a committed leftist, put it. She supported forest protection but was exasperated by the unfulfilled promises of the longtime PT-dominated government. It was time for a change, come what may. In this frustration, concerns about continued rural poverty and the inadequacy of government benefits could be subsumed into broader critiques of the status quo. For many reasons unrelated to the forest—a flagging economy, exhaustion with the PT, and concerns about corruption and crime among them—a higher percentage of the Acrean electorate (over 77 percent) voted for Bolsonaro than that in any other state in 2018.[15] In Acre that year, the PT also lost the governor's office for the first time in two decades, with voters electing Bolsonaro's ally Gladson Cameli, whose powerful political family had been linked to logging, harming Indigenous people and the environment, and corruption.[16] While Cameli's platform included "awareness, monitoring, and investment in environmentally sustainable activities," he criticized environmental agencies for "relentlessly persecuting rural producers."[17] Upon his 2018 election, he met with agribusiness representatives in the capital of Rondônia—the state to Acre's south, known for its rapid deforestation and associated violence against Indigenous and other rural people linked to the paving of the BR-364. There, he declared, "the economic salvation of Acre is agribusiness. Rondônia, our neighbor and brother, is the proof."[18] Upon becoming governor in 2019, he reportedly sought to eliminate the IMC, SISA's administrative body.[19]

The PT's 2018 loss in Acre and elsewhere in Brazil was part of a larger Latin American trend. As Kregg Hetherington describes it, "Throughout the region, the governments of the left turn fell apart, each more spectacularly than the last, ushering in regimes for whom environmental deregulation was an unquestionable good."[20] In the Amazon, these results echoed in "blood and fire."[21] In the years of the Bolsonaro administration (2019–2022), violence against Indigenous people and others working to protect the forest intensified.[22] So too did deforestation. In 2021, for example, Acre's deforestation rate was at its highest in almost two decades (though some Acrean government interlocutors told me that without SISA, deforestation would have increased even more).[23] Since forest emissions increased so substantially, Acre stopped receiving "payments for emissions reductions" via the German- and UK-funded REM Programme—a loss estimated at R$58 million.[24] The state-sponsored aquaculture complex Peixes da Amazônia—"the most modern aquaculture complex in Brazil"—shuttered, becoming, as one news source put it, a "cemetery."[25]

Things may now be changing. In a closely watched and closely divided election in 2022, Brazilian voters ousted Bolsonaro and voted back in PT leader Lula, under whose first two presidential administrations (2003–2010) Amazonian deforestation had fallen precipitously. However, the Acrean electorate doubled down: 70 percent voted for Bolsonaro while also reelecting Gladson Cameli over their previous governor Jorge Viana—the governor and political leader of the Government of the Forest.[26] Still, Lula—who promised to prioritize Amazonian conservation—became their president, and the Cameli administration, no doubt attuned to the changing federal political climate, recommitted to pursuing agribusiness-linked economic growth, but in ways that protected the forest.[27] The Cameli administration simultaneously publicized its efforts to "strengthen the transparency" of SISA governance and advertised Acre as "the new agricultural frontier of soy."[28] Reconciling these public commitments, Cameli adopted a productivist logic that recalled that of the early SISA administrators I knew: by improving "degraded land," he said publicly, "we don't need to cut down any trees in our forest to plant soybeans, corn, and a myriad of other agricultural crops." Moreover, he articulated the connection between agribusiness and climatic stability: "When the environment is treated with attention and care, agribusiness, which depends on rainfall and other natural factors, is also benefited."[29] Despite Acrean deforestation remaining high, in 2023, the REM Programme's European administrators came to an agreement with Acre's government to pay it millions of euros after it signed a pledge to reduce deforestation by at least 10 percent each year starting in 2024 and 50 percent total by 2027.[30] The state Minister of Planning called the agreement "a clear indication that the state has united to change the reality of Acre."[31] In 2023, the Acrean government also signed an agreement with the carbon market NGO Emergent, "pav[ing] the way for a binding Emissions Reduction Purchase Agreement (ERPA) to supply up to 10 million high-integrity forest carbon credits . . . for years 2023–2026" to a group whose members include Amazon, BlackRock, Walmart, and Airbnb.[32]

We shall see. As I write this in July 2023, deforestation has fallen in the Amazon overall and in Acre as compared to this point in 2022.[33] But there are many obstacles to keeping deforestation in check. The interplay of climate change with regional hydrological and deforestation dynamics makes it more difficult to contain fires.[34] Moreover, many of the underlying socioenvironmental relations driving deforestation, around cattle ranching and agribusiness, remain intact. These are still more politically powerful and better funded than those relations militating for forest carbon's continued sequestration. Carbon is still more monetarily valuable when extracted from the earth than when sequestered in it and the trees that grow from it. The cattle and crops that can be grazed and grown in deforested spaces, and the land claims that can be asserted there, are more valuable too. This remains no truer in the Amazon than in most other places, despite the special attention and critique the region and its residents receive.

Part of the reason that forest carbon's value remains low is because a key question of green capitalism remains unanswered. It is a question that a woman asked at the Acrean government meeting attended by the California Public Utilities Commission speaker: who will pay? The woman had been contracted by the IMC to help them develop another program under SISA, this one focused on water—the "hydrological services" that the forest "provided," she explained. The value of water is "immeasurable," she said, but it was hard to actually "valorize" it. Yet there were ways to assign it monetary worth—for example, the cost of bringing water from an origin point to where it is consumed (what she called the "cost of the trip"). The real challenge, she told us, was different. It was "finding the payers." "We are all potential payers because we use water," she said. "We are, all of us, beneficiaries of the system." But who would actually pay? It was the same question SISA's administrators were asking about forest carbon.

All of those trying to imagine and enact green capitalist solutions to environmental crises face some version of this question. This common conundrum points to some of the good reasons to be critical of forest carbon offsets and other iterations of green capitalism that continue to proliferate. It also points to the good faith efforts of many of those seeking to answer it. There were, no doubt, some in Acre seeking to enrich themselves through conjuring the forest's value internationally and in Rio Branco, even as the forest burned in rural areas. But most of those I knew who were working to valorize the living forest and its carbon were convinced that their work was important to protect the forest, combat climate change, and benefit smallholders, as well as Indigenous people and extractivists—all while working within a system of which they too were often critical. One IMC staff person told me that she thought that the criticisms of SISA were "very much based on a position I think is very ideologically anti-capitalist. . . . They are criticisms that have no response. . . . I am also [anti-capitalist], but unfortunately I have to live within it because [that is] the existing system in our Western culture. There is no way I can make a separate world to live in."[35]

People like her felt that they were doing what they could from their position as administrators in an under-resourced government of a small state as international demand for commodities like beef and soy continued, demand for forest carbon sputtered, and support from politicians flagged. When I questioned the governmental strategy of making forest carbon into a monetarily valuable asset, another state employee responded to me with some exasperation: what alternative did they have? "If you have a better idea to generate income," she said to me, "to make resources flow to these communities, come. You'd be welcome! . . . This romantic idea that we have to think of something outside the market—I really like it [but] I think it hasn't appeared yet. So it would be very welcome." "Maybe tomorrow it will start raining gold or diamonds will appear," she joked. "But still, you would have to take that diamond and sell it in the market, so the market would still be there. . . . If anyone has a better idea, they should let me know. I'm always looking for one."[36] I laughed, unable to offer her an alternative. To those of us who feel like we live life within it, capitalism can appear to have no alternative, beyond some more inclusive and greener version of itself.

Yet there are, of course, alternatives, even if they can be hard for some of us to envision and enact. I heard a version of one of them from a man who had conceptualized the term *florestania* back when he was connected to the Government of the Forest in its early years. When I asked about *florestania* during an interview, I expected him to reinforce my understanding of it as citizenship in the forest. Instead, he wistfully rejected its dominant definition and prominence, as if he regretted ever introducing the concept for the government to enlist. In fact, he said, *florestania* was "contrary" to rights-centered conceptions citizenship. Rather, it was "much more a set of responsibilities that one has with others" than "a set of rights I demand because I am a human being." It was:

> the existing relationships of the forest and the values that emanate from [them], which are not just social relationships between human beings. It is their relationship with nature, nature with nature, animals, water, birds, rivers, from the forests, from the entities and spirits existing in the forests and from the people of the communities that live [in them]. So in this system of life, certain values are born, certain knowledge and this knowledge moves the evolution of life. We can say this . . . is florestania.[37]

Florestania, he went on "was lost the moment it was called forest peoples' citizenship," as the Government of the Forest had done. Yet it "still exist[ed] in a few places hidden in the forest." Those places, though, needed to remain a secret: "When I encounter [it] . . . in some of these places, I won't even tell others that it exists." We laughed our way into silence, knowing, that I, a foreign researcher of forest carbon and efforts to make it valuable internationally, was exactly the type of person he would not tell.

NOTES

Preface

1   I use both *climate change* and *climate crisis* in this book to denote geophysical changes associated with the atmospheric buildup of greenhouse gases on the one hand and their broader effects on the other. Neither term, though, feels adequate. The climate is changing, but *climate change* fails to communicate its extent and urgency. *Climate crisis* seeks to convey that urgency and motivate action, yet it can also cultivate despair. As Joseph Masco writes, "crisis talk" can generate a paralyzing "radical presentism" that "can work to maintain a status quo" by undermining the imagination and enactment of "alternative futures." Masco, "Crisis in Crisis," S73, S75. Something similar could be said about the term *climate emergency*, which I do not use in this book to avoid a confusing proliferation of terms.

2   See Jasanoff, "New Climate for Society," for an analysis of climate change's epistemic challenges and opportunities; Chakrabarty, *Climate of History*, and Pandian, *Possible Anthropology*, for examinations of how it reworks fundamental conceptions about humanity that underwrite disciplines like history and anthropology; Ghosh, *Great Derangement*, for an account of the difficulties of writing about it within established literary forms; and Knox, *Thinking like a Climate*, 6, for an exploration of the difficulties posed by climate as a "form of thought."

3   Attention tends to coalesce around occurrences that are read as events in ways that it does not around more everyday and slower forms of violence. This makes the latter harder to perceive. See Ahmann, "'It's Exhausting to Create an Event out of Nothing'"; Nixon, *Slow Violence and the Environmentalism of the Poor*; and Sahlins, "Return of the Event, Again."

4   Frontline communities tend to face the worst impacts of climate change due to historical and ongoing forms of injustice that can also undermine their ability to recover from or adapt to climate impacts.

5   Lovejoy and Nobre, "Amazon Tipping Point"; Walker, "Collision Course." See also Science Panel for the Amazon, *Amazon Assessment Report 2021* (UN Sustainable Development Solutions Network), edited by Carlos Nobre et al., https://doi.org/10.55161/RWSX6527.

6   There are, of course, other definitions of green capitalism. For example, Scott Prudham defines it as "a set of responses to environmental change and environmentalism that relies on harnessing capital investment, individual choice, and entrepreneurial innovation to the green cause." Prudham, "Pimping Climate Change," 1595.

7   David McDermott Hughes, "Ayn Rand's Climate Movement: Libertarians, Juries, and the End of Fossil Fuels" (lecture, Dartmouth College, Hanover, NH, April 27, 2023).

8   See Tsing, Mathews, and Bubandt, "Patchy Anthropocene."

Introduction

1   The conference was the Thirteenth Session of the Conference of the Parties (COP) to the United Nations Framework Convention on Climate Change (UNFCCC), also known as COP 13. It took place in Bali, Indonesia.

2   Jens Stoltenberg, "Speech at UN Climate Change Conference in Bali," UNFCCC COP, December 12, 2007, https://www.regjeringen.no/en/historical-archive/Stoltenbergs-2nd-Government/Office-of-the-Prime-Minister/taler-og-artikler/2007/speech-at-un-climate-conference-in-bali/id493899/. Stoltenberg was not the first to propose this logic. While "avoided deforestation" was initially excluded from UNFCCC agreements, such as the 1997 Kyoto Protocol, the idea of integrating reductions from deforestation into UN climate mitigation had been under discussion in the UN negotiations and related policy spaces for several years—at least since COP 9 in 2003, at which Brazilian scientists presented the idea of "compensated reductions." Santilli et al., "Tropical Deforestation and the Kyoto Protocol." A new Coalition of Rainforest Nations led by Papua New Guinea and Costa Rica then formally presented the idea of an

international market-based mechanism for compensating for reductions in emissions from deforestation—then called RED—at COP 11 in 2005. Pistorius, "From RED to REDD+."

3   In the years after the 2007 conference, REDD+ was adopted into UN climate policy through the 2013 Warsaw Framework for REDD+ and Article 5 of the 2015 Paris Climate Accords.

4   In his 2007 speech, Stoltenberg himself went on to commit Norway to supplying over US$500 million per year to reducing carbon emissions from tropical forests. Stoltenberg, "Speech at UN Climate Change Conference." Norway has been a major financial contributor to REDD+ in the years since.

5   Concern about environmental destruction and efforts to address it have a long history in Brazil, as José Augusto Pádua shows in *Sopro de destuição*.

6   See Macedo et al., "Decoupling of Deforestation and Soy Production." Rapid deforestation continued in other Brazilian ecosystems, though, including the Cerrado savannah. Strassburg et al., "Moment of Truth for the Cerrado Hotspot." In this, the Cerrado could seem like a "sacrifice zone" enabling Amazonian conservation. Oliveira and Hecht, "Sacred Groves, Sacrifice Zones and Soy Production."

7   Nepstad et al., "End of Deforestation in the Brazilian Amazon"; Justin Gillis, "Restored Forests Breathe Life into Efforts against Climate Change," *New York Times*, December 24, 2014.

8   Malhi et al., "Regional Variation of Aboveground Live Biomass," 1120.

9   See, for example, Climate Focus, "Acre, Brazil: Subnational Leader in REDD+," May 2013, https://gcftf.org/wp-content/uploads/2020/12/Acre_Brazil_Subnational_Leader_in_REDD.pdf; Kelley Hamrick, "Acre and Goliath: One Brazilian State Struggles to End Deforestation," Ecosystem Marketplace, May 5, 2014, https://www.ecosystemmarketplace.com/articles/acre-and-goliath-one-brazilian-state-struggles-to-end-deforestation/; Slayde Hawkins, "Brazilian State Lays Foundation for Nature-Based Economy," Ecosystem Marketplace, December 7, 2010, https://www.ecosystemmarketplace.com/articles/brazilian-state-lays-foundation-for-nature-based-economy/.

10 KfW, "REDD Early-Movers Acre Fact Sheet," 2017, 10, accessed July 1, 2021, https://www.kfw-entwicklungsbank.de/PDF/Entwicklungsfinanzierung/Themen-NEU/REDD-Early-Movers-Acre-Fact-Sheet.pdf.

11 The 2010 SISA law lays out several ecosystem services other than carbon sequestration: natural beauty, socio-biodiversity, hydrological services, climate regulation, soil conservation and improvement, and traditional ecological knowledge. Government of Acre, "Lei N. 2.308," capítulo I.

12 For another anthropological analysis of REDD+ in Acre and elsewhere, see Rodrigues Machaqueiro, *Carbon Calculation*, which deftly analyzes REDD+ as advancing neoliberal transnational governance.

13 Graeber, *Toward an Anthropological Theory of Value*, 2.

14 In this, the Acrean effort to value forest carbon can be understood as part of long-standing rural development efforts critiqued by anthropologists and others. See, for example, Escobar, *Encountering Development*; Ferguson, *Anti-Politics Machine*; and Li, *Will to Improve*.

15 While not yet officially adopted as the geological name of this epoch, in the environmental humanities and humanistic social sciences, the term *Anthropocene* has been both generative and controversial, especially for implying species-wide culpability for environmental harm primarily wrought by the actions of a relatively small number of people, often as part of long-standing forms of colonization. See, for example, Danowski and Viveiros de Castro, *Ends of the World*; Davis and Todd, "On the Importance of a Date"; Haraway et al., "Anthropologists Are Talking"; Kawa, *Amazonia in the Anthropocene*; Marras and Taddei, *O Antropoceno*; Tsing, Mathews, and Bubandt, "Patchy Anthropocene"; Mathews, "Anthropology and the Anthropocene"; and Whyte, "Settler Colonialism, Ecology, and Environmental Injustice."

16 Appel, *Licit Life of Capitalism*, 2. See also, for example, Ekers and Prudham, "Socioecological Fix"; and McCarthy, "Socioecological Fix to Capitalist Crisis and Climate Change?"

17 Moore, "Rise of Cheap Nature." See also McCarthy and Prudham, "Neoliberal Nature."

18 On the role of states in carbon markets see, for example, Bryant, "Politics of Carbon Market Design"; and McElwee et al., "Payments for Environmental Services." Moreover, the categories of public and private and the distinction between them are themselves constructed. They are, Lisa Rofel and Sylvia Yanagisako show, "forged by historically specific processes, including the formation of differentiated transnational capitalist projects" in ways that position states as central to capitalism's varied development. Rofel and Yanagisako, *Fabricating Transnational Capitalism*, 10.

19 The influential economic theory here included the Coase theorem. Named for American economist Ronald Coase, the Coase theorem asserts that, in the presence of clearly defined property rights and the absence of transaction costs, parties can negotiate to reach an optimal outcome that will minimize the effects of negative externalities, like pollution. I read part of Coase's argument in an environment law course but had already heard his name: the law student listserv for buying and selling things was called Coase's List. Such was the broad cultural influence of law and economics

theory at NYU Law. Coase, "Problem of Social Cost." See also Dales, *Pollution, Property and Prices*; Demsetz, "Toward a Theory of Property Rights."

20 On the ways that economic theory can shape the world, see Appel, "Offshore Work"; Çalışkan and Callon, "Economization, Part 1," and "Economization, Part 2"; Callon, *Laws of the Markets*; MacKenzie, *An Engine, Not a Camera*; and Miyazaki, *Arbitraging Japan*.

21 The terms *carbon offset* and *carbon credit* are often used interchangeably, but they may also mean different things. Here is one commonly made distinction between them: offsets represent GHG reductions or removals bought to compensate for a polluter's GHG emissions whereas credits are quantifications of GHG reductions or removals. Credits may be traded within carbon markets, sold as offsets, or purchased to retire without being meant to compensate for pollution through payments for emissions reductions. I generally focus on offsets in this book, as well as payments for emissions reductions.

22 The percentage of their GHG emissions a polluter may cover with offset purchases in carbon markets varies significantly. In Aotearoa New Zealand, for example, regulated entities can cover 100 percent of their emissions with offset purchases. Ministry for the Environment, "New Zealand Emissions Trading Scheme," accessed July 22, 2022, https://environment.govt.nz/what-government-is-doing/areas-of-work/climate-change/ets/. In California, entities can only cover a much smaller percentage of their emissions (between 5 and 8 percent) with offsets, depending on the year. California Air Resources Board, "Compliance Offset Program," accessed July 22, 2022, https://ww2.arb.ca.gov/our-work/programs/compliance-offset-program/about. Yet, a CARB official told me that even this relatively small percentage is important to cost containment: it keeps the price of pollution allowances down for regulated emitters since offset carbon reductions and removals are often cheaper than reductions made by emitters themselves. This is one major reason that industry groups advocate for offsets within cap-and-trade systems.

23 BAU reference levels and baselines are among the many seemingly technical aspects of carbon markets that, in fact, contain many political choices. As Larry Lohmann, "Uncertainty Markets and Carbon Markets," 244, points out, they make something inherently uncertain and unknowable (what would have happened) seem technical and apolitical. Moreover, it creates an incentive to exaggerate how much you would have polluted without carbon financing. See Bumpus, "Matter of Carbon." I discuss these issues more in chapter 1.

24 United Nations Framework Convention on Climate Change, "Emissions Trading," accessed August 14, 2022, https://unfccc.int/process/the-kyoto-protocol/mechanisms/emissions-trading.

25   Fletcher et al., "Questioning REDD+ and the Future of Market-Based Conservation," 674.

26   Aguilar-Støen, "Better Safe than Sorry?"

27   As quoted in Svampa, "Resource Extractivism and Alternatives," 52, and Coronil, *Magical State*, 1, respectively. Fernando Coronil quotes Cabrujas in his explication of the Venezuelan state's use of and entanglement with oil.

28   See Forest Carbon Partnership Facility, "The Readiness Fund," accessed August 14, 2022, https://www.forestcarbonpartnership.org/readiness-fund.

29   The biggest contribution to the Amazon Fund as of this writing came from the Norwegian government, which paid US$1.2 billion into it between 2008 and 2018. After deforestation increased in the Amazon with Jair Bolsonaro's election as president, Norway and others suspended payments. Daniel Boffey, "Norway Halts Amazon Fund Donation in Dispute with Brazil," *The Guardian*, August 16, 2019.

30   This carbon was then retired within the Markit Global Carbon Index, so that it could not be credited again. As part of the deal, the Acrean government also retired the same amount of carbon, meaning, as one SISA administrator pointed out, the state government really received only US$2.5/ton of avoided emissions.

31   See Mauss, *The Gift*.

32   Newell and Paterson, *Climate Capitalism*; Goldstein, *Planetary Improvement*.

33   BlackRock, "Larry Fink's 2021 Letter to CEOs," accessed March 1, 2023, https://www.blackrock.com/us/individual/2021-larry-fink-ceo-letter.

34   BlackRock, "Larry Fink's 2022 Letter to CEOs: The Power of Capitalism," accessed March 1, 2023, https://www.blackrock.com/corporate/investor-relations/larry-fink-ceo-letter.

35   Redford, Padoch, and Sunderland, "Fads, Funding, and Forgetting"; Fletcher et al., "Questioning REDD+," 673; Angelsen et al., "Learning from REDD+."

36   See INPE, "TerraBrasilis," accessed November 13, 2023, http://terrabrasilis.dpi.inpe.br/app/dashboard/deforestation/biomes/legal_amazon/rates. See also Casado, Letícia, and Ernesto Londoño, "Under Brazil's Far-Right Leader, Amazon Protections Slashed and Forests Fall," *New York Times*, July 28, 2019; and Escobar, "Deforestation in the Amazon."

37   An increasing number of large companies are pursuing low-carbon or zero-carbon strategies, many of which rely upon offset purchases in the voluntary carbon market, as documented in Ecosystem Marketplace's

annual *State of the Voluntary Carbon Market* reports. Stephen Donofrio et al., "Voluntary Carbon and the Post-pandemic Recovery," Ecosystem Marketplace, September 21, 2020, https://www.forest-trends.org/publications/state-of-the-voluntary-carbon-markets-2020-2/; Stephen Donofrio et al., "Markets in Motion: State of the Voluntary Carbon Markets 2021," Ecosystem Marketplace, September 15, 2021, https://www.ecosystemmarketplace.com/publications/state-of-the-voluntary-carbon-markets-2021/; Stephen Donofrio, Christopher Daley, and Katherine Lin, "The Art of Integrity: State of the Voluntary Carbon Market 2022," Ecosystem Marketplace, accessed March 7, 2023, https://www.ecosystemmarketplace.com/publications/state-of-the-voluntary-carbon-markets-2022/. In an analysis of 482 large companies with some kind of carbon neutrality pledge, Kreibich and Hermwille found that "offsetting will be a key strategy" for these companies. Kreibich and Hermwille, "Caught in Between," 942. These companies' revenues totaled more than US$16 trillion, which is more than the GDP of China. The value of the voluntary carbon market increased to almost US$2 billion in 2021, almost four times its 2020 value. Donofrio, Daley, and Lin, "The Art of Integrity." About 30 percent of carbon offsets sold in the voluntary market in 2020 were from forest protection, up from 10 percent five years previously. World Economic Forum, "Nature and Net Zero," accessed July 1, 2022, https://www3.weforum.org/docs/WEF_Consultation_Nature_and_Net_Zero_2021.pdf.

New, and often linked, organizations and standards have been created to facilitate these transactions, including the REDD+ Environmental Excellence Standard. See Architecture for REDD+ Transactions, "Our Standard," accessed November 13, 2023, https://www.artredd.org/trees/; Emergent, "Protecting Forests, Beyond Net Zero," accessed November 13, 2023, https://www.emergentclimate.com; the LEAF Coalition, "Uniting to Protect Tropical Forests," accessed November 13, 2023, https://leafcoalition.org.

38 On "forest bonds," see Climate Bonds Initiative, "Forestry, Land Conservation and Forestry," accessed November 13, 2023, https://www.climatebonds.net/resources/publications/forest-bonds/.

At COP 26 in Glasgow, Scotland, in 2021, international leaders from twelve countries pledged US$12 billion in public funding to "support work to protect, restore and sustainably manage forests," to be delivered between 2021 and 2025 via the Glasgow Leaders' Declaration on Forests and Land Use. UN Climate Change Conference UK 2021, "The Global Forest Finance Pledge," February 11, 2021, https://ukcop26.org/the-global-forest-finance-pledge/. The "Leaders' Declaration" also included US$7.2 billion in investments from private corporate and philanthropic funds. Aruna Chandrasekhar and Giuliana Viglione, "COP26: Key Outcomes for

Food, Forests, Land Use and Nature in Glasgow," CarbonBrief, November 17, 2021, https://www.carbonbrief.org/cop26-key-outcomes-for-food-forests-land-use-and-nature-in-glasgow/. These COP 26 commitments echoed similar past pledges that largely went unfulfilled. Julia P. G. Jones, "Deforestation: Why COP26 Agreement Will Struggle to Reverse Global Forest Loss by 2030," *The Conversation*, November 2, 2021.

In 2019, California's Air Resources Board adopted the Tropical Forest Standard, creating a legal pathway for some tropical forest carbon offsets to potentially be admitted into the state's cap-and-trade market. California Air Resources Board, "California Tropical Forest Standard," September 19, 2019, https://ww2.arb.ca.gov/our-work/programs/california-tropical-forest-standard. The UNFCCC negotiations at COP 26 left open the possibility that REDD+ emissions reductions from 2021 onward could be included in UN-sanctioned carbon markets under Article 6 of the 2015 Paris Agreement. Chandrasekhar and Viglione, "COP26: Key Outcomes."

39 Emergent, Forest Trends, UN Environment Programme, and Environmental Defense Fund, "Why Large-Scale Forest Protection Must Urgently Be Part of Corporate Climate Mitigation Strategy: How the Jurisdictional Approach to Emission Reduction Crediting Unlocks Transformational and Systemic Change," July 2021, 5, https://www.emergentclimate.com/wp-content/uploads/2021/07/Jurisdictional-White-Paper-1.pdf.

40 Jessica Brice and Michael Smith, "The Amazon Is Fast Approaching a Point of No Return," *Bloomberg News*, July 29, 2021; Paulo Trevisani and Timothy Puko, "Brazil's Climate Overture to Biden: Pay Us Not to Raze Amazon," *Wall Street Journal*, April 21, 2021.

41 Chandrasekhar and Viglione, "COP26: Key Outcomes"; Jennifer Ann Thomas, "Analysis: With New Law, Brazil Seeks to Boost Payments for Protecting Nature," *Reuters*, March 25, 2021.

42 At COP 28, which took place in Dubai in 2023, for example, the Acrean government signed a "term sheet" with Emergent to "supply up to 10 million high-integrity forest carbon credits . . . for years 2023–2026." Emergent, "Acre Leads the Way in Brazil with the First LEAF Coalition Term Sheet," December 5, 2023, https://emergentclimate.com/wp-content/uploads/2023/12/Acre-is-the-first-Brazilian-State-to-sign-a-LEAF-Term-Sheet.pdf.

43 Callon, "Civilizing Markets."

44 Several critical scholars in Acre, including Maria de Jesus Morias, in *Acreanidade*, and Elder Andrade de Paula, in *(Des)envolvimento Insustentável na Amazônia Ocidental*, offer critical analyses of the Acrean government's forest-based governance strategies. Some nongovernmental groups and activists in Acre have also been highly critical. The 2021

"Letter in Defense of the Amazon and Mother Earth, against the Invasion of Capital, Extreme Violence and Green Scams," for example, is a letter signed by almost one hundred people from Acre and outside of it (and translated into English). The letter states that "programs and projects— of 'sustainable development' and of a 'green economy'—presented as though they are 'solutions' for us, for the forests and the world's climate, exert indirect, yet no less severe, violence [than the 'fascist policies of the current Brazilian president'] such that they restrict our traditional coexistence with the forest, placing at risk our cultural and spiritual survival and threatening our food sovereignty, our ways of life and our relation with the territories." "Letter in Defense of the Amazon and Mother Earth, against the Invasion of Capital, Extreme Violence and Green Scams," 2021, accessed November 1, 2023, https://wrm.org.uy/wp-content/uploads/2021/06/Carta-Defesa-Amazonia_EN.pdf. See also CIMI, "Do$$iê Acre: O Acre que os mercadores da natureza escondem," 2012, accessed November 1, 2023, http://www.cimi.org.br/pub/Rio20/Dossie-ACRE.pdf; "Open Letter to California," 2013, accessed November 1, 2023, https://1bps6437gg8c169ioy1drtgz-wpengine.netdna-ssl.com/wp-content/uploads/2018/10/Open_Letter_Acre_english_portugese_spanish.pdf; "Xapuri Declaration," May 28, 2017, https://wrm.org.uy/actions-and-campaigns/xapuri-declaration-may-28-2017/.

45 Even if seemingly technical processes of monitoring, reporting, and verification are reliable, deforestation halted in one place might just *leak* somewhere else. Then there is the issue of *additionality*: perhaps emissions would have declined regardless of any intervention, making any emissions reductions nonadditional. And what of the future—who is to say whether a forest protected today will burn tomorrow? This issue is known as *permanence*.

46 For example, Böhm, Misoczky, and Moog, "Greening Capitalism?"; and McAfee, "Selling Nature to Save It?" To the extent that REDD+ enables continued emissions either directly, via carbon offset sales, or indirectly, as a climate mitigation strategy, it is subject to the critique that it enables continued harms from fossil fuel extraction, transportation, and usage—harms that often disproportionately impact marginalized communities and have been the subject of environmental justice critiques and activism. See, for example, Tamra Gilbertson, "Carbon Pricing a Critical Perspective for Community Resistance," Climate Justice Alliance and the Indigenous Environmental Network, October 2017, https://www.ienearth.org/wp-content/uploads/2017/11/Carbon-Pricing-A-Critical-Perspective-for-Community-Resistance-Online-Version.pdf.

47 Fairhead, Leach, and Scoones, "Green Grabbing"; Lund et al., "Promising Change, Delivering Continuity"; Lyons and Westoby, "Carbon

Colonialism and the New Land Grab"; Wittman, Powell, and Corbera, "Financing the Agrarian Transition?" As Adam Bumpus and Diana Liverman write, "By enlisting the help of the developing world, international offsets not only provided a spatial fix for capital entities that were mandated to make emissions reductions, but also opened new channels of finance that allowed capital to create cheap carbon credits in the South and sell them into Northern markets where emissions reduction activities were more expensive." Bumpus and Liverman, "Accumulation by Decarbonization," 213. See also McAfee, "Contradictory Logic."

48 See Harvey, *New Imperialism*; Nichols, *Theft Is Property!*; West, *Dispossession and the Environment*; and Luxemburg, *Accumulation of Capital*. REDD+ has been charged with causing and/or exacerbating divisions between and within Indigenous and other forest communities, with some supporting and some opposing it. Relatedly, there is concern about consent processes undertaken as part of REDD+ development. See Lansing, "Realizing Carbon's Value"; Lyons and Westoby, "Carbon Colonialism and the New Land Grab"; McAfee, "Contradictory Logic"; and Myers et al., "Messiness of Forest Governance." See also Lisa Song and James Temple, "The Climate Solution Actually Adding Millions of Tons of CO_2 into the Atmosphere," *ProPublica*, April 29, 2021; Tamra Gilbertson, "Carbon Pricing."

49 Audre Lorde's famous warning that "the master's tools will never dismantle the master's house" could certainly be used in this critique of green capitalism. Lorde, *Master's Tools Will Never Dismantle the Master's House*.

50 Yanagisako, *Producing Culture and Capital*. See also Appel, *Licit Life of Capitalism*; Laura Bear et al., "Gens: A Feminist Manifesto for the Study of Capitalism," Society for Cultural Anthropology, March 30, 2015, https://culanth.org/fieldsights/gens-a-feminist-manifesto-for-the-study-of-capitalism; Gibson-Graham, *End of Capitalism*; Rofel and Yanagisako, *Fabricating Transnational Capitalism*; Tsing, *Friction*; and Tsing, *Mushroom at the End of the World*. The gens approach to studying capitalism, for example, is "to reveal the constructedness—the messiness and hard work involved in making, translating, suturing, converting, and linking diverse capitalist projects that enable capitalism to appear totalizing and coherent." Bear et al., "Gens: A Feminist Manifesto for the Study of Capitalism."

51 Çalışkan and Callon, "Economization, Part 1," 370.

52 See, for example, Appadurai, *Social Life of Things*; Cook, "Follow the Thing"; Guthman, "Unveiling the Unveiling"; Hartwick, "Geographies of Consumption"; Hudson and Hudson, "Removing the Veil?"; Marcus,

"Ethnography in/of the World System"; and West, *From Modern Production to Imagined Primitive*.

53   Documents themselves, of course, are material texts that can have important effects in the world. See Hetherington, *Guerrilla Auditors*; Hull, "Documents and Bureaucracy"; Hull, *Government of Paper*; and Riles, *Documents*.

54   Payments for emissions reductions are often retrospective and offer compensation for emissions reductions that have already occurred. Offset programs tend to require that carbon be stored for at least a set period of time into the future. Among the institutions the Acrean state worked with to monetarily valorize forest carbon was one created by the SISA law: the Company for the Development of Environmental Services, a public/private company in which the Acrean state was the majority owner. They also needed to work with external institutions like the Markit Global Carbon Index.

55   Harris et al., "Global Maps."

56   Other influential sources also advocated for REDD+ around the same time, including McKinsey consultants and the UK Government's Stern Review. Both similarly positioned it as cost-effective. See McKinsey & Company, *Pathways to a Low-Carbon Economy*; Nicholas Stern, "Stern Review on the Economics of Climate Change," Government of the United Kingdom, October 30, 2006, https://webarchive.nationalarchives.gov.uk/20100407172811/http://www.hm-treasury.gov.uk/stern_review_report.htm.

57   This perceived lack of labor may be part of why ecosystem services such as carbon sequestration, like other aspects of nature and undervalued forms of production and reproduction often performed by women and racialized groups, are often not valued or are undervalued within capitalist economies. See Battistoni, "Bringing in the Work of Nature"; Fraser and Jaeggi, *Capitalism*; Moore, *Capitalism in the Web of Life*; O'Connor, *Natural Causes*; Weeks, *Problem with Work*.

58   The climate crisis also means that even where people are not physically present, forests may be harmed through changing hydrological and other geophysical dynamics.

59   Friedlingstein et al., in "Global Carbon Budget 2022," report emissions from deforestation to be 1.8 ± 0.4 GtC per year between 2012 and 2021, with total anthropogenic emissions averaging 10.8 ± 0.8 GtC per year; Baccini et al., "Tropical Forests Are a Net Carbon Source."

60   Sivaramakrishnan, *Modern Forests*; Agrawal, *Environmentality*; Scott, *Art of Not Being Governed*; Mathews, *Instituting Nature*.

61   Campbell, *Conjuring Property*.

62 As quoted in *Globo Rural*, "Sinal de alerta na Amazônia: o desmatamento volta a crescer." *Globo Rural*, June 17, 2014. The man had received deforestation fines for R$30 million, about US$15 million at the time.

63 See Campbell, *Conjuring Property*; Hoelle, *Rainforest Cowboys*.

64 On the standardization of carbon and other commodities, see, for example, Cronon, *Nature's Metropolis*; Castree, "Commodifying What Nature?"; Dalsgaard, "Commensurability of Carbon"; Dempsey and Robertson, "Ecosystem Services"; Gifford, "'You Can't Value What You Can't Measure'"; Huber, "Resource Geographies I"; Knox-Hayes, "Spatial and Temporal Dynamics of Value in Financialization"; Lansing, "Realizing Carbon's Value"; Lohmann, "Marketing and Making Carbon Dumps"; Lovell, "Climate Change, Markets and Standards"; Lovell and Liverman, "Understanding Carbon Offset Technologies"; MacKenzie, "Making Things the Same"; Mathews, "Scandals, Audits, and Fictions"; Osborne, "Tradeoffs in Carbon Commodification"; Osborne and Shapiro-Garza, "Embedding Carbon Markets"; Robertson, "Nature That Capital Can See"; Robertson, "Measurement and Alienation"; Rodrigues Machaqueiro, *Carbon Calculation*; Whitington, "Carbon as a Metric of the Human."

65 MacKenzie, "Making Things the Same." There are six other greenhouse gases that can be converted into tCO$_2$e based on their "global warming potential." Deforestation can be translated into tCO$_2$e based on studies of how much carbon a given area of forest sequesters. This work is complex and entails making estimates about forests whose species composition and ecology is often not fully understood. For example, scientists from the academic and governmental institutions in Acre worked to estimate aboveground carbon stocks in the state based on studies conducted in a state forest not far from Rio Branco called Antimary. Using sampling, allometric and statistical methods, and tree species data, they estimated carbon sequestration throughout the state. For an article on this approach to estimating carbon stocks in Acre, see Salimon et al., "Estimating State-Wide Biomass Carbon Stocks." This work was largely beyond the scope of my research but is an important part of forest carbon's commodification.

66 Bridge, "Resource Geographies 1," 821; Whitington, "Carbon as a Metric of the Human"; Dalsgaard, "Commensurability of Carbon," 83.

67 See Dalsgaard, "Commensurability of Carbon," "Carbon Value between Equivalence and Differentiation," and "Carbon Valuation."

68 Marigo et al., "Carbon Star Formation."

69 Dalsgaard, "Commensurability of Carbon," 82. Dalsgaard goes on to argue that carbon's "commodification refers to the physical transformation from one form to another—typically fossil fuel to greenhouse gas."

70   Levi, *Complete Works of Primo Levi*, 940.

71   Popularized in the 1980s, the term *socioenvironmental* is often linked to the rubber tapper movement in Acre and then to the Government of the Forest that was elected in the state in 1998.

72   Besky and Blanchette, "Introduction," 11.

73   Following Polanyi, in *Great Transformation*, Osborne and Shapiro-Garza, in "Embedding Carbon Markets," 91, position forest carbon as a "fictitious commodity," like land, labor, and money, to reveal the contradictions inherent in efforts to commodify it. See also Brockington, "Ecosystem Services and Fictitious Commodities"; and Lohmann, "Uncertainty Markets and Carbon Markets."

74   Indigenous people and their territories have often been particularly effective at protecting forests, research shows, but they need more support and land because they are often impacted by violence and deforestation linked to mining and other incursions. See Barber et al., "Roads, Deforestation, and the Mitigating Effect"; Nepstad et al., "Inhibition of Amazon Deforestation"; Ricketts et al., "Indigenous Lands, Protected Areas"; Soares-Filho et al., "Role of Brazilian Amazon Protected Areas"; W. S. Walker et al., "Role of Forest Conversion, Degradation, and Disturbance."

75   Bear et al., "Gens." Marxist scholarship on commodity fetishism shows how commodification can mask capitalism's constitutive relations. Marx, *Capital*, chapter 1. See also Castree, "Commodity Fetishism." Feminist scholars have shown the broad range of relationships on which capitalism relies. See, for example, Battistoni, "Bringing in the Work of Nature"; Federici, "Social Reproduction Theory"; Fraser and Jaeggi, *Capitalism*; Weeks, *Problem with Work*. These relations often involve nonhuman species as well, as work in multispecies ethnography has explored, often drawing on Indigenous understandings of relations between humans and other-than-humans. See, for example, Besky and Blanchette, *How Nature Works*; Chao, *In the Shadow of the Palms*; de la Cadena, *Earth Beings*; Kirksey and Helmreich, "Emergence of Multispecies Ethnography"; Kohn, *How Forests Think*; Miller, *Plant Kin*; Ogden, Hall, and Tanita, "Animals, Plants, People, and Things"; D. B. Rose, *Wild Dog Dreaming*; Tsing, *Mushroom at the End of the World*; Tsing et al., *Arts of Living on a Damaged Planet*.

76   Other scholars have also highlighted the role of different types of relations, such as property relations, within REDD+ and/or have taken an ethnographic approach to studying forest carbon. See, for example, Mahanty et al., "Unravelling Property Relations around Forest Carbon"; McAfee and Shapiro, "Payments for Ecosystem Services in Mexico"; Milne, "Grounding Forest Carbon"; Milne and Adams, "Market Masquerades"; Milne et al., "Learning from 'Actually Existing' REDD+"; Osborne,

"Tradeoffs in Carbon Commodification"; and Osborne and Shapiro-Garza, "Embedding Carbon Markets."

77 Anthropologists and other scholars have examined infrastructure's (including roads') (in)visibility, temporality, affect, links with conceptions of progress, and exclusions. See, for example, Anand, *Hydraulic City*; Anand, Gupta, and Appel, *Promise of Infrastructure*; Dalakoglou and Harvey, *Roads and Anthropology*; Harvey and Knox, "Enchantments of Infrastructure"; Harvey and Knox, *Roads*; Hetherington, "Waiting for the Surveyor"; Hetherington, "Surveying the Future Perfect"; Hetherington, *Infrastructure, Environment, and Life in the Anthropocene*; Hetherington and Campbell, "Nature, Infrastructure, and the State"; Knox, "Affective Infrastructures and the Political Imagination"; Larkin, *Signal and Noise*; and Star, "Ethnography of Infrastructure."

78 Tropical forest regions, in other words, have been important in creating REDD+ and other forest policies, not just implementing ideas created outside of them, just as other locations in the Global South have been sites of science and technology innovation, creativity, and agency. See, for example, Cribelli, *Industrial Forests and Mechanical Marvels*; Greenleaf et al., "Forest Policy Innovation at the Subnational Scale"; Laveaga, *Jungle Laboratories*; Medina, Marques, and Holmes, "Introduction: Beyond Imported Magic"; and Pollock and Subramaniam, "Resisting Power, Retooling Justice."

79 Instituto Brasileiro de Geografia e Estatística (IBGE), "Cidades e Estados," accessed November 1, 2023, https://www.ibge.gov.br/cidades-e-estados/ac.html.

80 For more on the history and present of Indigenous people in Acre, see, for example, Apurinã, *Nos caminhos da BR-364*; Iglesias, *Kaxinawá de Felizardo*; and Matos, "A Comissão Pró-Índio do Acre e as línguas indígenas acreanas." Before colonization, there were about fifty Indigenous communities—belonging to the Pano, Aruak, and Arawá linguistic families (including the Huni kuĩ, Apurinã, and Yawanawa)—in what would become Acre.

That so many Acreans do not identify as Indigenous speaks to larger processes of erasure (*apagamento*) in Brazil. See Gabriel Andrade, "Precisamos falar sobre o apagamento de identidades indígenas no Brasil," *TODXS* (blog), March 21, 2021, https://medium.com/todxs/precisamos-falar-sobre-o-apagamento-de-identidades-ind%C3%ADgenas-no-brasil-2797bd37ce1f; and Miki, *Frontiers of Citizenship*. Interestingly, though, according to the census, the percentage of Acreans identifying as Indigenous has increased substantially in recent years. O Globo, "Censo 2022: Acre tem quase 32 mil indígenas e mais de 60% ainda vivem em áreas delimitadas," August 7, 2023, https://g1.globo.com/ac/acre/noticia

/2023/08/07/censo-2022-acre-tem-quase-32-mil-indigenas-e-mais-de-60percent-ainda-vivem-em-areas-delimitadas.ghtml.

81 For deforestation data in Acre, see INPE, "PRODES: Desmatamento no Municípios," accessed November 13, 2023, https://www.dpi.inpe.br/prodesdigital/prodesmunicipal.php.

82 In most cases, the term *state forest* does not refer to publicly owned forest, but rather to forest located within Acre. This forest may be in state-designated protected areas, Indigenous territories, privately owned lands, or land without clear ownership.

83 For discussions of resource frontiers see, for example, Cons and Eilenberg, *Frontier Assemblages*; and Rasmussen and Lund, "Reconfiguring Frontier Spaces." For discussions of how the Amazon has been imagined, see, for example, Bunker, *Underdeveloping the Amazon*; Slater, *Entangled Edens*; Slater, "Visions of the Amazon"; Viveiros de Castro, "Images of Nature and Society."

84 Krenak, *Ideas to Postpone the End of the World*.

85 Capitalism, of course, has always relied on regions figured as peripheral, many of them through ongoing forms of colonialism. See Mignolo, "Introduction."

86 In the hopes of securing a United States–controlled source of rubber, powerful American companies and the US government secretly allied with Bolivian interests against Brazil in the late nineteenth century. Hecht, *Scramble for the Amazon*. The implicit aim was for Acre "to become an American colony, in fact if not in name." Hecht and Cockburn, *Fate of the Forest*, 83.

87 Hecht and Cockburn, *Fate of the Forest*, 80.

88 Deforestation is not unique to Brazil, of course. While humans have long used trees for fuel and cleared forest to make space for cultivation, recent deforestation has been of a different scale and order often as a part of large-scale processes of colonialization and capitalism. See Williams, *Deforesting the Earth*.

89 For example, in the twentieth century, President Getúlio Vargas (1930–1945, 1951–1954) promoted a "March to the West" to consolidate rule in central Brazil. Garfield, *Indigenous Struggle at the Heart of Brazil*. The subsequent government constructed a fabricated and centrally located city, Brasília, and moved the federal capital there from coastal Rio de Janeiro. Holston, *Modernist City*.

90 This deforestation-dependent development entailed the creation of extensive infrastructure, property and tax policies, concessions, settlements, credits, and subsidies, much of which was administered by a troubled bureaucracy. Alston, Libecap, and Mueller, *Titles, Conflict, and Land*

Use; Becker, *Amazônia*; Bunker, *Underdeveloping the Amazon*; Costa, *Formação agropecuária da Amazônia*; Hecht, "Logic of Livestock and Deforestation"; Moran, *Developing the Amazon*; Hecht and Cockburn, *Fate of the Forest*; Schmink and Wood, *Contested Frontiers in Amazonia*. The Amazon region has been the focus of particular governmental angst in Brazil. Portuguese and Brazilian states gained official control of much of the region through treaties with other imperial governments and post-colonial states. But it was such a large area (almost 60 percent of Brazil's territory), so densely forested and so sparsely populated by Brazilians, that it was difficult to govern. The Amazon, and its abundance of species, was also the subject of significant and often avaristic interest from outsiders. Scientists, adventurers, and prospectors, as well as neighboring countries and competing imperial powers, sought to claim its resources and territory, causing officials and the military ongoing unease about Brazilian sovereignty there. Hecht, *Scramble for the Amazon*; Hecht and Cockburn, *Fate of the Forest*.

91 See, for example, Aldrich et al., "Contentious Land Change."

92 See Slater, *Entangled Edens*.

93 "Paradise Lost" is, of course, the name of John Milton's epic biblical poem about the "fall of man." As Susanna Hecht, in *Scramble for the Amazon and the "Lost Paradise" of Euclides da Cunha*, explores, "Lost Paradise" is also the name of the unfinished work about the Amazon by Brazilian journalist and writer Euclides da Cunha (1866–1909; he was killed before he could complete the piece). It was based in part on da Cunha's visit to Acre in the early 1900s on behalf of the Brazilian government, which wanted to justify its 1903 annexation of Acre against Peruvian ownership claims. Da Cunha wrote that Acre's Brazilian migrant rubber tappers "reclaimed their national heritage in a novel and heroic way, extending the fatherland to the new territories they occupied." As quoted in Hecht, *Scramble for the Amazon*, 446. As this indicates, the Amazonian frontier has served as an important space of Brazilian nation-building.

94 See Cons, "Ecologies of Capture."

95 Brum, in *Banzeiro Òkòtó*, describes the Amazon as the center of the world not only because of its centrality to climate stability but also because the Amazon is one of the frontlines of the climate crisis and those of us from outside of it have a lot to learn from the forest and those who live there.

96 Kainer et al., "Experiments in Forest-Based Development"; Morais, *Acreanidade*; Schmink, "Forest Citizens"; Schmink et al., "Forest Citizenship in Acre, Brazil."

97 As quoted in Kainer et al., "Experiments in Forest-Based Development," 870.

98   Schmink, "Forest Citizens," 151.

99   As quoted in Morais, *Acreanidade*, 202–3.

100 Hetherington, "Waiting for the Surveyor," 196.

101 The Government of the Forest was replaced with an allied state government in 2011.

102 See IMC, "Programa para pioneiros em REDD+ (REM)," accessed June 15, 2022, http://imc.ac.gov.br/programa-para-pioneiros-em-redd-rem/.

103 Roseneide Sena, "Acre celebra 10 anos do Programa REM, primeiro instrumento de pagamento por resultados e de repartição justa de benefícios - Programa REM Acre - Fase II," Government of Acre, July 2022, https://programarem.ac.gov.br/2022/07/01/acre-celebra-10-anos-do-programa-rem-primeiro-instrumento-de-pagamento-por-resultados-e-de-reparticao-justa-de-beneficios/. This number may include money raised via SISA more generally—a number that the former director of the IMC put at R$200 million. Brasilamaz, "Candidaturas aos governos na Amazônia Legal prometem inserção no mercado de carbono—Brasil Amazônia Agora," September 12, 2022, https://brasilamazoniaagora.com.br/2022/amazonia-mercado-de-carbono/.

104 To develop and implement SISA, the Acrean government worked with a number of outside institutions and actors, including transnational collaborations such as the Governors' Climate and Forest Taskforce, NGOs both large and small (such as the World Wildlife Fund), and carbon credit certification programs, like that offered by the Climate, Community, and Biodiversity Alliance.

105 See Greenleaf et al., "Forest Policy Innovation at the Subnational Scale."

106 Indigenous people are important to protecting the Amazonian forest and keeping its carbon sequestered, with deforestation often linked to violence against them. Walker and colleagues, for example, estimate that more than a third of the Amazon's aboveground carbon is located in Indigenous territories, where deforestation rates are often lower than outside of them. Walker et al., "The Role of Forest Conversion, Degradation, and Disturbance." Indigenous peoples have also been important players in the Acrean effort to value the forest. Some of them supported SISA, and some vehemently opposed it. "Letter in Defense of the Amazon and Mother Earth." This echoes a history of uneven Indigenous engagement with and benefits from the Acrean government. See Apurinã, *Nos caminhos da BR-364*. Some Indigenous Acreans received KfW-funded benefits, including stipends for Indigenous environmental agroforestry agents, training of new Indigenous agents, monetary grants to community organizations to, among other things, implement "ethno-management plans," "overall support" for many Indigenous territories,

and funding to develop an institutional structure within SISA that would enable Indigenous participation. KfW, "REDD Early-Movers Acre Fact Sheet." This structure included the Indigenous Peoples Working Group (GTI), composed of Indigenous Acreans. The group helped to design the ISA-Carbono Indigenous subprogram (which received €3 million of KfW funding) and provided other guidance about how to implement SISA. Maria DiGiano et al., "The Twenty-Year-Old Partnership between Indigenous Peoples and the Government of Acre, Brazil: Lessons for Realizing Climate Change Mitigation and Social Justice in Tropical Forest Regions through Partnerships between Subnational Governments and Indigenous Peoples," Earth Innovation Institute, September 2018, https://earthinnovation.org/uploads/2018/09/Acre_EN_online.pdf.

107   Duchelle et al., "Acre's State System of Incentives." The mucuna seeds were black mucuna (*Mucuna aterrima*).

108   See Hetherington, *Government of Beans*, for a discussion of the imagined role of cotton in securing rural campesino citizenship in twentieth century Paraguay, for example. As Hetherington writes, "Perhaps most importantly . . . [cotton] kept the promise itself alive that one day Paraguay's rural poor would be able to count on a state that provided for their welfare as citizens." Hetherington, *Government of Beans*, 20.

109   As quoted in Repórter Brasil 20 anos, "Acre against Chico Mendes," *Repórter Brasil*, October 26, 2017, https://reporterbrasil.org.br/2017/10/acre-against-chico-mendes/.

110   Rentier state theory posits that the ability to sell resources, like oil, means that states do not have to rely as much on domestic taxation and therefore do not need to promote economic development. See Beblawi and Luciani, *Rentier State*.

111   See Mathews, *Instituting Nature*; Sivaramakrishnan, *Modern Forests*.

112   Li, "Practices of Assemblage," 266.

113   Agrawal, *Environmentality*.

114   Amaral and Burity, *Inclusão social, identidade e diferença*; and Ansell, *Zero Hunger*. For a critical analysis of the use of the concept of social inclusion in Brazil, see Meyer et al., "Políticas Públicas." Inclusion was also emphasized by other left-wing Latin American governments that governed around the same time. Balán and Montambeault, *Legacies of the Left Turn in Latin America*.

115   Partido de Trabalhadores, "Lula: 'Nossa política de inclusão foi muito mais que o Bolsa Família,'" Partido dos Trabalhadores, August 11, 2021, https://pt.org.br/lula-nossa-politica-de-inclusao-foi-muito-mais-que-o-bolsa-familia/.

116 On petro-states, see Appel, Mason, and Watts, *Subterranean Estates*; Coronil, *Magical State*; Karl, *Paradox of Plenty*; Mitchell, *Carbon Democracy*; Lu, Valdivia, and Silva, *Oil, Revolution, and Indigenous Citizenship*; and Lyall and Valdivia, "Speculative Petro-State."

117 Ferguson, *Give a Man a Fish*, 3 (emphasis in the original).

118 Foucault, *"Society Must Be Defended,"* 241.

119 Mitchell, *Carbon Democracy*.

120 As Timothy Mitchell explores in *Carbon Democracy*, "carbon energy" production and circulation as fossil fuels has shaped democratic institutions and practices, social movements, and politics. The two kinds of carbon democracies are linked. The forest carbon form of carbon democracy formed as a market-based response to the fossil fuel type that Mitchell analyzes. Sequestered carbon only has monetary value because extracted carbon has been so valuable.

121 For critical engagement with resource determinism and "the resource curse," see Appel, "Toward an Ethnography of the National Economy"; Appel, Mason, and Watts, *Subterranean Estates*; Barry, *Material Politics*; Gilberthorpe and Rajak, "Anthropology of Extraction"; Watts, "Resource Curse?"; Mitchell, *Carbon Democracy*; and Weszkalnys, "Cursed Resources."

122 See Cavanagh and Benjaminsen, "Virtual Nature, Violent Accumulation"; Dehm, "Indigenous Peoples and REDD+ Safeguards"; Dunlap and Fairhead, "Militarisation and Marketisation of Nature"; Jodoin, *Forest Preservation in a Changing Climate*; Lyons and Westoby, "Carbon Colonialism and the New Land Grab"; Milne et al., "Learning from 'Actually Existing' REDD+"; Osborne, "Tradeoffs in Carbon Commodification"; Sarmiento Barletti and Larson, *Rights Abuse Allegations*; Sarmiento Barletti and Larson, "Environmental Justice in the REDD+ Frontier." As Juan Pablo Sarmiento Barletti and Anne Larson put it in "Environmental Justice in the REDD+ Frontier," 168, "Currently victories in terms of justice for indigenous and local populations in relation to REDD+ cannot be attributed to its design but rather occurred because social movements have been able to strategically and creatively make gains within the context of REDD+ negotiations, preparation or readiness, and implementation."

123 See Porter and Craig, "Third Way and the Third World"; McAfee, "Contradictory Logic."

124 Since the European invasion of the "New World," the Amazon has been a place for adventurers and prospectors to invoke and travel to for divergent reasons, including, for example, to seek territory to reestablish US Confederacy–style slavery, chase after redemptive adventure, and pursue

dreams of riches. See, for example, Millard, *River of Doubt*; Raffles, *In Amazonia*; Slater, *Entangled Edens*.

125 Haraway, "Situated Knowledges," 581.

126 Jennifer Watling and colleagues found that geoglyphs on approximately thirteen thousand km² in Acre were built "within anthropogenic forest that had been actively managed for millennia." Watling et al., "Impact of Pre-Columbian 'Geoglyph' Builders," 1868. For research on Indigenous Amazonian cultivation within and outside of forests, see, for example, Balée, "Indigenous Transformation of Amazonian Forests"; Glaser and Woods, *Amazonian Dark Earths*; Iriarte et al., "Origins of Amazonian Landscapes"; Neves and Heckenberger, "Call of the Wild"; Parker, "Forest Islands and Kayapó Resource Management in Amazonia"; and Posey, "Indigenous Knowledge, Biodiversity, and International Rights."

127 Raffles, *In Amazonia*, 34.

128 See Secretaria de Estado de Planejamento, "Evolução da população, taxa de urbanização e crescemento populacional—Acre," accessed October 30, 2021, https://seplan.ac.gov.br/evolucao-da-populacao-taxa-de-urbanizacao/. Many of these urban residents were migrants from rural areas or their descendants. Schmink and Cordeiro, *Rio Branco*.

129 For more detail on CIFOR's Global Comparative Study on REDD+, see CIFOR, "CIFOR's Comparative Study on REDD+," accessed October 30, 2021, https://www2.cifor.org/gcs/.

130 All names used in this book are pseudonyms, unless the person I am writing about is a public figure. Fernanda was tremendously helpful during this part of my research, assisting with asking questions and taking notes. I also hired someone to drive us in rural areas where I was advised that it was not safe for me to travel on my own. These concerns stemmed in part from the 2006 rape and murder of a female PhD student in the area.

131 Relatedly, I was sometimes told about Acre's heritage as being a balanced mixture of all Brazilians, and Acreans not having an accent because of it, echoing entrenched Brazilian tropes about the country as a "racial democracy." See Freyre, *Masters and the Slaves*. This myth mixed discourse positing racial "mixing" as a path to a superior tropical civilization with entrenched and violent racism that benefited white Brazilians and harmed racialized Brazilians, including Black and Indigenous people. See Miki, *Frontiers of Citizenship*; Mitchell, *Constellations of Inequality*; Nascimento, "Myth of Racial Democracy"; and Twine, *Racism in a Racial Democracy*.

132 The fact that smallholders were often current or former rubber tappers, or descendants of them, meant that they have been thought of

and organized as rural workers, rather than peasants. Martins, "Representing the Peasantry?" The figure of the worker is a socioculturally powerful category in Brazil in part because of the way that President Getúlio Vargas made formal, waged workers the centerpiece of early- to mid-twentieth-century welfare state policies. See Fischer, *Poverty of Rights*; Holston, *Insurgent Citizenship*; and Millar, *Reclaiming the Discarded*.

133 This might have been different in rubber tapper movement strongholds like Xapuri, the historic center of the rubber tapper movement, of which the Xapuri Rural Workers Union was an important part. For more on rural workers unions, see Welch and Sauer, "Rural Unions and the Struggle for Land in Brazil."

134 Hoelle, *Rainforest Cowboys*.

135 Adams et al., *Amazon Peasant Societies in a Changing Environment*; Nugent, *Amazonian Caboclo Society*; Pacheco, "Smallholder Livelihoods"; Vadjunec, Schmink, and Greiner, "New Amazonian Geographies."

136 See Rojas, Olival, and Olival, "Cultivating Alternatives."

137 On the anthropology of roads, see Dalakoglou and Harvey, *Roads and Anthropology*; Harvey and Knox, *Roads*; Jobson, "Road Work."

138 See, for example, Barber et al., "Roads, Deforestation, and the Mitigating Effect"; Fearnside, "Desmatamento na Amazônia"; Ferrante and Fearnside, "Amazon's Road to Deforestation"; Nelson and Hellerstein, "Do Roads Cause Deforestation?"

139 The idea of extending the BR-364 so that it continues all the way to Peru and the Pacific Ocean has circulated for decades and was revived under the presidency of Jair Bolsonaro, despite concerns that doing so would threaten local Indigenous and forest communities and the famed biodiversity of the Serra do Divisor National Park. See Fabiano Maisonnave, "Planned Brazil-Peru Highway Threatens One of Earth's Most Biodiverse Places," Mongabay.com, July 22, 2021, https://news.mongabay.com/2021/07/planned-brazil-peru-highway-threatens-one-of-earths-most-biodiverse-places/.

140 Communities like the Kaxarari of Vila Extrena are still waiting for compensation. Apurinã, *Nos caminhos da BR-364*.

141 See Hochstetler and Keck, *Greening Brazil*, 162–165; and Keck, "Social Equity and Environmental Politics in Brazil," 415–417.

142 Anticipating increased deforestation, the government initially made the land along the BR-364 in Feijó a "priority area" for a planned REDD+ project, with money set to flow there. These plans were replaced by SISA's statewide approach. See Duchelle et al., "Acre's State System of Incentives."

Chapter 1. Carbon Boom

1   Formed in the late 1960s to address air pollution, CARB is part of the California state government. More recently, CARB has been at the center of many of California's efforts to reduce GHG emissions and combat climate change. See California Air Resources Board, "History," accessed November 2, 2023, https://ww2.arb.ca.gov/about/history.

2   The California cap-and-trade program covers about 85 percent of California emissions and applies to about 450 businesses. Center for Climate and Energy Solutions, "California Cap and Trade," accessed August 19, 2022, https://www.c2es.org/content/california-cap-and-trade/. As of 2020, the program generated over US$8 billion in revenue that the state government spent to further reduce emissions, with over US$4 billion "benefiting priority populations." California Climate Investments, "Annual Report to the Legislature on California Climate Investments Using Cap-and-Trade Auction Proceeds," April 2021, https://ww2.arb.ca.gov/sites/default/files/auction-proceeds/2021_cci_annual_report.pdf. The percentage of their emissions that regulated entities are allowed to cover with carbon offsets depends on a few factors. These include the year of purchase: 8 percent through 2020, 4 percent in 2021–2025, and 6 percent in 2026–2030. Starting in 2021, at least half of these offsets had to come from projects that "provide direct environmental benefits in the state." To be admissible, offsets must comply with California's sector-by-sector offset protocols, and only half of an entity's offset purchases can come from international sources. As of this writing, the province of Québec's cap-and-trade program is linked with California's. On that linkage, see California Air Resources Board, "Program Linkage," accessed August 19, 2022, https://ww2.arb.ca.gov/our-work/programs/cap-and-trade-program/program-linkage.

3   This program was part of SISA.

4   The MOU anticipates the possibility of linking the signatory jurisdictions, which would enable forest offsets from Acre to be sold in the California carbon market. The MOU also includes Chiapas, Mexico. Like California, CARB notes, both Acre and Chiapas "have been implementing innovative strategies to address climate change." California's Air Resources Board, "Sector-Based Offset Credits," accessed May 1, 2022, https://ww2.arb.ca.gov/our-work/programs/compliance-offset-program/sector-based-offset-credits. During my fieldwork, however, I did not hear Chiapas talked about as a viable source of offsets in the near term. The MOU is available at https://ww2.arb.ca.gov/sites/default/files/cap-and-trade/sectorbasedoffsets/2010_mou_acre-california-chiapas.pdf.

A 2015 CARB report assesses Acre's forest carbon offsets as the only ones from tropical forests that were "technically capable of being considered for formal inclusion in the Cap-and-Trade Program" at that time, while they found that those from other developing jurisdictional programs "may be nearing readiness in the near future." California Air Resources Board, "Scoping Next Steps for Evaluating the Potential Role of Sector Based Offset Credits under the California Cap and Trade Program, Including from Jurisdictional Reducing Emissions from Deforestation and Forest Degradation Programs," October 19, 2015, ix, https://www.thecornerhouse.org.uk/sites/thecornerhouse.org.uk/files/ARB%20Staff%20White%20Paper%20Sector-Based%20Offset%20Credits.pdf.

5   For a discussion of the airline industry's Carbon Offset and Reduction Scheme for International Aviation in relationship to the Amazon, see Gonçalves, "Carbon Offset from the Amazon Forest to Compensate Aviation Emissions."

6   See Tsing, *Friction*.

7   See Greenleaf et al., "Forest Policy Innovation at the Subnational Scale," for a critique of "top-down" characterizations of REDD+.

8   Rofel and Yanagisako, in *Fabricating Transnational Capitalism*, 15, write that "the production of capitalist value is always a process of negotiation," rather than "simply a direct effect of capitalist investments or the result of a global stage of capitalism."

9   Rofel and Yanagisako, *Fabricating Transnational Capitalism*, 307–308. Such socioenvironmental relations are part of the "relations of interdependence" that Rofel and Yanagisako identify.

10 Laura Bear critiques "narrative economics" for keeping "the social relations and labour of speculation . . . entirely invisible." Bear, "Speculation," 4–5. In contrast, I seek to illuminate some of these relations here and in subsequent chapters.

11 California Air Resources Board, "Linkage Process and Acre, Brazil," April 28, 2016, 25.

12 Describing the rubber tapper movement as an "environmental justice" movement may have been meant to address criticism from environmental justice groups in California and elsewhere directed at California's cap-and-trade program. See, for example, Tamra Gilbertson, "Carbon Pricing a Critical Perspective for Community Resistance," Climate Justice Alliance and the Indigenous Environmental Network, October 2017, https://www.ienearth.org/wp-content/uploads/2017/11/Carbon-Pricing-A-Critical-Perspective-for-Community-Resistance-Online-Version.pdf.

13 In analyzing the speculative components of what she calls "future-oriented" commodities, Laura Bear defines speculation as "future-oriented affective, physical and intellectual labour that aims to accumulate capital for various ends." Bear, "Speculation," 2.

14 Tsing, *Friction*, 57.

15 See Sax and Tubb, "Buzz Phase of Resource Extraction"; Tsing, *Friction*; and Weszkalnys, "Geology, Potentiality, Speculation."

16 For a discussion of some of REDD+'s other stories, see Mathews, "Scandals, Audits, and Fictions"; and Mathews, "Imagining Forest Futures and Climate Change."

17 Bumpus and Liverman, "Accumulation by Decarbonization," 137. But see Boyd and Salzman, "Curious Case of Greening in Carbon Markets," for a discussion of "green differentiation" in UNFCCC carbon markets.

18 See Dalsgaard, "Carbon Value between Equivalence and Differentiation."

19 "Qualification," as Callon, Méadel, and Rabeharisoa explore in "Economy of Qualities," can matter as much as quantification. See Karen Hébert, "In Pursuit of Singular Salmon," for example, for an analysis of salmon and singularity—how wild salmon is made more valuable than farmed salmon, and Paige West, *From Modern Production to Imagined Primitive*, on the labor that goes into making coffee from Papua New Guinea valuable by making the people there seem primitive and poor. Standardization and singularization are also not mutually exclusive. As Sarah Besky in *Tasting Qualities*, 3, writes about the creation of quality in standardized black tea, "a product like, say, PG Tips tea bags, is paradoxically both 'singular' on the market and 'comparable' to other tea bags available, even at the same price point." See also Appadurai, "Introduction"; Baudrillard, *Simulacra and Simulation*; Debord, *Society of the Spectacle*; Igoe, *Nature of Spectacle*; and Kopytoff, "Cultural Biography of Things."

20 Although Amazonian rubber had long been traded widely, production expanded rapidly between the mid-1800s, when Charles Goodyear invented vulcanization, and the early 1910s. Many of the migrants that moved to Acre to tap rubber during this time were single men who came from the Northeast (particularly the state of Ceará) after a devastating El Niño–related drought in the late 1870s that killed an estimated four to five hundred thousand people and propelled many others to leave the area. See Dean, *Brazil and the Struggle for Rubber*; Patrick Gage, "'A Grande Seca': El Niño and Brazil's First Rubber Boom," Historical Climatology, February 6, 2017, https://www.historicalclimatology.com/features/a-grande-seca-el-nino-and-brazils-first-rubber-boom; Garfield, *In Search of the Amazon*; Hecht, *Scramble for the Amazon*; Nugent, *Rise and Fall of the Amazon Rubber Industry*; and Weinstein, *Amazon Rubber Boom*, for histories of rubber production in the Amazon. Women were also important to Acrean rubber extraction, as Cristina Scheibe Wolff shows, even though there were fewer of them. They were sometimes treated as a "luxury object" that could be bought (for 500 kg of rubber, for example), a valued possession considered to need extra protection,

or a prize worth murdering male competitors for. Wolff, *Mulheres da floresta*, 71. What she calls the "*nordestino* model" (from Brazil's northeast, where Ceará is located) shaped gender relations in rural Acre during the rubber boom, with some gendered customs and separation of spaces from the nineteenth-century Northeast apparent in contemporary rural life in the part of Acre she writes about (Alta Juruá). Indigenous women were also sometimes captured and forced to marry rubber tappers from the northeast. Wolff, *Mulheres da floresta*, chapter 3.

21   Acre was annexed into Brazil through the Treaty of Petrópolis. In exchange, Brazil gave Bolivia a small parcel of land elsewhere and two million British pounds, as well as committing to building a railroad through the Amazon that would bypass the treacherous Madeira rapids and link Bolivia to the Atlantic. Hecht, *Scramble for the Amazon*, 181.

22   For more on the history of Acre, see, for example, Albuquerque, "História e historiografia do Acre"; Hecht, *Scramble for the Amazon*; Hecht and Cockburn, *Fate of the Forest*; Souza, *História do Acre*; and Tocantins, *Formação histórica do Acre*.

23   See, for example, Apurinã, *Nos caminhos da BR-364*; Matos, "A Comissão Pró-Índio do Acre e as línguas indígenas acreanas"; and Wolff, *Mulheres da floresta*.

24   These rules were enforced by armed guards (*fiscais*). Rubber tappers in Acre tended to live and work alone. For about eight months per year, they worked fifteen-hour days, tapping rubber from far-flung rubber trees on prescribed, solitary routes. Bakx, "From Proletarian to Peasant," 146. In other parts of the Brazilian Amazon, rubber production sometimes worked differently. It might be extracted through familial, Indigenous clan-based, cooperative, or slavery-based relations of extraction. Hecht, *Scramble for the Amazon*, 265–68. In some areas, rubber tappers were more like autonomous petty commodity producers. Weinstein, *Amazon Rubber Boom*. Bakx, in "From Proletarian to Peasant," 145–49, argues that, while rubber tappers in the eastern Amazon may have had more autonomy, those in Acre subsisted in a stifling form of debt peonage. This analysis is in large part supported by the descriptions of Brazilian writer Euclides da Cunha, who documented labor conditions there in the early 1900s. See Hecht, *Scramble for the Amazon*.

25   During the Second World War, when Japanese control of Europe's Southeast Asian colonies cut off Allied powers from cheap plantation rubber, the Amazonian rubber economy had a brief resurgence. Supported by both the Brazilian and US governments, this resurgence precipitated another wave of Brazilian migration to Acre and elsewhere by so-called rubber soldiers. See Garfield, *In Search of the Amazon*.

26   Schmink et al., "Forest Citizenship in Acre, Brazil," 42.

27 Bakx, "From Proletarian to Peasant," 146.

28 Deforestation-dependent rubber extraction tended to center on a different species of tree—*Castilla*—and in the Amazon often entailed the violent exploitation of Indigenous people. *Castilla* trees could be tapped while standing, as they were in parts of the pre-conquest Americas, but generally were not during the rubber boom. Hecht, *Scramble for the Amazon*, 268–73.

29 As quoted in Hecht, *Scramble for the Amazon*, 443. Through an analysis of da Cunha's writing based on his state-sponsored visit to Acre in the early 1900s, Susanna Hecht, in *Scramble for the Amazon and the "Lost Paradise" of Euclides da Cunha*, shows how the contrast in the political ecologies of these two forms of rubber extraction were used to justify Brazilian control of Acre and invoked as a manifestation of a uniquely Brazilian civilization. According to da Cunha, the Peruvian *caucho* collection from *Castilla* trees entailed "killing trees and men" and was a "primitive activity," one of "barbarous individualism" (272, 416). In contrast, da Cunha argued that, in the way they tapped *Hevea* trees, Brazilian migrants were "reclaim[ing] their national heritage in a novel and heroic way, extending the fatherland to the new territories they occupied" (446). They had, after all, fought the Bolivians to make the area part of Brazil. Da Cunha argued that these rubber tappers possessed "the moral beauty of manly souls who have defeated the wilderness," creating "a newborn society" through their settlement and labor (449, 448). In their protective use of the forest, da Cunha saw the potential for, as Hecht puts it, "a new tropical civilization of mixed-blood pioneers, a New World counterweight to white European imperialism," an early version of Brazilian "Tropicalist" essentialism (12, 197–217). Gilberto Freyre, who articulated the idea of lusotropicalismo in *The Masters and the Slaves*, called da Cunha "the First Tropicalist" (421). Based on the large number of Brazilian migrants, Brazil also claimed that its annexation of Acre was justified by *uti possedeti*, the state-embraced principle that occupation and use bestows ownership in the absence of an agreement to the contrary (179).

30 Grandin, *Fordlandia*. Fordlândia also failed because workers did not want to live or work as prescribed.

31 As quoted in Gomes, Vadjunec, and Perz, "Rubber Tapper Identities," 260.

32 For more on Acre's rubber tapper movement, see, for example, Allegretti, "Extractive Reserves"; Allegretti, "A construção social de políticas públicas"; Almeida, "Politics of Amazonian Conservation"; Cardoso, *Extractive Reserves in Brazilian Amazonia*; and Keck, "Social Equity and Environmental Politics in Brazil." The rubber tapper movement was

one of many social and socioenvironmental movements advocating for democracy toward the end of the Brazilian military dictatorship. See Hochstetler and Keck, *Greening Brazil*.

33 By offering extractivists usufruct rights to land, the reserves can offer a meaningful improvement over the violent contestation that plagued parts of rural Acre and other parts of the Amazon. They also served as an influential model of use-based environmental protection in other parts of the world. The first reserve was the 930,000-hectare Chico Mendes Reserve in Acre, created in 1990, with many more created since. For more on extractive reserves, see Allegretti, "Extractive Reserves"; Almeida, "Politics of Amazonian Conservation"; Cardoso, *Extractive Reserves in Brazilian Amazonia*; Keck, "Social Equity and Environmental Politics in Brazil"; Almeida, Allegretti, and Postigo, "O legado de Chico Mendes"; Rich, *Mortgaging the Earth*; and Wallace et al., "Chico Mendes Extractive Reserve."

34 See, for example, Tsing, *Friction*, 230; and Nixon, *Slow Violence and the Environmentalism of the Poor*, 38–39, 135–36. The rubber tapper movement was attractive to outside environmentalists because, as Margaret Keck puts it, it "announced the possibility of an environmentalism that was not an amenity but rather was part and parcel of a struggle around basic rights to subsistence." Keck, "Social Equity and Environmental Politics in Brazil," 417. See also Hochstetler and Keck, *Greening Brazil*, chap. 4.

35 See, for example, Cronkleton et al., *Environmental Governance*; Hochstetler and Keck, *Greening Brazil*; Keck, "Social Equity and Environmental Politics in Brazil"; and Souza, *História do Acre*.

36 This international attention surprised many Brazilians, given that violence against activists, Indigenous people, and other rural people in the Amazon was, and remains in some places, quite routine. Keck, "Social Equity and Environmental Politics in Brazil," 410.

37 There were other politicians in Jorge Viana's family, including his brother and father. He and his brother Tião Viana, both members of the PT, swapped places as governor and senator of Acre during the two decades that the PT dominated state politics in Acre (1999–2018). Their father, Wildy Viana das Neves, was also a politician. He was elected to various positions as a member of a few different political parties, though he later joined the PT.

38 Jorge Viana was governor for the first two terms of the Government of the Forest. Binho Marques was governor for the third. Afterward, Tião Viana was elected governor, continuing, to some extent, forest-focused development policies.

39 For a discussion of other older efforts at environmental protection and socioeconomic reform in Brazil, see Pádua, *Um sopro de destruição*.

40 As quoted in Kainer et al., "Experiments in Forest-Based Development," 877.

41 This state law countered the elimination of a federal rubber subsidy in the early 1990s, which had prompted many rubber tappers to move to cities and threatened to decimate an already fragile sector. Kainer et al., "Experiments in Forest-Based Development," 876–77; Sills and Saha, "Subsidies for Rubber."

42 Schmink et al., "Forest Citizenship in Acre, Brazil"; Débora Almeida, Fernanda Basso Alves, and Liliana Pires, "Governança em cadeias de valor da sociobiodiversidade: Experiências e aprendizados de grupos multi-institucionais da Castanha-do-Brasil e Borracha-FDL no Acre," GIZ, Núcleo Maturi, IUCN, WWFBrazil, 2012, accessed November 5, 2023, https://d3nehc6yl9qzo4.cloudfront.net/downloads/livro_governanca_de_cadeias_de_valor.pdf.

43 The rubber narrative had local and domestic audiences too. Stories of extractivism and Chico Mendes motivated my research assistant, for example, to move to Acre, where she eventually began working for one of the new SISA-related institutions. It motivated other interlocutors of mine in Acre to study forest engineering, as I discuss in chapter 5.

44 See, for example, KfW, "REDD Early-Movers Acre Fact Sheet," 2017, accessed July 1, 2021, https://www.kfw-entwicklungsbank.de/PDF/Entwicklungsfinanzierung/Themen-NEU/REDD-Early-Movers-Acre-Fact-Sheet.pdf.2017. See also Environmental Defense Fund, "Ready for REDD: Acre's State Programs for Sustainable Development and Deforestation Control," 2, accessed July 2, 2022, https://www.edf.org/sites/default/files/Acre_Ready_for_REDD_EDF.pdf; and Greg Fishbein and Donna Lee, "Early Lessons from Jurisdictional REDD+ and Low Emissions Development Programs," The Nature Conservancy, Forest Carbon Partnership Facility, World Bank Group, January 2015, 9, https://www.nature.org/media/climatechange/REDD+_LED_Programs.pdf.

45 Revkin, *Burning Season*.

46 Tsing, *Friction*, 230.

47 See West, Igoe, and Brockington, "Parks and Peoples."

48 "Rio Branco Declaration," August 11, 2014, https://www.gcftf.org/wp-content/uploads/2020/12/Rio_Branco_Declaration_ENG.pdf. See also Stickler et al., "Rio Branco Declaration."

49 The GCF Task Force is an international collaboration composed of forty-three states and provinces that represent a third of the world's tropical

50 forests (as of this writing). Both the states of Acre and California are long-time members. See https://www.gcftf.org.

50 On "tournaments of value," see Appadurai, "Introduction," 27.
    For discussions of spectacle and conservation, see Igoe, *Nature of Spectacle*; and Igoe, Neves, and Brockington, "Spectacular Eco-Tour around the Historic Bloc." For analyses of the role of spectacle in the production of resource value and politics, see Apter, *Pan-African Nation*; and Coronil, *Magical State*.

51 In recounting the history of the rubber tapper movement, the rubber narrative also worked, to borrow from Verónica Gago's analysis of left-wing neoliberalism in Argentina, to "weave together that vitality of revolt with the categories of political economy." Gago, *Neoliberalism from Below*, 27.

52 Paxson, *Life of Cheese*, 14. This dynamic speaks to the ways that multiple forms of value, including but not limited to economic value, are constituted together. As Paxson writes, commodities that are unfinished, like artisanal cheese, can "call[] attention to the instability, and hence open promise, of its heterogeneous forms of value" (13).

53 Some climate science and policy literature relies on similar calculations: "Climate change impacts studies, by definition, focus on changes attributable to climate change. Most commonly, results [from climate change impact studies] are reported as changes in some outcome of interest relative to what would have happened in the absence of climate change. Such statements require developing a baseline point of comparison, and these hypothetical ultimately unobservable estimated trajectories pervade the climate science and policy literatures." Moore, Mankin, and Becker, "Challenges in Integrating," 177.

54 The PPCD used the average deforestation rate from 1996 to 2005 to set the reference level through 2010 (602 kms/year) and the average of 2001 to 2010 to set the 2010–2015 reference level (496 kms/year). Government of Acre, "Plano estadual de prevenção e controle de desmatamento no Acre," State Secretariat of Planning, State Secretariat of the Environment, January 7, 2010, 33, https://sema.ac.gov.br/wp-content/uploads/2020/05/PPCD.pdf. After that, a new baseline was to be adopted every five years based on the shifting, and hopefully declining, deforestation rates. Acre's baseline and reference level did not come up frequently in my ethnographic research, though I did ask about them in some interviews.

55 Government of Acre, "Jurisdictional Program Description: Jurisdictional Program of Incentives for Environmental Services Related to Carbon of the State of Acre, Brazil (IES Carbon Program of Acre)," Institute of Climate Change and Regulation of Environmental Services, August 13, 2013,

19–22. The Brazilian federal approach was to use the 1996–2005 average to set a baseline through 2020. Because the 1996–2005 emissions rates were higher than the later rates, the federal baseline was higher than the Acrean rate established by the state's PPCD after 2010, creating the potential to earn more money by using it. In this sense, the Acrean rate was more conservative than the federal rate. Some Acrean officials also considered the Acrean rate to be more conservative because of socioenvironmental dynamics that they believed would increase deforestation in the state after 2010, absent state interventions—in particular, the paving of the BR-364 and the BR-317, "which has opened new access to forests and consequently would accelerate deforestation and via the trend of rising commodities' prices (among them those of meat), the production of which represents an important regional driver of deforestation." Government of Acre, "Jurisdictional Program Description," 20.

56 See Moutinho and Escobar, "Falling Star." Starting in 1989, INPE began to monitor Amazonian deforestation using satellite imagery—the world's first space agency to do so. Analyzing primarily imagery from the NASA satellite Landsat, taken every sixteen days, PRODES detects deforestation at 6.25 hectares and larger. The Acrean state had its own deforestation monitoring that could detect clearing at a significantly smaller scale, but for the sake of consistency, I was told, SISA administrators used INPE data to determine the state's deforestation rate for SISA. INPE also offers daily alerts on Amazonian deforestation to aid in environmental law enforcement through a system called DETER.

57 Comitê Científico—SISA, "Resolução de conformidade 001/2013," August 22, 2013. Brazil's 2010 Climate Policy adopts 132.3 tons as the average total stored carbon per hectare of land in the Amazon and 3.67 tCO_2e (44/12) per ton of stored carbon released. Government of Brazil, "Decreto No. 7.390, de 9 de Dezembro de 2010," Presidência da República Casa Civil, December 9, 2010, https://climate-laws.org/documents/decree-no-7390-of-december-9-2010_ce2a?id=law-12-187-2009-establishing-the-national-policy-on-climate-change-npcc-regulated-by-decree-7-390-2010_dd44. World Wildlife Fund authors describe this latter conversion as "a standard figure based on the relative masses of carbon and CO_2 molecules." Karen Lawrence and Sarah Hutchison, "Estimating the Contribution of the Sky Rainforest Rescue Project to Reducing Deforestation and Carbon Dioxide Emissions in the State of Acre, Brazil," World Wildlife Fund, 2015, 12, accessed July 14, 2023, https://assets.wwf.org.uk/downloads/report_online_1.pdf. For more on Acrean forest carbon stocks, see Salimon et al., "Estimating State-Wide Biomass Carbon Stocks."

58 Evan Johnson, "California, Acre and Chiapas, Partnering to Reduce Emissions from Tropical Deforestation: Recommendations to Conserve

Tropical Rainforests, Protect Local Communities and Reduce State-Wide Greenhouse Gas Emissions," REDD Offset Working Group, 2013, 23, accessed November 5, 2023, https://ww2.arb.ca.gov/sites/default/files/cap-and-trade/sectorbasedoffsets/row-final-recommendations.pdf. The REDD Offset Working Group was established by the 2010 MOU between the governments of California, Chiapas, and Acre.

59   Bear, "Speculation," 8.

60   According to the REDD Offset Working Group, "the best way to estimate the RL [Reference Level] is usually as a continuation into the future of the historical emissions level averaged over a period that is long enough to capture this year-to-year variation. In some cases, it is appropriate to adjust the historical emissions average upward if there is a compelling, scientifically rigorous reason that the business-as-usual trend is moving toward a significantly higher rate of emissions. Major new investments in highways across remote forest regions that address critical transportation needs, or substantially higher profitability of forest conversion to crops or livestock because of higher market demand, are examples of possible reasons for upward adjustments of the RL. Similarly, downward adjustments may also be necessary under some circumstances, e.g. if a jurisdiction is simply running out of forests to clear or degrade." Johnson, "California, Acre and Chiapas," 24.

61   Government of Acre, "Jurisdictional Program Description," 19–22.

62   Preeminent among these strategies were increasing forms of production deemed sustainable (see chapter 2), building on the model of Acrean rubber extraction. Interlocutors told me that such production was a work in progress. Neither Acrean "rural producers" nor foreign buyers were, one official explained, sufficiently "consolidated" into these "sustainable production chains." State efforts to build these supply chains therefore needed to continue. This characterization of what is normal and will continue and what is a deviation resembles David McDermott Hughes's discussion of how his interlocutors used a selective and speculative interpretation of ecological and economic data to project a future of lucrative ecotourism that treated deviations as "abnormal" anomalies, rather than indications that their projections might be flawed. Hughes, "Third Nature," 171–72.

63   Mathews, "Imagining Forest Futures and Climate Change," 206.

64   The rubber narrative turned the reference level and departure from it into a kind of "drama" that Tsing describes as an effective way to attract investment. Tsing, *Friction*, 63.

65   For recent anthropological engagements with speculation, see, for example, Bear, "Speculation"; Bear, "Speculations on Infrastructure"; Campbell,

Conjuring Property; Ferry, "Speculative Substance"; Fortun, *Promising Genomics*; Ho, *Liquidated*; Morris, "Speculative Fields"; Weszkalnys, "Geology, Potentiality, Speculation"; and Zaloom, "How to Read the Future." As Weszkalnys and others show, these practices of speculation are about resources' and land's materiality, as well as financial practices.

66 Bear, "Speculation," 3. Given the climate crisis and market-oriented efforts to address it, future carbon sequestration in living forests becomes an additional imagined component of what David McDermott Hughes, writing about wildlife conservation in southern Africa, calls "third nature," or "the potential of landforms in a given area to support specific types of wildlife communities" in the future. Hughes, "Third Nature," 158. In this case, what is being envisioned is the potential of a forest to support carbon sequestration into the future.

67 Rasmussen and Lund, "Reconfiguring Frontier Spaces," 23.

68 Tsing, *Friction*, 68.

69 The value of global consumer commodities, as Paige West shows for coffee from Papua New Guinea, can also rely on relatedly standardized (and inaccurate) stories and images of poverty and primitivity. West, *From Modern Production to Imagined Primitive*.

70 Guthman, *Agrarian Dreams*; Besky, *Darjeeling Distinction*.

71 See, for example, Besky, *Darjeeling Distinction*; Bryant and Goodman, "Consuming Narratives"; Guthman, *Agrarian Dreams*; Jaffee, *Brewing Justice*; Méndez et al., "Agrobiodiversity and Shade Coffee Smallholder Livelihoods"; Moberg, *Fair Trade and Social Justice*; Robbins, "Coffee, Fair Trade, and the Commodification of Morality"; and West, *From Modern Production to Imagined Primitive*.

72 MacKenzie, "Making Things the Same."

73 Discussing environmental conservation and "traditional people" in the Amazon, Manuela Carneiro da Cunha and Mauro W. B. de Almeida, in "Indigenous People, Traditional People, and Conservation in the Amazon" (315), argue, for example, that it would be a "misunderstanding" to "assert[] that 'foreign' nongovernmental organizations and ideologies were responsible for the connection made between conservation of biological diversity and traditional people of the Amazon." Instead, they argue, "rubber-tappers took the lead in establishing a link between their struggle and ecological concerns" (321). Similarly, the use of the rubber narrative to make forest carbon valuable was not a straightforward imposition of external ideas or modes of governance.

74 On the TFS, see California Air Resources Board, "California Tropical Forest Standard," September 19, 2019, https://ww2.arb.ca.gov/our-work

/programs/california-tropical-forest-standard. On carbon offset standards, see, for example, Corbera and Martin, "Carbon Offsets"; Gupta et al., "In Pursuit of Carbon Accountability"; Lansing, "Performing Carbon's Materiality"; and Lovell, Bulkeley, and Liverman, "Carbon Offsetting."

75 Hughes, "Third Nature," 173.
76 Gomes et al., "Extractive Reserves in the Brazilian Amazon."
77 Smallholders on the BR-364 in Feijó also often offered rubber's low price to me as the reason that they had stopped tapping it.
78 Gomes et al., "Extractive Reserves in the Brazilian Amazon."
79 Some academic research also questioned extractivism's economic viability. See Jaramillo-Giraldo et al., "Is It Possible?"; and Salisbury and Schmink, "Cows versus Rubber."
80 Hoelle, *Rainforest Cowboys*.
81 Vadjunec, Schmink, and Gomes, "Rubber Tapper Citizens."
82 Kröger, "Deforestation, Cattle Capitalism and Neodevelopmentalism"; Salisbury and Schmink, "Cows versus Rubber"; Toni et al., *Expansão e trajetórias da pecuária na Amazônia*.
83 Schmink et al., "Forest Citizenship in Acre, Brazil," 39; Repórter Brasil 20 anos, "Acre against Chico Mendes," *Repórter Brasil*, October 26, 2017, https://reporterbrasil.org.br/2017/10/acre-against-chico-mendes/.
84 See Debord, *Society of the Spectacle*; and Taussig, "Beach (a Fantasy)."
85 As quoted in Repórter Brasil 20 anos, "Acre against Chico Mendes."
86 Jorge Viana, "O governo federal e o governo estadual recuperando a BR-364 (Rio Branco a Sena)," Facebook, July 28, 2014, https://www.facebook.com/jorgevianaacre/posts/pfbid0MqqvMHmopSYXvdsL9wS0Er30gEBfmHRpzydu1EWFAZz7MEMewsKQiC499XnynVVLl.

Interlude I. Highway Landscapes

1 See Rojas, Olival, and Olival, "Cultivating Alternatives."
2 Under the Federal Forest Code, private landowners in the Amazon can legally deforest up to 20 percent of their land (with some exceptions). Moreover, a 2012 court ruling permitted smallholders in Acre to burn a few hectares of land each year for subsistence purposes, provided they had the requisite permits. The case, known as *Fogo Zero* (Zero Fire), was brought by the Ministerio Público of the state of Acre against the state government after extensive fires in 2005. Ministerio Públicos are designed to be independent, meant to represent the public interest, and considered a "virtual fourth branch of government." Hochstetler and Keck, *Greening Brazil*, 56. In 2009, the court supported the Ministerio

Público's action and required the state to ban all burning and to increase assistance to smallholders to help them transition away from fire usage. The state government appealed, arguing that clearing and burning were "traditional" land use practices of Acrean smallholders and were necessary for their food security. In 2012, another court overturned the initial ruling and allowed smallholders to obtain permits to burn a few hectares to grow food for their own sustenance. Some government interlocutors tied the state government's emphasis on increasing production without the use of fire to these court cases. The governor at the time brought together relevant government agencies and actors and directed them to provide rural people with viable land use alternatives. Out of this came the 2008 State Policy for the Valorization of Forest Environmental Assets (Política estadual de valorização do ativo ambiental floresta), often seen as the precursor to SISA.

3   See Ferguson, "Bovine Mystique," for a discussion of livestock as savings.

4   See Bowman et al., "Persistence of Cattle Ranching in the Brazilian Amazon"; Skidmore et al., "Cattle Ranchers and Deforestation in the Brazilian Amazon"; and Walker, Moran, and Anselin, "Deforestation and Cattle Ranching in the Brazilian Amazon." On cattle in Acre, see Toni et al., *Expansão e trajetórias da pecuária na Amazônia*. There have been extensive academic discussions about the causes of deforestation. In an influential paper, Geist and Lambin, for example, distinguish between proximate causes and underlying drivers in their analysis of causes of tropical deforestation. They call cattle ranching a proximate cause in the sense that at the local level, it "originate[s] from intended land use and directly impact[s] forest cover." Geist and Lambin, "Proximate Causes and Underlying Driving Forces," 143. Underlying drivers include the market for beef at various scales, government incentives for cattle ranching, and cultural valuation of cattle.

5   See, for example, Cortner et al., "Perceptions of Integrated Crop-Livestock Systems"; Strassburg et al., "Moment of Truth for the Cerrado Hotspot"; and Valentim and Andrade, "Tendências e perspectivas da pecuária bovina na Amazônia Brasileira."

6   Brazil nut trees were often saved when the rest of the forest was felled around them. It was, I was told, illegal to cut them down and you could be charged hefty fines for felling them. The nuts they made were also economically valuable, yet they often made few of them when growing in open fields, without the forest's microclimate. Smoke too can make them less productive, I was told.

7   See Hoelle, *Rainforest Cowboys*, 16.

8   Hecht, "Logic of Livestock and Deforestation"; Hoelle, *Rainforest Cowboys*.

9   Hoelle, *Rainforest Cowboys*.
10 As Jeff Hoelle shows in *Rainforest Cowboys*, *fazendeiros* dominated much of the Acrean upper class.
11 The enforcement of deforestation policies starting in the mid-2000s may have made smallholders' land a more valuable part of cattle ranching. A 2014 study found that larger properties (more than 500 hectares) accounted for almost half of Amazonian deforestation between 2004 and 2011. Godar et al., "Actor-Specific Contributions." These properties were obvious targets of early anti-deforestation enforcement. Moreover, because they tended to have larger swaths of deforested land, they were easier for satellite-based monitoring to flag. The same study found that the portion of Amazonian deforestation attributable to smallholders increased in the late 2000s.
12 Bowman et al., "Persistence of Cattle Ranching in the Brazilian Amazon"; Gomes, Perz, and Vadjunec, "Convergence and Contrasts in the Adoption of Cattle Ranching"; Hoelle, *Rainforest Cowboys*; Kröger, "Deforestation, Cattle Capitalism and Neodevelopmentalism"; Walker, Moran, and Anselin, "Deforestation and Cattle Ranching in the Brazilian Amazon." Changes in monetary policy in the late 1990s lowered the cost of Brazilian exports, encouraging growth in export-oriented commodities including cattle, while banks began to loan more money for Amazonian cattle production in tandem with the development of cattle vaccination and hygiene programs. These policies and practices contributed to cattle-linked deforestation.

Chapter 2. Producing the Forest

Some of the ideas in chapter 2 first appeared in "Rubber and Carbon: Opportunity Costs, Incentives and Ecosystem Services in Acre, Brazil," *Development and Change* 51, no. 1 (2020): 51–72.

1   Açaí is a staple food for at least three million people in Pará and elsewhere in eastern Amazônia. IPAM, "Desafios para a sustentabilidade na cadeia do açaí: Subsídios para a iniciativa açaí sustentável," November 7, 2018, https://blog.institutoterroa.org/wp-content/uploads/2020/11/Desafios-para-a-Sustentabilidade-na-Cadeia-do-Acai.pdf.

    For further discussion, see Eduardo Brondízio's *Amazonian Caboclo and the Açaí Palm*, which is focused on the state of Pará and the berry's growth into a trendy food in Brazil and internationally.

2   Souza and Souza, "Crescimento da produção de açaí e castanha-do-Brasil no Acre," 162. These numbers are not certain, however; a 2019 article on the state government's website said that were no "consolidated numbers

on the production of açaí in the state." Samuel Bryan, "Seminário entre técnicos e produtores discute formas de expandir o mercado do açaí no Acre," *Notícias do Acre*, September 26, 2019, https://agencia.ac.gov.br/seminario-entre-tecnicos-e-produtores-discute-formas-de-expandir-o-mercado-do-acai-no-acre/.

3   Lopes et al., "Mapping the Socio-Ecology of Non Timber Forest Products (NTFP) Extraction in the Brazilian Amazon," 112.

4   Souza and Souza, "Crescimento da produção de açaí e castanha-do-Brasil no Acre."

5   See Gupta et al., "In Pursuit of Carbon Accountability"; Kosoy and Corbera, "Payments for Ecosystem Services as Commodity Fetishism"; Leach and Scoones, "Carbon Forestry in West Africa"; Turnhout et al., "Envisioning REDD+ in a Post-Paris Era"; and Vijge and Gupta, "Framing REDD+ in India."

6   Regarding individual productivity, entrepreneurship, and neoliberalism, see, for example, Elyachar, *Markets of Dispossession*.

7   Dove, "Revisionist View of Tropical Deforestation and Development," 23. NTFPs became important to global forest conservation starting in the 1990s. See Arnold and Pérez, "Can Non-timber Forest Products Match?"; Neumann and Hirsch, *Commercialisation of Non-timber Forest Products*; and Ticktin, "Ecological Implications of Harvesting Non-timber Forest Products." As practiced in Acrean forests, rubber can be considered an NTFP.

8   Some land-use change scientists argue that intensification leads to increased supply and therefore decreased price, which in turn leads to decreased production, less pressure for ecosystem conversion, and even a *transition* back to forest. This causal chain is often called land-sparing and is linked to the Borlaug hypothesis, named for Norman Borlaug, the "father" of the Green Revolution. Scott Kilman and Roger Thurow, "Father of 'Green Revolution' Dies," *Wall Street Journal*, September 13, 2009. The hypothesis argues that increased crop yields lead to decreased deforestation and other benefits. See Borlaug, "Feeding a Hungry World." However, others argue that intensification is not associated with forest protection. Rudel et al., "Agricultural Intensification and Changes in Cultivated Areas." Or they have found that intensification can contribute to more deforestation since it makes cleared land more valuable. Kaimowitz and Angelsen, "Will Livestock Intensification Help Save Latin America's Tropical Forests?" This analysis sometimes draws on the Jevons Paradox, which posits that increased efficiency leads to increased consumption. Alcott, "Jevons' Paradox." See Pelletier et al., in "Does Smallholder Maize Intensification Reduce Deforestation?"

for a discussion of the links between smallholder intensification and reduced deforestation in Zambia. On soy and forest sparing, see Phalan et al., "Reconciling Food Production and Biodiversity Conservation"; Thaler, "Land Sparing Complex"; and Garrett et al., "Intensification in Agriculture-Forest Frontiers."

As LaShandra Sullivan has argued about another part of Brazil, the "implicit indexing of whiteness through its association with such modern land use practices facilitates . . . land grab[s]." Sullivan, "Identity, Territory and Land Conflict in Brazil," 455. See also Baletti, "Saving the Amazon?"; Rojas, Olival, and Olival, "Cultivating Alternatives"; Sullivan, "Overseen and Unseen"; and Thaler, "Land Sparing Complex."

9   See Neumann, *Imposing Wilderness*; and West, Igoe, and Brockington, "Parks and Peoples." While in some places protected areas have been created and governed as the "antithesis of production," often a significant amount of labor is involved. Sodikoff, *Forest and Labor in Madagascar*, 445; Münster, "Working for the Forest."

10 On the inadequacies of some community-focused conservation efforts, see, for example, Brosius, Tsing, and Zerner, *Communities and Conservation*; Charnley and Poe, "Community Forestry in Theory and Practice"; Dressler et al., "From Hope to Crisis and Back Again?"; and Hughes, *From Enslavement to Environmentalism*.

11 The Acrean strategy more closely aligned with the "land sharing" approach that is increasingly advocated for as an alternative to "land sparing." See Fischer et al., "Land Sparing versus Land Sharing"; and Law and Wilson, "Providing Context for the Land-Sharing and Land-Sparing Debate." But neither were terms that I heard used in interviews, workshops, or everyday discussions.

12 Susanna Hecht, in "From Eco-Catastrophe to Zero Deforestation?," 7, uses the term "politics of agreement" to explicate the decrease in Amazonian deforestation between 2005 and 2012.

13 Mathews, *Instituting Nature*; Tsing, *Friction*.

14 See Hugh Raffles's *In Amazonia*, chapter 3, for a beautiful account of hard work, desire, imagination, and entanglement with international political economy that went into placemaking in another part of the Brazilian Amazon.

15 McKinsey & Company, *Pathways to a Low-Carbon Economy*, 120. The McKinsey consultants write that the land uses practiced in tropical forest areas (including "slash-and-burn agriculture and conversion to pasture") are worth below €2 per ton of avoided carbon dioxide emissions (120, 121). Critics of REDD+ and related programs to pay for ecosystem services have argued that this cheapness is based on, and reinforces,

global inequities that keep many people in places like the Brazilian Amazon poor. See McAfee, "Contradictory Logic of Global Ecosystem Services Markets."

In fact, though, REDD+ has not been a cheap way to reduce emissions. The McKinsey report (121) offers one clue as to why in its list of costs not included in their estimates: "While these costs include ongoing . . . monitoring and management of preserved forests, they do not include transaction costs, the cost of building new infrastructure, or the capacity-building cost necessary to set up the monitoring and management infrastructure, which itself could account for a reasonably large portion of the total cost in tropical countries. Also, the costs of avoiding leakage and insuring the permanence of carbon stocks against natural disturbance events are not included." The term "transaction costs" does a lot of work. Greenleaf, "Rubber and Carbon."

16   McKinsey & Company, *Pathways to a Low-Carbon Economy*, 118.

17   See Alatas, *Myth of the Lazy Native*, for a discussion of this trope in the colonial period, and Li and Semedi, *Plantation Life*, and Li, "Dynamic Farmers, Dead Plantations," for discussions of its ongoing impacts in Indonesia's oil palm plantations.

18   Li, "Fixing Non-market Subjects," 45–46.

19   Hughes, *From Enslavement to Environmentalism*, 185 (emphasis in the original).

20   See Li and Semedi, *Plantation Life*, for a discussion of this logic of efficiency and the resulting harmful "plantation life" and "corporate occupation" in and around Indonesian oil palm plantations.

21   See Li, "Fixing Non-market Subjects," for a Foucauldian analysis of how, since colonialism, certain rural land and people (often Indigenous) have been slotted into "a non-market niche" (34).

22   The cash payments included the rubber subsidy I describe in chapter 1, a "certification" program that paid a bonus to smallholders who adopted sustainable agriculture plans, and specific monetary grants for some Indigenous communities.

23   See Greenleaf, "Rubber and Carbon."

24   Keeping poor rural people from moving to cities—where jobs were scarce, food was sometimes difficult to access, and crime was a consistent concern—was something of a secondary goal of Acrean rural development, something mentioned to me offhandedly by some government and NGO interlocutors.

25   The Acrean approach to valorizing the forest embraced the fundamental productivist idea that, as James Ferguson puts it, "altering a person's relation to the system of production" by incorporating them into it would

"produce[] a transformation far more profound, more 'structural'" than simply distributing money for forgoing deforestation. Ferguson, *Give a Man a Fish*, 39. As I explore in chapters 3 and 4, however, Acrean forest carbon valorization entailed its own politics of distribution.

26 See Tsing, Mathews, and Bubandt, "Patchy Anthropocene."

27 However, as I examine in chapters 3 and 4, it was not always the case that forest beneficiaries wanted to become more productive market participants.

28 Açaí often entailed gendered and generationally structured work done within families. Young men and children did most of the collecting. Women tended to do the at-home processing. See Lopes et al., "Mapping the Socio-Ecology of Non Timber Forest Products (NTFP) Extraction in the Brazilian Amazon."

29 Small churches, many of them Pentecostal, also dotted the landscape.

30 For a discussion of the importance of açaí in Feijó, see Lopes et al., "Mapping the Socio-Ecology of Non Timber Forest Products (NTFP) Extraction in the Brazilian Amazon."

31 Something similar happened in the floodplains of Pará, where smallholders increased açaí production in response to market demand based on existing methods. Brondízio, *Amazonian Caboclo and the Açaí Palm*; Brondízio and Siquera, "From Extractivists to Forest Farmers."

32 See Harrison, *Forests*.

33 In recent years, the plantation form has been a generative and controversial concept within the humanities and humanistic social sciences, pioneered by Black feminist scholars. In particular, Katherine McKittrick, in "Plantation Futures," 4, analyzes the use of the plantation as a way to "think[] through long-standing and contemporary practices of racial violence," drawing on Sylvia Wynter's work. Wynter, "Novel and History, Plot and Plantation." Speaking to its importance, the plantation has also, controversially, been used to offer an alternative epochal name to the Anthropocene: the "Plantationocene." See Davis et al., "Anthropocene, Capitalocene, . . . Plantationocene?"; Haraway, "Anthropocene, Capitalocene, Plantationocene, Chthulucene"; Jegathesan, "Black Feminist Plots"; and Wolford, "Plantationocene."

Anna Tsing, Andrew Mathews, and Nils Bubandt, in "Patchy Anthropocene," offer the plantation as a prototypical example of one of the "dominant landscape forms in the contemporary world." It is a "modular simplification" that violently "attempt[s] to reduce the number of living things in an area to just one kind." Forest-based *nativo* açaí, though, differs both from plantations and the "feral proliferations" that can escape them, which Tsing, Mathews, and Bubandt also highlight (S189).

34   See Nogueira and Santana, "Benefícios socioeconômicos da adoção de novas tecnologias," for an economic analysis of the benefits of the "technicization" of açaí production in Pará, the adoption of "industrial management" within areas of native açaí palm cultivation in wet *várzea* areas, and the introduction of irrigated açaí cultivation on dry land.

35   Maria do Socorro Padilha de Oliveira and João Tomé de Farias Neto, "Cultivar brs-Pará: Açaizeiro para produção de frutos em terra firme," Embrapa, December 2004, https://www.infoteca.cnptia.embrapa.br/bitstream/doc/382295/1/com.tec.114.pdf; Eliane Silva, "Açaí de terra firme avança no Pará e ganha espaço no interior de São Paulo," *O Globo Rural*, June 7, 2021, https://globorural.globo.com/Noticias/Agricultura/noticia/2021/06/acai-de-terra-firme-avanca-no-para-e-ganha-espaco-no-interior-de-sao-paulo.html.

36   Embrapa was a representative on one of sisa's key institutional bodies: the Commission for Validation and Accompaniment, which was meant to advise and monitor sisa's implementation and was composed of representatives from four governmental and four nongovernmental institutions. Additionally, the first director of the imc had been a researcher at Embrapa and returned to it after he stepped down from the imc in 2013.

37   Roberto Vaz, "Acreano investe mais de R$6 milhões em plantação de açaí," *Ac24horas*, January 31, 2013, https://ac24horas.com/2013/01/31/acreano-investe-mais-de-r-6-milhoes-em-plantacao-de-acai/. See also Amazon Sat, "Projeto para exportar açaí produzido no Acre," YouTube, October 5, 2013, video, 8:01, https://www.youtube.com/watch?v=2fRJDv3YjBw.

38   In Brazilian Portuguese, the term *plantation* does not automatically carry the same scalar, labor, or racial implications as it does in English. You can have a small *plantação* in your backyard, for example, for growing vegetables. In this context though, plantations indicated large-scale açaí planting.

39   Elton Disner and Da Redação, "Rei do açaí irrigado e réu na Madeira Limpa, Eloy Vaccaro morre nos eua," Jeso Carneiro, August 29, 2021, https://www.jesocarneiro.com.br/para/rei-do-acai-irrigado-e-reu-na-madeira-limpa-eloy-vaccaro-morre-nos-eua.html; Evilásio Cosmiro, "Principal 'Pirata' da exploração de madeira ilegal na Amazônia recebeu prêmio de Sebastião Viana," *Site Cultural de Feijó* (blog), August 2015, http://acrefeijonew.blogspot.com/2015/08/principal-pirata-da-exploracao-de.html.

40   As quoted in *Notícias do Acre*, "'Acre está pronto para empreender o açaí', afirma maior produtor do país," February 19, 2015, https://agencia.ac.gov.br/acre-esta-pronto-para-empreender-o-acai-afirma-maior-produtor

-do-pais/. In 2015, Notícias do Acre, a state government news site, published an article publicizing Eloy Pedro Vaccaro's visit to the state. According to the article, Acre's governor said that the state government would continue to invest in açaí research, beginning by visiting Vaccaro's plantations in the state of Pará, to learn how to promote plantations in Acre's açaí development.

41 As quoted in Eliane Silva, "Açaí de terra firme avança no Pará e ganha espaço no interior de São Paulo."

42 Hecht and Cockburn, *Fate of the Forest*, 65.

43 Bunker, *Underdeveloping the Amazon*; Moran, *Developing the Amazon*; Schmink and Wood, *Contested Frontiers in Amazonia*; Smith, *Rainforest Corridors*.

44 This transformation happened in part through the conversion of a lot of the central savanna ecosystem—the Cerrado—into a space of intensive agribusiness. How did the country so quickly convert this ecosystem—which Norman Borlaug said had such poor soils that "nobody thought . . . [it would] ever . . . be productive"—into an agribusiness center? As quoted in Larry Rohter, "Scientists Are Making Brazil's Savannah Bloom," *New York Times*, October 2, 2007. *The Economist* offers an answer: "Embrapa, Embrapa, Embrapa." Rather than import models of agricultural growth developed elsewhere, Embrapa researchers successfully advanced tropical agricultural research, transforming Cerrado agriculture. "The Miracle of the Cerrado," *The Economist*, August 26, 2010. Embrapa has also long been active in the Amazon, including in Acre. Its sprawling campus just outside Rio Branco on the BR-364 employed dozens of agricultural and forestry researchers, drawing talent from around Brazil to work on topics like sustainable timber extraction and intensive rubber production.

45 As quoted in Bakx, "From Proletarian to Peasant," 153.

46 Gustavo Oliveira and Susanna Hecht describe similar contrasting production styles in the context of soy: soy "carries the visual imprint of order and efficiency, the reality of a highly technified, 'fordist' style of production rather than the scruffy, atavistic and externally unintelligible features of locally complex agriculture that include substantial areas with woody species and differentiated landscape management." Oliveira and Hecht, "Sacred Groves, Sacrifice Zones and Soy Production," 252.

47 On the appeal of plantation agriculture in Mozambique, for example, see Wolford, "Plantationocene."

48 As quoted in Bruno Taitso, "Sustainable Production Chains Gain Strength in Acre," World Wildlife Fund, June 28, 2011, https://www.wwf.org.br/?29103/Sustainable-production-chains-gain-strength-in-Acre.

49 As quoted in Samuel Bryan, "Seminário entre técnicos e produtores discute formas de expandir o mercado do açaí no Acre," *Notícias do Acre*, September 26, 2019, https://agencia.ac.gov.br/seminario-entre-tecnicos-e-produtores-discute-formas-de-expandir-o-mercado-do-acai-no-acre/.

50 See Braverman, *Planted Flags*, chapter 1.

51 Eliane Silva, "Açaí de terra firme avança no Pará e ganha espaço no interior de São Paulo."

52 See Kaimowitz and Angelsen, "Will Livestock Intensification Help Save Latin America's Tropical Forests?," for a discussion of this dynamic around livestock.

53 Weinstein and Moegenburg, "Açaí Palm Management in the Amazon Estuary," 337. See Weinstein and Moegenburg's discussion about this dynamic in Pará's floodplains (*várzea*). Recent years have seen efforts to avoid açaí becoming another monocrop there. Embrapa discourages this kind of planting, instead recommending that açaí be planted as part of agroforestry systems, alongside cocoa and bananas, for example. See IPAM, "Desafios para a sustentabilidade na cadeia do açaí"; Eliane Silva, "Açaí de terra firme avança no Pará e ganha espaço no interior de São Paulo." Pará's government established a maximum number of açaí trees that can be cultivated per area unit. Mongabay Brasil, "O impacto da demanda do açaí nas florestas de várzea da Amazônia," October 4, 2021, https://conexaoplaneta.com.br/blog/o-impacto-da-crescente-demanda-global-do-acai-nas-florestas-de-varzea-da-amazonia/. Moreover, *várzea* is not supposed to be cleared for planted açaí. Elisa Vaz, "Sazonalidade do açaí diminui produção e eleva preços," *O Liberal*, January 14, 2022, https://www.oliberal.com/economia/sazonalidade-do-acai-diminui-producao-e-eleva-precos-1.483440. Yet many do not follow these rules, resulting in negative ecological impacts in the *várzea* landscapes where traditional açaí production takes place. Most of the research on açaí has taken place in Pará, and differences in hydrology and ecology in Acre may shape its impacts there in ways that are not accounted for in research to date.

54 On the Green Revolution agriculture, see, for example, Patel, "Long Green Revolution." On its exclusions, see, for example, Hetherington's discussion in *Government of Beans* of how soy production in Paraguay often violently excludes smallholders.

55 IPAM, "Desafios para a sustentabilidade na cadeia do açaí"; Eliane Silva, "Açaí de terra firme avança no Pará e ganha espaço no interior de São Paulo."

56 IPAM, "Desafios para a sustentabilidade na cadeia do açaí." In Pará, IPAM reports that high açaí prices have a number of negative impacts: they un-

dermine local consumers' ability to eat this staple and culturally important food, and they increase land grabs and speculation in areas where rural people lack secure land tenure. Additionally, large-scale açaí processing threatens those who traditionally process açaí in Belém (*batedores*). IPAM, "Desafios para a sustentabilidade na cadeia do açaí," 27, 29, 31.

57 The way that açaí is sometimes extracted from its traditional floodplain *várzea* landscapes in Pará can have negative impacts on biodiversity. IPAM, "Desafios para a sustentabilidade na cadeia do açaí." Moreover, the irrigation demands of planted açaí could potentially reduce water availability for other species during Acre's dry summer months, particularly during climate-linked droughts. However, it appears that little research has been published on intensified açaí's impact on water availability.

58 Eliane Silva, "Açaí de terra firme avança no Pará e ganha espaço no interior de São Paulo."

59 Li, *Will to Improve*.

60 See, for example, Dove, "Theories of Swidden Agriculture"; Li, *Will to Improve*; Li and Semedi, *Plantation Life*; Mertz et al., "Swidden Change in Southeast Asia"; and Peluso and Vandergeest, "Genealogies of the Political Forest and Customary Rights." For example, in 1957, staff of the Food and Agricultural Organization described swidden agriculture as "the greatest obstacle not only to the immediate increase of agricultural production, but also to the conservation of the production potential for the future, in the form of soils and forests," and as "not only a backward type of agricultural practice . . . [but] also a backward stage of culture in general." As quoted in Mertz et al., "Swidden Change in Southeast Asia," 259.

61 Dove, "Plants, Politics, and the Imagination"; Li and Semedi, *Plantation Life*; Li, "Dynamic Farmers, Dead Plantations, and the Myth of the Lazy Native."

62 Elisa Vaz, "Sazonalidade do açaí diminui produção e eleva preços."

Interlude II. The Flood

1 To the outrage of many Acreans, a Rondônian politician named Roberto Dorner ensured that, for many years, no bridge crossed the Rio Madeira. Instead, all the traffic importing goods to Acre had to ride on ferries he owned—a lucrative racket that positioned him as the "owner" of the river. Altino Machado, "Deputado ganha R$1,8 milhão por mês com transporte no Rio Madeira," *Terra Magazine*, August 30, 2011, https://www.rondoniagora.com/politica/deputado-ganha-r-1-8-milhao-por-mes-com-transporte-no-rio-madeira. Those ferries could not run when the river was too high, as in the 2014 flood. They also had to be docked when

the river was too low, as could happen during dry summers. At both times, Acre was left temporarily isolated. Eventually, a bridge was built and opened in 2021. "Ponte do Abunã sobre o rio Madeira é inaugurada na sexta-feira (7) e cria expectativa para o desenvolvimento da região," *O Globo*, May 6, 2021, https://g1.globo.com/ro/rondonia/noticia/2021/05 /06/ponte-do-abuna-sobre-o-rio-madeira-e-inaugurada-na-sexta-feira -7-e-cria-expectativa-para-o-desenvolvimento-da-regiao.ghtml.

Chapter 3. Robin Hood in the Untenured Forest

A different version of chapter 3 appeared as "The Value of the Untenured Forest: Land Rights, Green Labor, and Forest Carbon in the Brazilian Amazon," *Journal of Peasant Studies* 47, no. 2 (February 2020): 286–305.

1 "The traditional power structure fragmented in most of the state," as the rubber economy declined. Bakx, "From Proletarian to Peasant," 151. The dictatorship did not promote the state as a replacement power structure in some parts of the Amazon. As David Cleary put it in the early 1990s, "Wherever one looks in the Amazonian economy, the state is in retreat: unable to finance tax breaks or build highways without the aid of multilateral banks, unable to include more than one per cent of the rural population in official colonisation schemes, unable to control land titling or land conflicts, unable to register or tax the greater part of the Amazonian economy, unable to enforce federal law on more than a sporadic basis." Cleary, "After the Frontier," 344. The transition to democracy brought a devolution of power from the federal to state and municipal governments. Hochstetler and Keck, *Greening Brazil*. But through the 1990s, and beyond, the state government remained quite absent in much of Acre. See Kainer et al., Experiments in Forest-Based Development," 874; and Schmink and Cordeiro, *Rio Branco*.

2 See, for example, Dean, *With Broadax and Firebrand*.

3 Bolsa Família is a conditional cash transfer program that was formulated (based on four existing government programs) and expanded under President Lula's administration. It is credited with significantly reducing poverty in Brazil. It pays low-income households (with money usually distributed to women) based on conditionalities like children's school attendance and vaccination, though it is questionable the extent to which such conditionalities are enforced. It also includes payments to extremely poor households without children. See, for example, Soares, Ribas, and Osório, "Evaluating the Impact of Brazil's Bolsa Família"; Cotta and Machado, "Programa Bolsa Família e segurança alimentar e nutricional no Brasil"; and Neves et al., "Brazilian Cash Transfer Program (Bolsa Família)."

4 See, for example, Luttrell et al., "Who Should Benefit from REDD+?"; Dunlop and Corbera, "Incentivizing REDD+"; Pham et al., "Approaches to Benefit Sharing"; Weatherley-Singh and Gupta, "Drivers of Deforestation and REDD+ Benefit-Sharing"; Schroeder and McDermott, "Beyond Carbon"; and Wong et al., "Narratives in REDD+ Benefit Sharing."

5 Landowners could then decide whether and how to redistribute money received from offset sales. Getting carbon offsets certified by institutions like the Climate, Community, and Biodiversity Alliance could require a certain degree of benefit sharing. Private REDD+ projects could also be run by NGOs, businesses, or other entities.

6 Mendoza, Greenleaf, and Thomas, "Green Distributive Politics"; Ferguson, *Give a Man a Fish*.

7 Hecht, "From Eco-Catastrophe to Zero Deforestation?," 5.

8 See, for example, Rofel and Yanagisako, *Fabricating Transnational Capitalism*.

9 Anthropologists have long studied property as a nexus of social relations with respect to things. Campbell, *Conjuring Property*; Gluckman, *Ideas in Barotse Jurisprudence*; Hann, *Property Relations*; Hetherington, *Guerrilla Auditors*; Humphrey and Verdery, *Property in Question*; Morris, "Speculative Fields"; Verdery, *Vanishing Hectare*. For legal scholarship taking a relational approach, see, for example, Geisler and Daneker, *Property and Values*; and Rose, *Property and Persuasion*. Meghan Morris, in "Property's Relations," offers a linked history that examines anthropology's influence on conceptions of property.

10 Dehm, "Indigenous Peoples and REDD+ Safeguards," 191–92.

11 Eliasch, *Climate Change*, 35.

12 Baker McKenzie and Covington and Burling LLP, "Background Analysis of REDD Regulatory Frameworks," May 2009, https://www.un-redd.org/sites/default/files/2021-10/Background_Analysis_of_REDD_Regulatory_Frameworks_9%20June%202009.pdf; as cited in Dehm, "Indigenous Peoples and REDD+ Safeguards," 192.

13 Forest Carbon Partnership Facility, "Carbon Fund Methodological Framework," World Bank, April 2020, 23, https://www.forestcarbonpartnership.org/system/files/documents/fcpf_carbon_fund_methodological_framework_revised_2020_final_posted_1.pdf. A study of "Readiness Plan Idea Notes" that countries prepared to obtain funding from the World Bank's Forest Carbon Partnership Facility found that "almost all . . . recognize the need to clarify land tenure systems in the context of REDD implementation." Crystal Davis et al., "A Review of 25 Readiness Plan Idea Notes from the World Bank Forest Carbon Partnership Facility," World Resources Institute, February 1, 2009, 2, https://www.wri

.org/research/review-25-readiness-plan-idea-notes-world-bank-forest-carbon-partnership-facility.

14 CCBA, "Climate, Community and Biodiversity Standards, Version 3.1," June 21, 2017, 21, https://verra.org/wp-content/uploads/2017/12/CCB-Standards-v3.1_ENG.pdf.

15 Börner et al., "Direct Conservation Payments in the Brazilian Amazon," 1273, 1275. To make REDD+ both effective and equitable, other REDD+ scholarship emphasizes the importance of, as Duchelle and colleagues put it, "clarifying and securing tenure rights—before REDD+ begins—is thus needed for the application of both regulatory and incentive-based REDD+ mechanisms." Duchelle et al., "Linking Forest Tenure Reform," 54. See also Corbera et al., "Rights to Land, Forests and Carbon in REDD+"; Greenleaf, "Using Carbon Rights"; Larson et al., "Land Tenure and REDD+"; Sunderlin et al., "Creating an Appropriate Tenure Foundation for REDD+."

16 John Locke positioned private property ownership as a key to proper citizenship and land use. Locke, *Second Treatise of Civil Government*. As economist Hernando de Soto puts it, "legal property"—via the state issuance of individual titles—gives citizens of "the advanced nations of the West . . . the means to discover . . . the most potentially productive qualities of their resources." It is the "key to modern development." Soto, *Mystery of Capital*, 51.

17 Hardin, in "Tragedy of the Commons," framed environmental degradation, like resource depletion and pollution, as the result of rational individual overuse of unowned commons. Landowners with secure property rights are positioned as perhaps the best resource stewards in this line of thought. Hardin and many others did not recognize that many "commons" throughout the world are effectively governed through community rules and relationships, as Elinor Ostrom and others have since documented. See, for example, Ostrom, *Governing the Commons*; and Ostrom et al., "Revisiting the Commons."

18 See, for example, Coase, "Problem of Social Cost"; Dales, *Pollution, Property and Prices*; Demsetz, "Toward a Theory of Property Rights."

19 West, *From Modern Production to Imagined Primitive*, 27.

20 Depending on the legislation or other governing rules, carbon rights can also be separated from land rights.

21 Harvey, *Brief History of Neoliberalism*, 165; Castree, "Commodifying What Nature?"

22 Bumpus and Liverman, for example, write that "a key concern, therefore, is who defines ownership over carbon, over what scales and with what outcomes." Bumpus and Liverman, "Accumulation by Decarbonization,"

218. See also Boykoff et al., "Theorizing the Carbon Economy"; Bridge, "Resource Geographies 1"; Bryant, "Politics of Carbon Market Design"; Kongsager and Corbera, "Linking Mitigation and Adaptation in Carbon Forestry Projects"; Osborne, "Tradeoffs in Carbon Commodification"; Wang and Corson, "Making of a 'Charismatic' Carbon Credit."

23 Rofel and Yanagisako, in *Fabricating Transnational Capitalism*, also show how the meaning of "privatization" and the line between public and private are themselves historically and culturally constituted.

24 Tsing, *Mushroom at the End of the World*, 267.

25 Jeferson Almeida et al., "Leis e práticas de regularização fundiaria no estado do Acre," Imazon, March 2021, 9, https://imazon.org.br/wp-content/uploads/2021/03/LeisRegularizacaoFundiaria_Acre.pdf.

26 A study by the Brazilian research organization Imazon reported that 14 percent of Acre is composed of private property, 32 percent is in conservation areas, and 15 percent is designated as Indigenous Territories. Jeferson Almeida et al., "Leis e práticas de regularização fundiaria no estado do Acre."

27 Holston, *Insurgent Citizenship*, 121.

28 Bakx, "From Proletarian to Peasant," 145. See also Ribeiro, *Os índios e a civilização*.

29 What counts as productive use has changed somewhat over time, but cultivation has been consistently valued as evidence. Deforestation has been seen as cultivation's precursor and therefore a first step in land claiming, and even as productive use in and of itself. Alston, Libecap, and Mueller, *Titles, Conflict, and Land Use*; Araujo et al., "Property Rights and Deforestation"; Bunker, *Underdeveloping the Amazon*; Hecht, "Soybeans, Development and Conservation"; Schmink and Wood, *Contested Frontiers in Amazonia*. Following John Locke and other thinkers, Brazilian scholars and law have articulated a connection between cultivation, property, and moral character. For example, Clóvis Beviláqua, the Brazilian jurist and author of the 1916 Civil Code, explained that: "With the cultivation of land, the sentiment of individual property became accentuated, because productive work—creating, regularly, utilities corresponding to the effort employed—stabilized man and, tying him strongly to the generous soil, gave him a special personality." As quoted in Holston, *Insurgent Citizenship*, 116.

30 This process is often called *grilagem*, or land fraud. See Campbell, *Conjuring Property*. See also Holston, *Insurgent Citizenship*.

31 Campbell, *Conjuring Property*, 5.

32 Bakx, "From Proletarian to Peasant," 189.

33 Bakx, "From Proletarian to Peasant"; Ribeiro, *Os índios e a civilização*.

34 Bakx, "From Proletarian to Peasant," 146.

35 As quoted in Hecht, *Scramble for the Amazon*, 369.

36 Private REDD+ projects are permitted by SISA, and their forest carbon is deducted from the state's tally.

37 Diego's was one of five private forest carbon projects started by a few large Acrean landholders in the early to mid-2010s. They ranged in size from twenty-two thousand to two hundred thousand hectares and were developed in partnership with two small US-based companies. These projects aimed to generate carbon offsets to sell to companies and individuals seeking to compensate for their emissions for reputational reasons in the so-called voluntary carbon market. Some of the proceeds funded benefits for those *posseiros* who lived within the project boundaries—benefits that were not dissimilar from those funded through SISA, but which seemed, at least, to be smaller and less comprehensive than state benefits distributed along the BR-364 in Feijó. At least one of the projects was also controversial, with some outside observers accusing it of violating *posseiro* rights. See, for example, World Rainforest Movement, "Envira REDD+ Project in Acre, Brazil: Gold Certificate from Carbon Certifiers for Empty Promises," April 29, 2018, https://www.wrm.org.uy/bulletin-articles/envira-redd-project-in-acre-brazil-gold-certificate-from-carbon-certifiers-for-empty-promises.

38 To gain permission to visit their carbon projects, the carbon offset company had me sign an agreement requiring that that they review anything I write about my visits. Nothing I write here is based on visiting their projects.

39 Lack of formal title did not always mean insecurity. For example, a smallholder I interviewed, who lived down a newly constructed dirt side road off the BR-364 in Feijó, did not have title to his land but also said that he felt secure in his occupation because the landowner (*dono*) did not have all the documents to claim secure tenure either. Neither had a formalized or an easily formalizable land claim. This generated a certain version of security. The man and his wife had recently built a new house, and the couple had plans for the future: a new aluminum roof to replace the leaky thatched one; electricity—since the government had started to string wires down some side roads off the BR-364; and large trees they would plant for future logging. They planned to stay on this land and yet did not have land title to it. Theoretically, those who lack secure land title are not supposed to make such investments and take such care.

40 "Traditional people and communities" is a capacious category in Brazil that includes not only Indigenous people but also extractivists like

rubber tappers, *quilombolas* (descendants of escaped enslaved people), *ribeirinhos* (riverine people), and others. See Baletti, *Ordenamento Territorial*; Carneiro da Cunha and de Almeida, "Indigenous People, Traditional People, and Conservation in the Amazon"; Chagas, "A política do reconhecimento dos 'remanescentes das comunidades dos quilombos'"; Garfield, *Indigenous Struggle at the Heart of Brazil*; Little, "Territórios sociais e povos tradicionais no Brasil"; Mitchell, *Constellations of Inequality*; and Soares et al., "Fatores explicativos das demarcações de terras indígenas."

41 Article 231 of the 1988 Brazilian Constitution recognizes Indigenous peoples' "rights to the lands they traditionally occupy" and obligates the government to demarcate "Indigenous lands," for example. Some Indigenous lands have been demarcated. In Acre, 15 percent of the land is part of Indigenous territories. Jeferson Almeida et al., "Leis e práticas de regularização fundiaria no estado do Acre." However, many Indigenous territorial claims have not been demarcated—especially those in urban areas and other spaces occupied by politically powerful non-Indigenous communities. See, for example, Campbell, "Indigenous Urbanization in Amazonia"; Carneiro da Cunha et al., "Indigenous Peoples Boxed in by Brazil's Political Crisis"; and Chiavari and Lopes, "Indigenous Land Rights in Brazil." Some have undertaken other strategies, like "autodemarcation," in response. Vega et al., "Those Who Live like Us." On Quilombola land rights and antiblack racism, see, for example, Arregui, "Quilombola Movement"; French, "Dancing for Land"; Mitchell, *Constellations of Inequality*; and Perry, "Resurgent Far Right and the Black Feminist Struggle."

42 See chapter 4 for a discussion of rural producers' associations.

43 This requirement was out of step with the important role of rubber tapping in Acrean identity—a role that was reinforced by the rubber tapper movement and multiple state government administrations. For some, being a rubber tapper remained important to their identity even if they no longer actually tapped rubber. Hoelle, *Rainforest Cowboys*; Gomes, Vadjunec, and Perz, "Rubber Tapper Identities"; Vadjunec, Schmink, and Gomes, "Rubber Tapper Citizens."

44 These include both traditional settlements, which have encouraged deforestation, and newer forms, such as agro-extractivist settlements and sustainable development projects, which include more stringent forest conservation requirements than those of the Federal Forest Code.

45 In some other parts of Brazil, some social movements, such as the Landless Workers Movement (the Movimento dos Trabalhadores Rurais Sem Terra, or MST), have sometimes successfully pushed for land rights through use and occupation. See Caldar, "O MST e a formação dos sem

terra"; Carter, *Challenging Social Inequality*; Ondetti, *Land, Protest, and Politics*; Wolford, *This Land Is Ours Now*; and Wolford, "Participatory Democracy by Default." While there were some land occupations in Acre, I did not encounter such social movements in Feijó and was told that the MST was not very active in the state.

46 See Campbell, *Conjuring Property*, 179.

47 McKinsey & Company, *Pathways to a Low-Carbon Economy*.

48 This could include state regularization efforts. For example, the NGO Imazon found that Terra Legal, a federal regularization program for the Amazon, not only issued fewer titles than planned but has also favored wealthier landholders and agribusiness and exacerbated land speculation and land grabbing. Brenda Brito and Paulo Barreto, "A regularização fundiária avançou na Amazônia?," Imazon, January 2011, https://imazon.org.br/publicacoes/1404-2/. See also Campbell, *Conjuring Property*, 180–85; and Oliveira, "Land Regularization in Brazil."

49 Property size was also relevant: small plots, like Angelo and Lucia's, did not contain enough carbon to make them a viable source of carbon offsets on their own. Without land tenure, though, they had not yet confronted that obstacle.

50 Interview, August 2014, Rio Branco, Acre, Brazil.

51 The SISA program was considered a pioneering jurisdictional REDD+ program. See, for example, Palmer, Taschini, and Laing, "Getting More 'Carbon Bang' for Your 'Buck.'" It was an approach that some other governments were adopting as well. Dehm, "Indigenous Peoples and REDD+ Safeguards." Jurisdictional REDD+ programs are sometimes considered to be superior to private REDD+ projects for a number of reasons, including that they can be aligned with and integrated into other state land use policies and have the backing of governmental authority, enabling a more "landscape" level approach. See, for example, William Boyd et al., "Jurisdictional Approaches to REDD+ and Low Emissions Development: Progress and Prospects," World Resources Institute, June 2018, https://wriorg.s3.amazonaws.com/s3fs-public/ending-tropical-deforestation-jurisdictional-approaches-redd.pdf; Brandão et al., "Lessons for Jurisdictional Approaches"; Greenleaf et al., "Forest Policy Innovation at the Subnational Scale"; Ros-Tonen, Reed, and Sunderland, "From Synergy to Complexity"; and Turnhout et al., "Envisioning REDD+ in a Post-Paris Era." For a political ecology critique of jurisdictional REDD+ in Acre, see Santos Rocha da Silva and Correia, "Political Ecology of Jurisdictional REDD+."

52 Government of Acre, Lei N. 2.308, art 23, art 18, art 22.

53 Government of Acre, Lei N. 2.308, art 23.

54 Aggregating emissions reductions across the state was meant to make carbon calculations more robust. This is because efforts to reduce deforestation in one area can push deforestation into another area outside of a REDD+ project's boundaries. Since SISA included the entire state, it would theoretically capture any such leakage, as it is called, within state boundaries. In this way, the Acrean state became a kind of insurer of environmental quality. See Mathews, "Imagining Forest Futures and Climate Change," for a discussion of how the Mexican state took on this role in the context of a program with individual land ownership.

55 The two KfW contacts specified that a significant percentage (70 percent of the first payment and 90 percent of the second) "directly benefit actors at the local level." KfW, "REDD Early-Movers Acre Fact Sheet," 2017, 6, accessed July 1, 2021, https://www.kfw-entwicklungsbank.de/PDF/Entwicklungsfinanzierung/Themen-NEU/REDD-Early-Movers-Acre-Fact-Sheet.pdf.

56 Mathews, "Scandals, Audits, and Fictions," 99.

57 KfW, "REDD Early-Movers Acre Fact Sheet."

58 See also Environmental Defense Fund, "Ready for REDD: Acre's State Programs for Sustainable Development and Deforestation Control," 12, accessed July 2, 2022, https://www.edf.org/sites/default/files/Acre_Ready_for_REDD_EDF.pdf.

59 For a version of this argument, see Greenleaf, "Value of the Untenured Forest."

60 Government of Acre 2010, Section III, Art. 4.

61 For a discussion of how, in Brazil, Lockean views of labor undergird both neoliberal land reform and social movements that reject it and claim land directly, see Wolford, "Land Reform in the Time of Neoliberalism."

62 Ferguson, *Give a Man a Fish*, 39.

63 See Fischer, *Poverty of Rights*; Holston, *Insurgent Citizenship*; and Millar, *Reclaiming the Discarded*. Formal benefits for formal workers date particularly to the mid-twentieth-century administrations of President Getúlio Vargas.

64 Bakx, "From Proletarian to Peasant"; Martins, "Representing the Peasantry?"; Welch and Sauer, "Rural Unions and the Struggle for Land in Brazil."

65 See Hoelle, *Rainforest Cowboys*.

66 This analysis also aligns with research finding that the conditionality component of forest carbon and other PES programs often is not enforced, turning what are purportedly conditional environmental payments into something akin to subsidies. Fletcher and Breitling, "Market

Mechanism or Subsidy in Disguise?"; Lansing, "Understanding Linkages"; McElwee et al., "Payments for Environmental Services"; McElwee, Huber, and Nguyễn, "Hybrid Outcomes of Payments"; Shapiro-Garza, "Contesting the Market-Based Nature of Mexico's National Payments for Ecosystem Services Programs"; Van Hecken et al., "Silencing Agency in Payments for Ecosystem Services (PES) by Essentializing a Neoliberal 'Monster' into Being."

67   In the Amazon, the Vargas administration briefly emphasized, but did not fulfill, rubber tapper rights. Garfield, *In Search of the Amazon*, chapter 3.

68   For discussions of this kind of extractivism in Latin America, see, for example, Burchardt and Dietz, "(Neo-)Extractivism"; Gudynas, *Extractivisms*; Kingsbury, "Latin American Extractivism"; Mendoza, *Patagonia Sublime*; Morais and Saad-Filho, "Da economia política à política econômica"; Riofrancos, *Resource Radicals*; and Svampa, *Neo-Extractivism in Latin America*. These and other scholars and activists have argued that neoextractivism is both continuous with and disjunctive from past extractive economies. While extraction from Latin America and elsewhere to feed the appetites of the Global North has long been a mainstay of colonial and neoliberal exploitation, the term *neoextractivism* recognizes how leftist and left-of-center Latin American governments have also promoted extractive capitalism, sometimes to finance some of their redistributive policies, bolster state authority, and promote resource sovereignty.

69   The reasons for the decline in Amazonian deforestation at that time have been debated. See, for example, Assunção, Gandour, and Rocha, "Deforestation Slowdown in the Brazilian Amazon"; and Nepstad et al., "Slowing Amazon Deforestation through Public Policy."

70   Mitchell, "Society, Economy, and the State Effect."

71   See chapter 5 for further discussion of this form of state-building in Rio Branco.

72   See also Maria DiGiano et al., "The Twenty-Year-Old Partnership between Indigenous Peoples and the Government of Acre, Brazil: Lessons for Realizing Climate Change Mitigation and Social Justice in Tropical Forest Regions through Partnerships between Subnational Governments and Indigenous Peoples," Earth Innovation Institute, September 2018, 5, https://earthinnovation.org/uploads/2018/09/Acre_EN_online.pdf; and Duchelle et al., "Acre's State System of Incentives," 33–34. For a critical analysis of ZEE elsewhere, see Baletti, "*Ordenamento Territorial.*"

73   These fines could be high, over R$100,000, one official noted in a presentation I attended; landholders could sell their land many times over and not have enough money to pay them.

74 On forest carbon and territorialization, see Nel, "Contested Carbon." For analysis of resources and territorialization, see, for example, Bridge, "Resource Geographies 1"; Burchardt and Dietz, "(Neo-)Extractivism"; Rasmussen and Lund, "Reconfiguring Frontier Spaces"; and Vandergeest and Peluso, "Territorialization and State Power in Thailand." As Rasmussen and Lund (394) put it, "the ability to territorialize does not derive from a pre-existing authority. Rather, the successful territorializations of an extractive zone, a conservation area, or a community forest are co-productive of authority."

75 While I was in Acre, IMC officials were developing "social and environmental safeguards" meant to ensure the quality of both private forest carbon projects and state actions under SISA.

76 Early in REDD+'s development, scholars expressed concern about the possibility that it would lead to a new consolidation of state claims in and power over forested regions, including in places where significant decentralization had occurred in recent years. See Osborne, "Tradeoffs in Carbon Commodification"; and Phelps, Webb, and Agrawal, "Does REDD+ Threaten to Recentralize Forest Governance?"

77 Mathews, "Imagining Forest Futures and Climate Change," 217.

78 Government of Acre, Lei N. 2.308, art 4.

79 This is not unique to Acre. For example, in the 1960s, legal scholar Charles Reich traced how the distribution of government "largess" in the United States expanded the state's power through an "underlying philosophy . . . that the wealth that flows from government is held by its recipients conditionally." Reich, "New Property," 733, 768.

80 State-building efforts under SISA aligned with that of some other forest carbon programs, even though the mechanisms it adopted surrounding forest benefit distribution may have been more inclusive. See Osborne, "Tradeoffs in Carbon Commodification," for a discussion of how forest carbon's commodification in Chiapas, Mexico has undermined common property relations there via, among other dynamics, a centralization of forest governance.

81 Gago, *Neoliberalism from Below*, 19. Analysis of neoliberalism in the Global North has sometimes highlighted state retreat, the dissolution of social welfare programs and environmental protections, and the rise of volunteerism and NGOs. See, for example, Barry, Osborne, and Rose, *Foucault and Political Reason*; Muehlebach, *Moral Neoliberal*; and Pierson, *Dismantling the Welfare State?*

82 James Ferguson, in *Give a Man a Fish*, and Stephen Collier, in *Post-Soviet Social*, have explored some of the redistributive logics and policies that can develop from or adjacent to neoliberalism.

83   Land rights are a goal, for example, of the MST.

84   Recognizing the centrality of land rights to political power and sovereignty—as well as the importance of land and territory to many Indigenous communities—Indigenous communities and allied scholars have also emphasized securing land rights (often in forms other than individual titles) as a precondition to REDD+. For them, forest carbon's valorization is both a threat to hard-fought and still insecure land rights and an opportunity to secure new land rights for forest people and communities. See, for example, Larson et al., "Land Tenure and REDD+"; and Sunderlin et al., "Creating an Appropriate Tenure Foundation for REDD+."

85   Campbell, *Conjuring Property*; Holston, *Insurgent Citizenship*.

86   Campbell, *Conjuring Property*, 198.

87   See Carter, *Challenging Social Inequality*; Medeiros, "Social Movements"; Ondetti, *Land, Protest, and Politics*; Robles, "Revisiting Agrarian Reform in Brazil"; and Wolford, *This Land Is Ours Now*. See also Jo Guldi's account, in *The Long Land War*, of the fraught struggle for occupancy rights in many parts of the world.

Interlude III. The Rural Road, Part 1

1   Ogden, *Loss and Wonder at the World's End*, 45. See also Knox, "Affective Infrastructures and the Political Imagination."

Chapter 4. Beneficiaries and Forest Citizenship

A different version of chapter 4 was published as "Beneficiaries of Forest Carbon: Precarious Inclusion in the Brazilian Amazon," *American Anthropologist* 123, no. 2 (2021): 305–17.

1   Part of a program called ProAcre, the meetings took place in rural areas within a few hours of the state capital, Rio Branco.

2   See Aragão et al., "Interactions between Rainfall, Deforestation and Fires"; and Brondízio and Moran, "Human Dimensions of Climate Change."

3   The figure of the beneficiary appears in various contexts. Isabelle Strengers, for example, brings up beneficiaries in connection with capitalism: "More and more people . . . become defined as 'beneficiaries' . . . of something made possible 'elsewhere,' out of their reach." Latour et al., "Anthropologists Are Talking," 590. Even workers, she says, are now "said to enjoy the 'benefit' of a job, which they have to 'deserve.'" These beneficiaries, she says, are seen as distinct from those deemed to be "productive." Writing about a very different type of beneficiary—the

privileged beneficiary of global capitalism—Bruce Robbins argues that what beneficiaries receive is never "in fair exchange for services rendered or indeed recompenses you in any way for anything." Robbins, *Beneficiary*, 5.

4   The ways that exchange of various sorts can create relationships is, of course, a classic focus of economic anthropology. See, for example, Malinowski, *Argonauts of the Western Pacific*; and Mauss, *The Gift*.

5   Hetherington, *Government of Beans*, 114.

6   See Morais, *Acreanidade*; Schmink, "Forest Citizens"; Schmink et al., "Forest Citizenship in Acre, Brazil"; and Vadjunec, Schmink, and Gomes, "Rubber Tapper Citizens."

7   Forest citizenship is more akin to what Kregg Hetherington terms "regulatory citizens" meant to "continually make the state present by participating in the regulation of environment and health." Hetherington, *Government of Beans*, 114. Through the concepts of "environmentality" and "environmental rule," Arun Agrawal, in *Environmentality*, and Pamela McElwee, in *Forests Are Gold*, have also explored how governing the environment can also be a way to govern people.

8   See Han, "Precarity, Precariousness, and Vulnerability"; Millar, "Toward a Critical Politics of Precarity"; Molé, "Precarious Subjects"; Muehlebach, "On Precariousness and the Ethical Imagination"; and Thorkelson, "Precarity Outside."

9   Anand, *Hydraulic City*, 8. Other anthropologists have relatedly examined citizenship as uneven and contingent membership and political belonging instantiated and negotiated through quotidian state-citizen interactions, rather than as a legal status enacted through formalized rights and responsibilities. See, for example, Lazar, *Anthropology of Citizenship*.

10 Ferguson, *Give a Man a Fish*.

11 See, for example, Hunter, *Transformation of the Workers' Party in Brazil*, for a discussion of PT efforts to counter clientelist politics.

12 Benefit negotiations were often gendered, since both association presidents and technicians were usually male. Male association members also spoke more at association meetings I attended. However, women led or held important positions in many urban environmental agencies and NGOs involved in forest protection.

13 Both Gilberto Freyre, in *Sobrados e mucambos*, and Roberto DaMatta, in *Casa e a rua*, highlight the importance of the street, "*a rua*," as a contrast to "*a casa*"—the house. Whereas the latter symbolizes the private relations and responsibilities of family, the former is a public place characterized by things like struggle, pollution, and danger. Yet in rural Feijó, "the street" was also the town, where rural residents went to access

goods and services many wanted or needed—education, health care, and consumer goods. Moreover, when association members described Manuel going there, it was with approval. It showed that he was working hard to access benefits for them, and it signaled the possibility that more benefits might be forthcoming. In this, it could be a space of possibility.

14 As Kregg Hetherington has explored, clientelism can be "a form of accusation" that "has always tended to be used by educated urbanites against the rural poor." Hetherington, *Government of Beans*, 93.

15 For discussions of the entanglement of patronage and democratic politics in Latin America see, for example, Ansell, *Zero Hunger*. See also Ansell and Mitchell, "Models of Clientelism and Policy Change"; Auyero, *Poor People's Politics*; Auyero, Lapegna, and Poma, "Patronage Politics and Contentious Collective Action"; Eiró and Koster, "Facing Bureaucratic Uncertainty in the Bolsa Família Program"; Gay, "Rethinking Clientelism"; Goldman, *Como funciona a democracia*; Holston, *Insurgent Citizenship*; Koster, "Mediating and Getting 'Burnt' in the Gap"; Palmeira, "Voto"; and Tarlau, "Coproducing Rural Public Schools in Brazil." As technicians distributed benefits via associations and other groups (e.g., cooperatives) and pushed forest beneficiaries to work together to become independent, they promoted the kind of social solidarity and "horizontal" ties that patronage politics are sometimes said to undermine. Roniger, *Hierarchy and Trust in Modern Mexico and Brazil*.

16 These meetings were one type of intermittent interaction between technicians and beneficiaries, the frequency and intimacy of which depended on both the particular relationships of those involved and how easy it was for beneficiaries to visit urban headquarters. For many, visiting urban areas was prohibitively expensive and time-consuming.

17 Mucuna was popularized and promoted by Embrapa in southern Brazil, with its drier climate and more mechanized agriculture. In Acre, it was included in the Certification for Sustainable Rural Properties Project, adopted into law in 2008, that provided a small payment ("bonus") of US$250 per year, technical assistance, and agricultural equipment and inputs to smallholders who adopted agricultural plans and practices deemed sustainable. Among the inputs provided was mucuna. Schmink et al., "Forest Citizenship in Acre, Brazil," 36.

18 Weber, *Economy and Society*, 92.

19 Mucuna has been viewed as a problem in other areas where it has been incorporated into agricultural production systems. For example, Sophie Chao writes about a strain of mucuna with a conflicted status in West Papua oil palm plantations: "The problem was *Mucuna bracteata*, a leguminous cover plant introduced by the company for its soil moisture

retention capacities. In the absence of natural predators, however, *Mucuna* had multiplied uncontrollably and been reclassified by plantation management as an invasive species." Chao, "The Beetle or the Bug?," 486.

20 While many rural producers I spoke with shared these negative critiques, a few reported that they liked mucuna because it allowed them to grow more crops without deforesting. Among them were the mucuna family discussed in the introduction.

21 See Hecht, "Factories, Forests, Fields, and Family," for an analysis of the gendered dynamics of labor in extractive reserves.

22 The government already bought up to R$4500 per year worth of produce per family through a federally funded program. Discussions of a guaranteed price came up in other contexts as well. A workshop on native rubber production, for example, included an ongoing debate between rubber tappers, who talked about wanting a price that was not only higher but also guaranteed, and an economist and SEAPROF representative, who emphasized that the price could not be guaranteed because it was set by "the market." This reflected different ideas about what prices are and how they should be set, as well as about "the market." For the latter group, "the market" was something that functions externally over which neither they nor the state have control. For rubber tappers, it was something the government should establish.

23 See Scott, *Moral Economy of the Peasant*.

24 The idea of benefits as an exchange also echoed the rubber economy that had dominated Acre. In it, rubber bosses often enforced hierarchal relationships of exchange in which rubber tappers accessed material goods and services via the rubber they tapped. See, for example, Almeida, "Rubber Tappers of the Upper Juruá River, Brazil."

25 See Campbell, *Conjuring Property*, 46–51, for a discussion of how the term *os grandes* was used in Acre's neighboring state of Amazonas.

26 Such assertions of rights could be common in other contexts. See Dagnino, "Citizenship in Latin America"; Dagnino, "'We All Have Rights'"; and Holston, *Insurgent Citizenship*.

27 For discussions of exchange as part of conservation relationships elsewhere, see, for example, Cepek, "Foucault in the Forest"; and West, *Conservation Is Our Government Now*, chapter 7.

28 In other contexts, anthropologists have similarly found that benefits are not necessarily depoliticizing. Ferguson, *Give a Man a Fish*; Morton, "Managing Transience."

29 Ferguson, *Give a Man a Fish*, 162.

30 See Anand, *Hydraulic City*, for a discussion of water and citizenship.

31 Ferguson, *Give a Man a Fish*, 162.
32 Mitchell, *Carbon Democracy*.

Interlude IV. The Rural Road, Part 2

1   See Francisco Apurinã, *Nos caminhos da* BR-364, for an analysis of the BR-364's effects on Indigenous communities in Acre.
2   Jorge Viana, "As eleições e o risco da Acreanidade," Facebook, August 9, 2014, https://www.facebook.com/jorgevianaacre/posts/pfbid02BaDBdUujDrT2WqeyEAid7WpEYXjeWvzYbTfNraSf7TexaS C8AvJFhCGvYxWA5QVyl.

Chapter 5. The Urban Forest

Some of the ideas in chapter 5 first appeared in "Forest Urbanism: Urban Forest Life and Rural Forest Death in the Brazilian Amazon," Society for Cultural Anthropology, July 27, 2021, https://culanth.org/fieldsights/forest-urbanism-urban-forest-life-and-rural-forest-death-in-the-brazilian-amazon.

1   See Yanagisako, *Producing Culture and Capital*. The storytelling, productivity, and politics that I explore in previous chapters as part of the Acrean effort to valorize the forest are, of course, cultural as well.
2   Becker, "Geopolítica da Amazônia."
3   See Secretaria de Estado de Planejamento, "Evolução da população, taxa de urbanização e crescimento populacional—Acre," accessed October 30, 2021, https://seplan.ac.gov.br/evolucao-da-populacao-taxa-de-urbanizacao/. See also Becker, *Amazônia*; Browder and Godfrey, *Rainforest Cities*; Schmink, "Forest Citizens"; and Schmink and Cordeiro, *Rio Branco*.
4   See Morais, *Acreanidade*; Schmink, "Forest Citizens"; Schmink and Cordeiro, *Rio Branco*; and Vadjunec, Schmink, and Gomes, "Rubber Tapper Citizens."
5   Morais, *Acreanidade*, xxi.
6   Dagnino, "Citizenship in Latin America," 213.
7   Hecht and Rajão, "From 'Green Hell' to 'Amazonia Legal'"; Hoelle, *Rainforest Cowboys*; Horta Duarte, "Estado, sociedade e meio ambiente no Brasil em 200 anos de independência"; Pádua, "Natureza e território na construção do Brasil"; Rangel, *Inferno verde*.
8   For discussions of cultural citizenship in Latin America, see, for example, Alvarez, Dagnino, and Escobar, *Cultures of Politics/Politics of*

Cultures; Dagnino, "Citizenship in Latin America"; Dagnino, "'We All Have Rights'"; Holston, *Insurgent Citizenship*.
9 Williams, *Country and the City*.
10 Bunce, *Countryside Ideal*, 2.
11 *Mapinguaris* are large hairy beings with mouths in their stomach. In Amazonian cosmology, they are protectors of the forest. Slater, *Entangled Edens*, 250.
12 The concept drew on older discussions, such as the work by Acrean journalist and historian Leandro Tocantins, *Formação histórica do Acre*. For a careful and critical exposition of Acreanidade, see Morais, *Acreanidade*.
13 Schmink and Cordeiro, *Rio Branco*.
14 The gang was led by an elected federal deputy and retired military police officer named Hildebrando Pascoal (the "Chainsaw Deputy"), and Jorge received death threats for his government's successful work prosecuting it. Malu Gaspar, "Testemunha diz que havia planos para matar Viana," *Folha de S. Paulo*, December 3, 2000, https://www1.folha.uol.com.br/fsp/brasil/fc0312200002.htm.
15 Schmink, "Forest Citizens"; Schmink and Cordeiro, *Rio Branco*.
16 In the 2010s, crime increased in Acre, becoming a factor in the 2018 elections. Fernando Cymbaluk, "PT perde liderança de 20 anos no governo do Acre com vitória de Cameli," *UOL*, October 7, 2018; Fabiano Maisonnave and Fábio Pontes, "Ex-reduto do PT, Acre dá a Jair Bolsonaro maior votação relativa," *Folha de São Paulo*, October 30, 2018. Moreover, fear of crime remained a socially important component of life, as Theresa Caldeira elucidates in *City of Walls*.
17 Economic growth and the prioritization of public administration meant that, in its early years, the Government of the Forest significantly reduced state debt. This improved the state's credit and allowed it to borrow more from international and domestic institutions. Schmink and Cordeiro, *Rio Branco*, 59.
18 See Maron Greenleaf, "Forest Urbanism: Urban Forest Life and Rural Forest Death in the Brazilian Amazon," Society for Cultural Anthropology, July 27, 2021, https://culanth.org/fieldsights/forest-urbanism-urban-forest-life-and-rural-forest-death-in-the-brazilian-amazon.
19 Morais, *Acreanidade*, 242. See Morais, *Acreanidade*, chapter 5 for a detailed exposition of this urban renovation.
20 A historian who worked for Jorge Viana emphasized to me how the government had worked to identify and celebrate the sixteen different Indigenous cultures of Acre, rather than a singular generic "indian" (*índio*) culture. They had done this by regularizing more Indigenous

territories (*terras indigenous*) (though others pointed out to me that the process had stalled), by financially supporting the rejuvenation of languages and cultural practices, by creating Indigenous education and agroforestry programs, and through public celebrations like those at the Forest Library.

21 Portal Amazônia, "Calçadão da Gameleira," June 15, 2021, https://portalamazonia.com/amazonia-az/calcadao-da-gameleira.

22 In "Rubber Tapper Citizens," Vadjunec, Schmink, and Gomes detail some of these cultural invocations of the forest.

23 The column's author explained the title *História das margins* this way: Acre was located at the margin of Brazil, but the margin was still part of the whole.

24 Government of Acre, "Floresta Digital: Um novo tempo no Acre," *Notícias do Acre*, February 4, 2010, https://agencia.ac.gov.br/floresta-digital-um-novo-tempo-no-acre/.

25 Schmink, "Forest Citizens," 154.

26 Anderson, *Imagined Communities*.

27 As quoted in Morais, *Acreanidade*, 202–3.

28 Schmink, "Forest Citizens," 152.

29 Those with whom I spoke tended to be the ones who persevered through what they described as a difficult program—students who were in some way "conquered" by forestry. Many were relatively privileged, with resources that enabled them to attend required trips and weekend workshops, rather than working other jobs.

30 This understanding of government employment resembled that in other places; as Hetherington notes, "structurally, the project to reform the civil service tends to turn government employment into a site for the reproduction of the middle class." Hetherington, *Government of Beans*, 92. In its early days, the Government of the Forest worked to improve the stability of public sector employment by paying millions in back pay, reliably paying government workers' salaries on time, and giving them a standard bonus. See Schmink and Cordeiro, *Rio Branco*.

31 Being *concursado*—that is, having gone through a purportedly merit-based screening process and test called a *concurso*—was particularly desired. *Concursos* were given to hire a diverse array of government employees, from bus drivers to police officers to university professors. Some positions only required the submission of basic documentation, while others included selective tests. *Concursos* were advertised everywhere. Filling out an online *concurso* application was listed as a service available at tiny internet shops, sometimes listed first, above services like internet

use and printing. Even some of those I knew who disliked the government regularly sat for hours-long *concursos* on weekends, attracted by the security, pay, and benefits of being *concursado*.

32   Evelina Dagnino, in "Choices of the Left," argues that the PT's emphasis on the state's role in social transformation ended up undercutting more participatory forms of democracy. For a discussion of similar dynamics in Brazil and elsewhere in Latin America, see, for example, Druck, "Os sindicatos, os movimentos sociais e o governo Lula"; Hetherington, *Government of Beans*; Junge, "NGOs as Shadow Pseudopublics"; Nogueira, "Institutionalization of Rural Social Movements in the Lula Government and the Decline of Land Reform in Brazil"; and Prevost, Campos, and Vanden, *Social Movements and Leftist Governments in Latin America*.

33   Though the ease and pace of driving on the road was novel, there are long histories of travel within the Amazon, including between rural and urban areas. See, for example, Cleary, "After the Frontier"; Browder and Godfrey, *Rainforest Cities*; Becker, "Geopolítica da Amazônia"; Nugent and Harris, *Some Other Amazonians*; Padoch et al., "Urban Forest and Rural Cities"; Schmink and Cordeiro, *Rio Branco*; and Vadjunec, Schmink, and Greiner, "New Amazonian Geographies."

34   See Caldeira, *City of Walls*.

35   As quoted in Daiene Cardoso, "Viúva de Chico Mendes faz campanha por Marina," *Estadão*, September 3, 2010, https://politica.estadao.com.br/noticias/geral,viuva-de-chico-mendes-faz-campanha-por-marina,605003.

36   CIMI, "Do$$iê Acre: O Acre que os mercadores da natureza escondem," 2012, accessed November 1, 2023, http://www.cimi.org.br/pub/Rio20/Dossie-ACRE.pdf; "Letter in Defense of the Amazon and Mother Earth, against the Invasion of Capital, Extreme Violence and Green Scams," 2021, accessed November 1, 2023, https://wrm.org.uy/wp-content/uploads/2021/06/Carta-Defesa-Amazonia_EN.pdf; "Open Letter to California," 2013, accessed November 1, 2023, https://1bps6437gg8c169i0y1drtgz-wpengine.netdna-ssl.com/wp-content/uploads/2018/10/Open_Letter_Acre_english_portugese_spanish.pdf; Paula, *(Des)envolvimento insustentável na Amazônia Ocidental*; "Xapuri Declaration," May 28, 2017, https://wrm.org.uy/actions-and-campaigns/xapuri-declaration-may-28-2017/.

37   Morais, *Acreanidade*.

38   Whereas in the city, the urban forest was about the value of the living forest to the city, here the urban forest was something more akin to what Bertha Becker describes—a rural area transformed by urban values.

39   See Hoelle, *Rainforest Cowboys*.

40   Dean, *With Broadax and Firebrand*; Fearnside, "Desmatamento na Amazônia"; Ostos, "O Brasil e suas naturezas possíveis (1930–1945)"; Pádua, "Natureza e território na construção do Brasil."

41   Hoelle, *Rainforest Cowboys*, 4.

42   Hoelle, *Rainforest Cowboys*, 36.

43   Those who had more money than Edivan often hired lawyers to challenge their fines in courts, which was often an effective delay strategy.

44   *Floresta* was the term used most frequently when talking about the forest as something to be protected, as something with value. In contrast, *mata* was the term used when talking about forest as wasted space, something that had been or needed to be replaced, or a place to be avoided. Smallholders often used the term *mata* to dismissively describe the forest around them to me. It could index a neighbor's laziness, signaling that they did not want to do the hard work of keeping the land cleared. It was a place to put the trash, a local historian explained to me. It could signal the danger of the unknown and uninhabited as well as the unimportant landscapes of daily life. And it could carry a sense of backwardness. Nancy Scheper-Hughes, in *Death Without Weeping*, 89, writes about it in Brazil's Northeast, where the derivative word *matutos* has been "used to label the poor, the illiterate, the humble, the devout, the 'backward' and rural." Candice Slater, in *Entangled Edens*, 158–59, 178, puzzles over the distinction between *floresta* and *mata*, which she translates as "bush." She quotes an interlocutor as telling her that they are not different: "It's just that the rain forest [*floresta*] is something about which outsiders care. If I say 'the bush,' you might not listen. But if I say 'the rain forest,' then everybody wants to hear." But the *mata*, Slater notes, "also tends to be a forest full of landmarks readily familiar to the community" (178).

45   Schmink and Cordeiro, *Rio Branco*; Schmink, "Forest Citizens"; Vadjunec, Schmink, and Gomes, "Rubber Tapper Citizens."

46   Schmink, "Forest Citizens."

47   See Liz Kimbrough, "In Brazil's Acre, Smoke from Fires Threatens Health, Could Worsen COVID-19," Mongabay, August 18, 2021, https://news.mongabay.com/2021/08/in-brazils-acre-state-smoke-from-fires-affects-health-could-worsen-covid-19/.

Afterword

1   INPE, PRODES (Desmatamento), accessed April 14, 2023, http://terrabrasilis.dpi.inpe.br/app/dashboard/deforestation/biomes/legal_amazon/rates.

2   Samual Bryan, "Seminário entre técnicos e produtores discute formas de expandir o mercado do açaí no Acre," *Notícias do Acre*, September 26, 2019, https://agencia.ac.gov.br/seminario-entre-tecnicos-e-produtores-discute-formas-de-expandir-o-mercado-do-acai-no-acre/.

3   Elaine Gan, Heather Swanson, Anna Tsing, and Nils Bubandt describe "ghosts" as "vestiges and signs of past ways of life still charged in the present" that can "haunt" landscapes. They are often unnoticed or forgotten. Paying attention to them is a way, the authors suggest, to reveal hidden landscape histories that shape the present. Gan et al., "Introduction," G1, G2.

4   Morton, "Dark Ecology of Elegy," 255. Thanks to Kregg Hetherington for pointing me toward Morton's work on ecological elegy.

5   On enchantment and ethnography see, Stainova, "Enchantment as Method."

6   Tsing, *Mushroom at the End of the World*.

7   See Ricardo Abramovay, Joice Ferreira, Francisco de Assis Costa, Marco Ehrlich, Ana Margarida Castro Euler, Carlos Eduardo F. Young, David Kaimowitz, et al., "The New Bioeconomy in the Amazon: Opportunities and Challenges for a Healthy Standing Forest and Flowing Rivers," in Science Panel for the Amazon, *Amazon Assessment Report 2021*, ed. Carlos Nobre, Andrea Encalada, Elizabeth Anderson, Fernando Hector Roca Alcazar, Mercedes Bustamante, Carlos Mena, Marielos Peña-Claros, et al., UN Sustainable Development Solutions Network (SDSN), 2021, https://doi.org/10.55161/UGHK1968.

8   Stephen Donofrio, Christopher Daley, and Katherine Lin, "The Art of Integrity: State of the Voluntary Carbon Market 2022," Ecosystem Marketplace, accessed March 7, 2023, https://www.ecosystemmarketplace.com/publications/state-of-the-voluntary-carbon-markets-2022/.

9   See, for example, Patrick Greenfield, "Revealed: More than 90% of Rainforest Carbon Offsets by Biggest Certifier Are Worthless, Analysis Shows," *The Guardian*, January 18, 2023; and West et al., "Action Needed."

10 As quoted in Morais, *Acreanidade*, 202–3.

11 On the illusive appearance of state power and coherence, see, for example, Coronil, *Magical State*; Das and Poole, *Anthropology in the Margins of the State*; Hansen and Stepputat, *States of Imagination*; Mitchell, "Society, Economy, and the State Effect"; and Taussig, *Magic of the State*. See Mathews, *Instituting Nature*, for a discussion of the limits of state power over forests in Mexico, and Hetherington, *Government of Beans*, on a left-of-center administration's struggles to effectively regulate soy in Paraguay.

12   Angelsen et al., "Learning from REDD+"; Corbera and Martin, "Carbon Offsets"; Norman and Nakhooda, "State of REDD+ Finance," Center for Global Development Working Paper no. 378, May 2015, https://ssrn.com/abstract=2622743; Seymour and Busch, *Why Forests? Why Now?*; and Streck, "Financing REDD+."

13   Interview, May 2014, Rio Branco, Acre, Brazil.

14   As Laura Bear writes, "the labour of speculation connects directly to the recreation of inequalities of race, nation, sexuality and gender as well as class." Bear, "Speculation," 3. See also Rofel and Yanagisako, *Fabricating Transnational Capitalism*.

15   UOL, "Eleições 2018," accessed April 14, 2023, https://placar.eleicoes.uol.com.br/2018/2turno/ac/. There was a backlash against the PT (*antipetismo*) in many parts of the country in the 2018 election. In Acre, the PT's loss was in part attributed to voters being fed up with Jorge and Tião Viana "taking turns between congress and state government"; David Fleischer, a political scientist at the University of Brasília likened it to the end of a family dynasty. Fernando Cymbaluk, "PT perde liderança de 20 anos no governo do Acre com vitória de Cameli," UOL, October 7, 2018, https://noticias.uol.com.br/politica/eleicoes/2018/noticias/2018/10/07/pt-perde-lideranca-de-20-anos-no-governo-do-acre-com-vitoria-de-cameli-pp.htm.

Journalists writing about the PT's 2018 loss in Acre also cited the "explosion of violence" there. Cymbaluk, "PT perde liderança de 20 anos no governo do Acre com vitória de Cameli." This increase may have made it "the most violent state in the country." Fabiano Maisonnave and Fábio Pontes, "Ex-reduto do PT, Acre dá a Jair Bolsonaro maior votação relativa," *Folha de São Paulo*, October 30, 2018.

16   Chico Araújo and Montezuma Cruz, "Ex-governador do Acre é condenado por crime ambiental," *Tudo Rondônia - Independente!* (blog), accessed July 28, 2023, http://www.tudorondonia.com.br/noticias/ex-governador-do-acre-e-condenado-por-crime-ambiental-,2483.shtml; Pepita Ortega, "Diálogos de pai do governador do Acre indicam 'instrumentalização' de órgãos públicos para interesses privados da família Cameli, diz PF," *Estadão* (blog), March 22, 2023, https://www.estadao.com.br/politica/blog-do-fausto-macedo/dialogos-instrumentalizacao-orgaos-publicos-interesses-privados-da-familia-cameli-diz-pf/; Mariana Franco Ramos, "De Chico Mendes a Gladson Cameli, Acre abriga o agronegócio e as resistências," *De Olho nos Ruralistas*, December 23, 2021, https://deolhonosruralistas.com.br/2021/12/23/de-chico-mendes-a-gladson-cameli-acre-abriga-o-agronegocio-e-as-resistencias/; Xico Sá, "'Cartel de Manaus' dribla concorrências," *Folha de São Paulo*, May 19, 1997.

17　O Globo, "Gladson Cameli, do pp, é eleito governador do Acre," *O Globo*, October 7, 2018; Senado Notícias, "Gladson Cameli critica ação de órgãos de proteção do meio ambiente no Acre," March 16, 2017, https://www12.senado.leg.br/noticias/materias/2017/03/16/gladson-cameli-critica-acao-de-orgaos-de-protecao-do-meio-ambiente-no-acre.

18　As quoted in Alexander Zaitchik, "On the Front Lines of Bolsonaro's War on the Amazon, Brazil's Forest Communities Fight against Climate Catastrophe," *The Intercept*, July 6, 2019.

19　Brasilamaz, "Candidaturas aos governos na Amazônia Legal prometem inserção no mercado de carbono," September 12, 2022, https://brasilamazoniaagora.com.br/2022/amazonia-mercado-de-carbono/; Fabio Pontes, "Após gestão desmatadora, governador do Acre defende compromisso com meio ambiente," ((o))eco, October 17, 2022, https://oeco.org.br/noticias/apos-gestao-desmatadora-governador-do-acre-defende-compromisso-com-meio-ambiente/.

20　Hetherington, *Government of Beans*, 219.

21　This socioenvironmental violence might be seen as a version of the larger process of enclosure and dispossession that Marx, writing about this as "primitive accumulation" in England, described as "written in the annals of mankind in letters of blood and fire." Marx, *Capital*, chapter 26,

22　See Rapozo, "Necropolitics, State of Exception, and Violence."

23　inpe, "prodes (Desmatamento)." See also imazon, "Desmatamento na Amazônia cresce 29% em 2021 e é o maior dos últimos 10 anos," *Imazon* (blog), January 17, 2022, https://imazon.org.br/imprensa/desmatamento-na-amazonia-cresce-29-em-2021-e-e-o-maior-dos-ultimos-10-anos/.

24　Pontes, "Após gestão desmatadora, governador do Acre defende compromisso com meio ambiente."

25　Redação Notícias da Hora, "Um Cemitério Chamado Peixes da Amazônia," notícias da hora, March 5, 2023, https://www.noticiasdahora.com.br/politica/um-cemiterio-chamado-peixes-da-amazonia.html.

26　uol, "Presidente por Estado," accessed April 14, 2023, https://noticias.uol.com.br/eleicoes/2022/apuracao/2turno/votos-por-estado/presidente/.

27　Lisandra Paraguassu, "Brazil's Lula Unveils Plan to Stop Deforestation in Amazon by 2030 | Reuters," *Reuters*, June 5, 2023; Pontes, "Após gestão desmatadora, governador do Acre defende compromisso com meio ambiente." Cameli made this commitment while at a 2022 gcf Task Force meeting in Peru.

28　Ângela Rodrigues, "Instituto de Mudanças Climáticas fortalece a transparência das ações de governança do Sisa," *Noticias do Acre*,

May 9, 2023, https://agencia.ac.gov.br/instituto-de-mudancas-climaticas-fortalece-a-transparencia-das-acoes-de-governanca-do-sisa/; Wesley Moraes, "Acre: a nova fronteira agrícola da soja." *Noticias do Acre*, May 22, 2023, https://agencia.ac.gov.br/acre-a-nova-fronteira-agricola-da-soja/.

29 As quoted in Pontes, "Após gestão desmatadora, governador do Acre defende compromisso com meio ambiente."

30 Elynalia Lima, "Estado do Acre, Reino Unido e Alemanha pactuam nova etapa da Fase II do Programa REM Acre," *Noticias do Acre*, July 6, 2023, https://agencia.ac.gov.br/estado-do-acre-reino-unido-e-alemanha-pactuam-nova-etapa-da-fase-ii-do-programa-rem-acre/. This commitment aligns with the 2018 State Plan to Prevent and Combat Deforestation and Burning.

31 As quoted in Lima, "Estado do Acre, Reino Unido e Alemanha pactuam nova etapa da Fase II do Programa REM Acre."

32 Emergent, "Acre Leads the Way in Brazil with the First LEAF Coalition Term Sheet," December 5, 2023, https://emergentclimate.com/wp-content/uploads/2023/12/Acre-is-the-first-Brazilian-State-to-sign-a-LEAF-Term-Sheet.pdf.

33 Terrence McCoy, "Under Lula, Amazon Deforestation Is Declining. Can He Keep It Up?," *Washington Post*, July 17, 2023.

34 See Walker, "Collision Course."

35 Interview, February 2014, Rio Branco, Acre, Brazil.

36 Interview, August 2014, Rio Branco, Acre, Brazil.

37 Interview, July 2014, Rio Branco, Acre, Brazil.

BIBLIOGRAPHY

Adams, Cristina, Rui Murrieta, Walter Neves, and Mark Harris, eds. *Amazon Peasant Societies in a Changing Environment: Political Ecology, Invisibility and Modernity in the Rainforest*. New York: Springer, 2008.
Agrawal, Arun. *Environmentality: Technologies of Government and the Making of Subjects*. Durham, NC: Duke University Press, 2005.
Aguilar-Støen, Mariel. "Better Safe than Sorry? Indigenous Peoples, Carbon Cowboys and the Governance of REDD in the Amazon." *Forum for Development Studies* 44, no. 1 (2017): 91–108. https://doi.org/10.1080/08039410.2016.1276098.
Ahmann, Chloe. "'It's Exhausting to Create an Event out of Nothing': Slow Violence and the Manipulation of Time." *Cultural Anthropology* 33, no. 1 (2018): 142–71. https://doi.org/10.14506/ca33.1.06.
Alatas, Hussein. *The Myth of the Lazy Native: A Study of the Image of the Malays, Filipinos and Javanese from the 16th to the 20th Century and Its Function in the Ideology of Colonial Capitalism*. London: Frank Cass, 1977.
Albuquerque, Gerson Rodrigues de. "História e historiografia do Acre: Notas sobre os silêncios e a lógica do progresso" [History and historiography of Acre: Notes on silences and the logic of progress]. *Tropos: Comunicação, sociedade, e cultura* 1, no. 4 (2015): 1–19.
Alcott, Blake. "Jevons' Paradox." *Ecological Economics* 54, no. 1 (2005): 9–21.
Aldrich, Stephen, Robert Walker, Cynthia Simmons, Marcellus Caldas, and Stephen Perz. "Contentious Land Change in the Amazon's Arc of Deforestation." *Annals of the Association of American Geographers* 102, no. 1 (2012): 103–28.

Allegretti, Mary. "A construção social de políticas públicas: Chico Mendes e o movimento dos seringueiros" [The social construction of public policies: Chico Mendes and the rubber tappers' movement]. *Desenvolvimento e meio ambiente* 18 (2008): 39–59.

Allegretti, Mary. "Extractive Reserves: An Alternative for Reconciling Development and Environmental Conservation in Amazonia." In *Alternatives to Deforestation: Steps toward Sustainable Use of the Amazon Rain Forest*, edited by Anthony B. Anderson, 252–64. New York: Columbia University Press, 1990.

Almeida, Mauro Barbosa de. "The Politics of Amazonian Conservation: The Struggles of Rubber Tappers." *Journal of Latin American Anthropology* 7, no. 1 (2002): 170–219. https://doi.org/10.1525/jlca.2002.7.1.170.

Almeida, Mauro Barbosa de. "Rubber Tappers of the Upper Juruá River, Brazil: The Making of a Forest Peasant Economy." PhD diss., University of Cambridge, 1993. https://doi.org/10.17863/CAM.16267.

Almeida, Mauro Barbosa de, Mary Allegretti, and Augusto Postigo. "O legado de Chico Mendes: Êxitos e entraves das reservas extrativistas" [The legacy of Chico Mendes: Successes and obstacles of extractive reserves]. *Desenvolvimento e meio ambiente* 48 (2018): 25–55. https://doi.org/10.5380/dma.v48i0.60499.

Alston, Lee J., Gary D. Libecap, and Bernardo Mueller. *Titles, Conflict, and Land Use: The Development of Property Rights and Land Reform on the Brazilian Amazon Frontier*. Ann Arbor: University of Michigan Press, 1999.

Alvarez, Sonia E., Evelina Dagnino, and Arturo Escobar, eds. *Cultures of Politics/Politics of Cultures: Re-visioning Latin American Social Movements*. Boulder, CO: Westview Press, 1998.

Amaral, Aécio, and Joanildo A. Burity. *Inclusão social, identidade e diferença: Perspectivas pós-estruturalistas de análise social* [Social inclusion, identity and difference: Post-structuralist perspectives on social analysis]. São Paulo: Annablume, 2006.

Anand, Nikhil. *Hydraulic City: Water and the Infrastructures of Citizenship in Mumbai*. Durham, NC: Duke University Press, 2017.

Anand, Nikhil, Akhil Gupta, and Hannah Appel, eds. *The Promise of Infrastructure*. Durham, NC: Duke University Press, 2018.

Anderson, Benedict. *Imagined Communities: Reflections on the Origin and Spread of Nationalism*. New York: Verso Books, 2016.

Angelsen, Arild, Maria Brockhaus, Amy E. Duchelle, Anne Larson, Christopher Martius, William D. Sunderlin, Louis Verchot, Grace Wong, and Sven Wunder. "Learning from REDD+: A Response to Fletcher et al." *Conservation Biology* 31, no. 3 (2017): 718–20. https://doi.org/10.1111/cobi.12933.

Ansell, Aaron. *Zero Hunger: Political Culture and Antipoverty Policy in Northeast Brazil*. Chapel Hill: University of North Carolina Press, 2014.

Ansell, Aaron, and Ken Mitchell. "Models of Clientelism and Policy Change: The Case of Conditional Cash Transfer Programmes in Mexico and Brazil."

Bulletin of Latin American Research 30, no. 3 (2011): 298–312. https://doi.org/10.1111/j.1470-9856.2010.00497.x.

Appadurai, Arjun. "Introduction: Commodities and the Politics of Value." In *The Social Life of Things: Commodities in Cultural Perspective*, edited by Arjun Appadurai, 3–63. Cambridge: Cambridge University Press, 1986.

Appadurai, Arjun. *The Social Life of Things: Commodities in Cultural Perspective*. Cambridge: Cambridge University Press, 1986.

Appel, Hannah. *The Licit Life of Capitalism: US Oil in Equatorial Guinea*. Durham, NC: Duke University Press, 2019.

Appel, Hannah. "Offshore Work: Oil, Modularity, and the How of Capitalism in Equatorial Guinea." *American Ethnologist* 39, no. 4 (2012): 692–709. https://doi.org/10.1111/j.1548-1425.2012.01389.x.

Appel, Hannah. "Toward an Ethnography of the National Economy." *Cultural Anthropology* 32, no. 2 (2017): 294–322. https://doi.org/10.14506/ca32.2.09.

Appel, Hannah, Arthur Mason, and Michael Watts, eds. *Subterranean Estates: Life Worlds of Oil and Gas*. Ithaca, NY: Cornell University Press, 2015.

Apter, Andrew. *The Pan-African Nation: Oil and the Spectacle of Culture in Nigeria*. Chicago: University of Chicago Press, 2005.

Apurinã, Francisco. *Nos caminhos da BR-364: Povo Huni Kui e a Terra Indígena Colônia 27* [The paths of the BR-364: The Huni Kui people and the Colony 27 Indigenous Land]. Curitiba: Editora Prismas, 2015.

Aragão, Luiz Eduardo O. C., Yadvinder Malhi, Nicolas Barbier, Andre Lima, Yosio Shimabukuro, Liana Anderson, and Sassan Saatchi. "Interactions between Rainfall, Deforestation and Fires during Recent Years in the Brazilian Amazonia." *Philosophical Transactions of the Royal Society B: Biological Sciences* 363, no. 1498 (2008): 1779–85. https://doi.org/10.1098/rstb.2007.0026.

Araujo, Claudio, Catherine Araujo Bonjean, Jean-Louis Combes, Pascale Combes Motel, and Eustaquio J. Reis. "Property Rights and Deforestation in the Brazilian Amazon." *Ecological Economics* 68, no. 8 (2009): 2461–68. https://doi.org/10.1016/j.ecolecon.2008.12.015.

Arnold, J. E. Michael, and M. Ruiz Pérez. "Can Non-timber Forest Products Match Tropical Forest Conservation and Development Objectives?" *Ecological Economics* 39, no 3 (2001): 437–47. https://doi.org/10.1016/S0921-8009(01)00236-1.

Arregui, Aníbal G. "The Quilombola Movement: Sensing Futures in Afroindigenous Amazonia." *Ethos* 48, no. 3 (2021): 336–56. https://doi.org/10.1111/etho.12280.

Assunção, Juliano, Clarissa Gandour, and Rudi Rocha. "Deforestation Slowdown in the Brazilian Amazon: Prices or Policies?" *Environment and Development Economics* 20, no. 6 (2015): 697–722. https://doi.org/10.1017/S1355770X15000078.

Auyero, Javier. *Poor People's Politics: Peronist Survival Networks and the Legacy of Evita*. Durham, NC: Duke University Press, 2001.

Auyero, Javier, Pablo Lapegna, and Fernanda Page Poma. "Patronage Politics and Contentious Collective Action: A Recursive Relationship." *Latin American Politics and Society* 51, no. 3 (2009): 1–31. https://doi.org/10.1111/j.1548-2456.2009.00054.x.

Baccini, Alessandro, Wayne Walker, Luis Carvalho, Mary Farina, Damien Sulla-Menashe, and Richard A. Houghton. "Tropical Forests Are a Net Carbon Source Based on Aboveground Measurements of Gain and Loss." *Science* 358, no. 6360 (2017): 230–34. https://doi.org/10.1126/science.aam5962.

Bakx, Keith. "From Proletarian to Peasant: Rural Transformation in the State of Acre, 1870–1986." *Journal of Development Studies* 24, no. 2 (1988): 141–60. https://doi.org/10.1080/00220388808422060.

Balán, Manuel, and Françoise Montambeault, eds. *Legacies of the Left Turn in Latin America: The Promise of Inclusive Citizenship*. Notre Dame, IN: University of Notre Dame Press, 2020.

Balée, William. "Indigenous Transformation of Amazonian Forests: An Example from Maranhão, Brazil." *L'Homme* 33, no. 126/128 (1993): 231–54.

Baletti, Brenda. "*Ordenamento Territorial*: Neo-Developmentalism and the Struggle for Territory in the Lower Brazilian Amazon." *Journal of Peasant Studies* 39, no. 2 (2012): 573–98. https://doi.org/10.1080/03066150.2012.664139.

Baletti, Brenda. "Saving the Amazon? Sustainable Soy and the New Extractivism." *Environment and Planning A: Economy and Space* 46, no. 1 (2014): 5–25. https://doi.org/10.1068/a45241.

Barber, Christopher P., Mark A. Cochrane, Carlos M. Souza, and William F. Laurance. "Roads, Deforestation, and the Mitigating Effect of Protected Areas in the Amazon." *Biological Conservation* 177 (2014): 203–9. https://doi.org/10.1016/j.biocon.2014.07.004.

Barry, Andrew. *Material Politics: Disputes along the Pipeline*. Hoboken, NJ: John Wiley & Sons, 2013.

Barry, Andrew, Thomas Osborne, and Nikolas Rose, eds. *Foucault and Political Reason: Liberalism, Neo-Liberalism, and Rationalities of Government*. Chicago: University of Chicago Press, 1996.

Battistoni, Alyssa. "Bringing in the Work of Nature: From Natural Capital to Hybrid Labor." *Political Theory* 45, no. 1 (2017): 5–31. https://doi.org/10.1177/0090591716638389.

Baudrillard, Jean. *Simulacra and Simulation*. Ann Arbor: University of Michigan Press, 1994.

Bear, Laura. "Speculation: A Political Economy of Technologies of Imagination." *Economy and Society* 49, no. 1 (2020): 1–15. https://doi.org/10.1080/03085147.2020.1715604.

Bear, Laura. "Speculations on Infrastructure: From Colonial Public Works to a Post-colonial Global Asset Class on the Indian Railways 1840–2017." *Economy and Society* 49, no 1. (2020): 45–70. https://doi.org/10.1080/03085147.2020.1702416.

Beblawi, Hazem, and Giacomo Luciani, eds. *The Rentier State*. New York: Routledge, 2016.

Becker, Bertha K. *Amazônia*. São Paulo: Editora Atica, 1990.

Becker, Bertha K. "Geopolítica da Amazônia." *Estudos avançados* 19, no. 53 (2005): 71–86. https://doi.org/10.1590/S0103-40142005000100005.

Besky, Sarah. *The Darjeeling Distinction: Labor and Justice on Fair-Trade Tea Plantations in India*. Berkeley: University of California Press, 2013.

Besky, Sarah. *Tasting Qualities: The Past and Future of Tea*. Berkeley: University of California Press, 2020.

Besky, Sarah, and Alex Blanchette, eds. *How Nature Works: Rethinking Labor on a Troubled Planet*. Albuquerque: University of New Mexico Press, 2019.

Besky, Sarah, and Alex Blanchette. "Introduction: The Fragility of Work." In *How Nature Works: Rethinking Labor on a Troubled Planet*, edited by Sarah Besky and Alex Blanchette, 1–22. Albuquerque: University of New Mexico Press, 2019.

Böhm, Steffen, Maria Ceci Misoczky, and Sandra Moog. "Greening Capitalism? A Marxist Critique of Carbon Markets." *Organization Studies* 33, no. 11 (2012): 1617–38. https://doi.org/10.1177/0170840612463326.

Borlaug, Norman. "Feeding a Hungry World." *Science* 318, no. 5849 (2007): 359. https://doi.org/10.1126/science.1151062.

Börner, Jan, Sven Wunder, Sheila Wertz-Kanounnikoff, Marcos Rügnitz Tito, Ligia Pereira, and Nathalia Nascimento. "Direct Conservation Payments in the Brazilian Amazon: Scope and Equity Implications." *Ecological Economics* 69, no. 6 (2010): 1272–82. https://doi.org/10.1016/j.ecolecon.2009.11.003.

Bowman, Maria S., Britaldo S. Soares-Filho, Frank D. Merry, Daniel C. Nepstad, Hermann Rodrigues, and Oriana T. Almeida. "Persistence of Cattle Ranching in the Brazilian Amazon: A Spatial Analysis of the Rationale for Beef Production." *Land Use Policy* 29, no. 3 (2012): 558–68. https://doi.org/10.1016/j.landusepol.2011.09.009.

Boyd, William, and James Salzman. "The Curious Case of Greening in Carbon Markets." *Environmental Law* 41 (2011): 73–94.

Boykoff, Maxwell T., Adam Bumpus, Diana Liverman, and Samual Randalls. "Theorizing the Carbon Economy: Introduction to the Special Issue." *Environment and Planning A: Economy and Space* 41, no. 10 (2009): 2299–304. https://doi.org/10.1068/a42196.

Brandão, Frederico, Marie-Gabrielle Piketty, René Poccard-Chapuis, Brenda Brito, Pablo Pacheco, Edenise Garcia, Amy E. Duchelle, Isabel Drigo, and Jacqueline Carvalho Peçanha. "Lessons for Jurisdictional Approaches from Municipal-Level Initiatives to Halt Deforestation in the Brazilian Amazon." *Frontiers in Forests and Global Change* 3 (2020): 1–14. https://doi.org/10.3389/ffgc.2020.00096.

Braverman, Irus. *Planted Flags: Trees, Land, and Law in Israel/Palestine*. Illustrated edition. Cambridge: Cambridge University Press, 2014.

Bridge, Gavin. "Resource Geographies 1: Making Carbon Economies, Old and New." *Progress in Human Geography* 35, no. 6 (2010): 820–34. https://doi.org/10.1177/0309132510385524.

Brockington, Dan. "Ecosystem Services and Fictitious Commodities." *Environmental Conservation* 38, no. 4 (2011): 367–69. https://doi.org/10.1017/S0376892911000531.

Brondízio, Eduardo S. *The Amazonian Caboclo and the Açaí Palm: Forest Farmers in the Global Market.* Bronx, NY: The New York Botanical Garden Press, 2008.

Brondizio, Eduardo S., and Emilio F. Moran. "Human Dimensions of Climate Change: The Vulnerability of Small Farmers in the Amazon." *Philosophical Transactions of the Royal Society B: Biological Sciences* 363, no. 1498 (2008): 1803–9. https://doi.org/10.1098/rstb.2007.0025.

Brondízio, Eduardo S., and Andrea Siqueira. "From Extractivists to Forest Farmers: Changing Concepts of Caboclo Agroforestry in the Amazon Estuary." *Research in Economic Anthropology* 18 (1997): 233–79.

Brosius, J. Peter, Anna Lowenhaupt Tsing, and Charles Zerner, eds. *Communities and Conservation: Histories and Politics of Community-Based Natural Resource Management.* Walnut Creek, CA: AltaMira Press, 2005.

Browder, John O., and Brian J. Godfrey. *Rainforest Cities: Urbanization, Development, and Globalization of the Brazilian Amazon.* New York: Columbia University Press, 1997.

Bryant, Gareth. "The Politics of Carbon Market Design: Rethinking the Techno-Politics and Post-politics of Climate Change." *Antipode* 48, no. 4 (2016): 877–98. https://doi.org/10.1111/anti.12237.

Bryant, Raymond L., and Michael K. Goodman. "Consuming Narratives: The Political Ecology of 'Alternative' Consumption." *Transactions of the Institute of British Geographers* 29, no. 3 (2004): 344–66. https://doi.org/10.1111/j.0020-2754.2004.00333.x.

Bumpus, Adam G. "The Matter of Carbon: Understanding the Materiality of tCO$_2$e in Carbon Offsets." *Antipode* 43, no. 3 (2011): 612–38. https://doi.org/10.1111/j.1467-8330.2011.00879.x.

Bumpus, Adam G., and Diana M. Liverman. "Accumulation by Decarbonization and the Governance of Carbon Offsets." *Economic Geography* 84, no. 2 (2008): 127–55. https://doi.org/10.1111/j.1944-8287.2008.tb00401.x.

Bunce, Michael F. *The Countryside Ideal: Anglo-American Images of Landscape.* New York: Taylor and Francis, 1994.

Bunker, Stephen G. *Underdeveloping the Amazon: Extraction, Unequal Exchange, and the Failure of the Modern State.* Chicago: University of Chicago Press, 1988.

Burchardt, Hans-Jürgen, and Kristina Dietz. "(Neo-)Extractivism—a New Challenge for Development Theory from Latin America." *Third World Quarterly* 35, no. 3 (2014): 468–86. https://doi.org/10.1080/01436597.2014.893488.

Brum, Eliane. *Banzeiro Òkòtó: Uma viagem à Amazônia centro do mundo* [Banzeiro Òkòtó: A trip to the Amazon center of the world]. São Paulo: Companhia das Letras, 2021.

Cadena, Marisol de la. *Earth Beings: Ecologies of Practice across Andean Worlds.* Durham, NC: Duke University Press, 2015.

Caldar, Roseli Salete. "O MST e a formação dos sem terra: O movimento social como princípio educativo" [The MST and the formation of the landless: The social movement as an educational principle]. *Estudos avançados* 15 (2001): 207–24. https://doi.org/10.1590/S0103-40142001000300016.

Caldeira, Teresa P. R. *City of Walls: Crime, Segregation, and Citizenship in São Paulo.* Berkeley: University of California Press, 2001.

Çalışkan, Koray, and Michel Callon. "Economization, Part 1: Shifting Attention from the Economy towards Processes of Economization." *Economy and Society* 38, no. 3 (2009): 369–98. https://doi.org/10.1080/03085140903020580.

Çalışkan, Koray, and Michel Callon. "Economization, Part 2: A Research Programme for the Study of Markets." *Economy and Society* 39, no. 1 (2010): 1–32. https://doi.org/10.1080/03085140903424519.

Callon, Michel. "Civilizing Markets: Carbon Trading between in Vitro and in Vivo Experiments." *Accounting, Organizations and Society* 34, nos. 3–4 (2009): 535–48. https://doi.org/10.1016/j.aos.2008.04.003.

Callon, Michel. *The Laws of the Markets.* Sociological Review Monograph Series. Oxford: Blackwell Publishers, 1998.

Callon, Michel, Cécile Méadel, and Vololona Rabeharisoa. "The Economy of Qualities." *Economy and Society* 31, no. 2 (2002): 194–217. https://doi.org/10.1080/03085140220123126.

Campbell, Jeremy M. *Conjuring Property: Speculation and Environmental Futures in the Brazilian Amazon.* Culture, Place, and Nature. Seattle: University of Washington Press, 2015.

Campbell, Jeremy M. "Indigenous Urbanization in Amazonia: Interpretive Challenges and Opportunities." *Journal of Latin American and Caribbean Anthropology* 20, no. 1 (2015): 80–86. https://doi.org/10.1111/jlca.12136.

Cardoso, Catarina A. S. *Extractive Reserves in Brazilian Amazonia: Local Resource Management and the Global Political Economy.* London: Routledge, 2018.

Carneiro da Cunha, Manuela, and Mauro W. B. de Almeida. "Indigenous People, Traditional People, and Conservation in the Amazon." *Daedalus* 129, no. 2 (2000): 315–38.

Carneiro da Cunha, Manuela, Ruben Caixeta, Jeremy M. Campbell, Carlos Fausto, José Antonio Kelly, Claudio Lomnitz, Carlos D. Londoño Sulkin, Caio Pompeia, and Aparecida Vilaça. "Indigenous Peoples Boxed in by Brazil's Political Crisis." *HAU: Journal of Ethnographic Theory* 7, no. 2 (2017): 403–26. https://doi.org/10.14318/hau7.2.033.

Carter, Miguel, ed. *Challenging Social Inequality: The Landless Rural Workers Movement and Agrarian Reform in Brazil*. Durham, NC: Duke University Press, 2015.

Castree, Noel. "Commodifying What Nature?" *Progress in Human Geography* 27, no. 3 (2003): 273–97. https://doi.org/10.1191/0309132503ph4280a.

Castree, Noel. "Commodity Fetishism, Geographical Imaginations and Imaginative Geographies." *Environment and Planning A: Economy and Space* 33, no. 9 (2001): 1519–25. https://doi.org/10.1068/a3464.

Cavanagh, Connor, and Tor A. Benjaminsen. "Virtual Nature, Violent Accumulation: The 'Spectacular Failure' of Carbon Offsetting at a Ugandan National Park." *Geoforum* 56 (September 2014): 55–65. https://doi.org/10.1016/j.geoforum.2014.06.013.

Cepek, Michael L. "Foucault in the Forest: Questioning Environmentality in Amazonia." *American Ethnologist* 38, no. 3(2011): 501–15. https://doi.org/10.1111/j.1548-1425.2011.01319.x.

Chagas, Miriam de Fátima. "A política do reconhecimento dos 'remanescentes das comunidades dos quilombos'" [The policy of recognition of the 'remnants of quilombo communities']. *Horizontes Antropológicos* 7, no. 15 (2001): 209–35. https://doi.org/10.1590/S0104-71832001000100009.

Chakrabarty, Dipesh. *The Climate of History in a Planetary Age*. Chicago: University of Chicago Press, 2021.

Chao, Sophie. "The Beetle or the Bug? Multispecies Politics in a West Papuan Oil Palm Plantation." *American Anthropologist* 123, no. 3 (2021): 476–89. https://doi.org/10.1111/aman.13592.

Chao, Sophie. *In the Shadow of the Palms: More-than-Human Becomings in West Papua*. Durham, NC: Duke University Press, 2022.

Charnley, Susan, and Melissa R. Poe. "Community Forestry in Theory and Practice: Where Are We Now?" *Annual Review of Anthropology* 36, no. 1 (2007): 301–36. https://doi.org/10.1146/annurev.anthro.35.081705.123143.

Chiavari, Joana, and Cristina Leme Lopes. "Indigenous Land Rights in Brazil: Challenges and Barriers to Land Demarcation." In *Indigenous Amazonia, Regional Development and Territorial Dynamics: Contentious Issues*, edited by Walter Leal Filho, Victor T. King, and Ismar Borges de Lima, 39–59. The Latin American Studies Book Series. New York: Springer International Publishing, 2020. https://doi.org/10.1007/978-3-030-29153-2_3.

Cleary, David. "After the Frontier: Problems with Political Economy in the Modern Brazilian Amazon." *Journal of Latin American Studies* 25, no. 2 (1993): 331–49.

Coase, Ronald H. "The Problem of Social Cost." *Journal of Law and Economics* 56, no. 4 ([1960] 2013): 837–77.

Collier, Stephen. *Post-Soviet Social: Neoliberalism, Social Modernity, Biopolitics*. Princeton, NJ: Princeton University Press, 2011.

Cons, Jason. "Ecologies of Capture in Bangladesh's Sundarbans." *American Ethnologist* 48, no. 3 (2021): 245–59. https://doi.org/10.1111/amet.13022.

Cons, Jason, and Michael Eilenberg. *Frontier Assemblages: The Emergent Politics of Resource Frontiers in Asia*. Hoboken, NJ: John Wiley & Sons, 2019.

Cook, Ian. "Follow the Thing: Papaya." *Antipode* 36, no. 4 (2004): 642–64. https://doi.org/10.1111/j.1467-8330.2004.00441.x.

Corbera, Esteve, Manuel Estrada, Peter May, Guillermo Navarro, and Pablo Pacheco. "Rights to Land, Forests and Carbon in REDD+: Insights from Mexico, Brazil and Costa Rica." *Forests* 2, no. 1 (2011): 301–42. https://doi.org/10.3390/f2010301.

Corbera, Esteve, and Adrian Martin. "Carbon Offsets: Accommodation or Resistance?" *Environment and Planning A: Economy and Space* 47, no. 10 (2015): 2023–30. https://doi.org/10.1177/0308518X15611666.

Coronil, Fernando. *The Magical State: Nature, Money, and Modernity in Venezuela*. Chicago: University of Chicago Press, 1997.

Cortner, Owen, Rachael D. Garrett, Judson F. Valentim, Joice Ferreira, Meredith T. Niles, Júlio Reis, and Juliana Gil. "Perceptions of Integrated Crop-Livestock Systems for Sustainable Intensification in the Brazilian Amazon." *Land Use Policy* 82 (2019): 841–53. https://doi.org/10.1016/j.landusepol.2019.01.006.

Costa, Francisco de Assis. *Formação agropecuária da Amazônia: Os desafios do desenvolvimento sustentável* [Agricultural training in the Amazon: The challenges of sustainable development]. Belém: NAEA, 2000.

Cotta, Rosângela Minardi Mitre, and Juliana Costa Machado. "Programa Bolsa Família e segurança alimentar e nutricional no Brasil: Revisão crítica da literatura" [Bolsa Família Program and food and nutritional security in Brazil: Critical literature review]. *Revista Panamericana de salud pública* 33, no. 1 (2013): 54–60. https://doi.org/10.1590/S1020-49892013000100008.

Cribelli, Teresa. *Industrial Forests and Mechanical Marvels: Modernization in Nineteenth-Century Brazil*. Cambridge: Cambridge University Press, 2016.

Cronkleton, Peter, Peter Leigh Taylor, Deborah Barry, Samantha Stone-Jovicich, and Marianne Schmink. *Environmental Governance and the Emergence of Forest-Based Social Movements*. Bogor, Indonesia: Center for International Forestry Research (CIFOR), 2008. https://doi.org/10.17528/cifor/002348.

Cronon, William. *Nature's Metropolis: Chicago and the Great West*. New York: W. W. Norton, 1992.

Dagnino, Evelina. "The Choices of the Left: The Paradox of the New Developmentalist State." In *The Brazilian Left in the 21st Century: Conflict and Conciliation in Peripheral Capitalism*, edited by Vladimir Puzone and Luis Felipe Miguel, 69–90. New York: Springer International Publishing, 2019.

Dagnino, Evelina. "Citizenship in Latin America: An Introduction." *Latin American Perspectives* 30, no. 2 (2003): 211–25.

Dagnino, Evelina. "'We All Have Rights, but . . .' Contesting Concepts of Citizenship in Brazil." In *Inclusive Citizenship: Meanings and Expressions*, edited by Naila Kabeer, 149–63. London: Zed, 2005.

Dalakoglou, Dimitris, and Penny Harvey. *Roads and Anthropology: Ethnography, Infrastructures, (Im)Mobility*. New York: Routledge, 2015.

Dales, John H. *Pollution, Property and Prices*. Toronto: University of Toronto Press, 1968.

Dalsgaard, Steffen. "Carbon Valuation: Alternatives, Alternations and Lateral Measures?" *Valuation Studies* 4, no. 1 (2016): 67–91. https://doi.org/10.3384/VS.2001-5992.164167.

Dalsgaard, Steffen. "Carbon Value between Equivalence and Differentiation." *Environment and Society* 5, no. 1 (2014). https://doi.org/10.3167/ares.2014.050106.

Dalsgaard, Steffen. "The Commensurability of Carbon: Making Value and Money of Climate Change." *HAU: Journal of Ethnographic Theory* 3, no. 1 (2013): 80–98. https://doi.org/10.14318/hau3.1.006.

DaMatta, Roberto. *A casa e a rua: Espaço, cidadania, mulher e morte no Brasil* [The house and the street: Space, citizenship, women, and death in Brazil]. Rio de Janeiro: Editora Rocco, 1997.

Danowski, Déborah, and Eduardo Viveiros de Castro. *The Ends of the World*. Translated by Rodrigo Nunes. New York: Polity, 2016.

Das, Veena, and Deborah Poole, eds. *Anthropology in the Margins of the State*. Santa Fe, NM: School of American Research Press, 2004.

Davis, Heather, and Zoe Todd. "On the Importance of a Date, or, Decolonizing the Anthropocene." *ACME: An International Journal for Critical Geographies* 16, no. 4 (2017): 761–80.

Davis, Janae, Alex A. Moulton, Levi Van Sant, and Brian Williams. "Anthropocene, Capitalocene, . . . Plantationocene?: A Manifesto for Ecological Justice in an Age of Global Crises." *Geography Compass* 13, no. 5 (2019): e12438. https://doi.org/10.1111/gec3.12438.

Dean, Warren. *Brazil and the Struggle for Rubber: A Study in Environmental History*. Studies in Environment and History. Cambridge: Cambridge University Press, 1987.

Dean, Warren. *With Broadax and Firebrand: The Destruction of the Brazilian Atlantic Forest*. Berkeley: University of California Press, 1997.

Debord, Guy. *The Society of the Spectacle*. Detroit, MI: Black & Red, [1967] 2014.

Dehm, Julia. "Indigenous Peoples and REDD+ Safeguards: Rights as Resistance or as Disciplinary Inclusion in the Green Economy?" *Journal of Human Rights and the Environment* 7, no. 2 (2016): 170–217. https://doi.org/10.4337/jhre.2016.02.01.

Dempsey, Jessica, and Morgan M. Robertson. "Ecosystem Services: Tensions, Impurities, and Points of Engagement within Neoliberalism." *Progress in Human Geography* 36, no. 6 (2012): 758–79. https://doi.org/10.1177/0309132512437076.

Demsetz, Harold. "Toward a Theory of Property Rights." *American Economic Review* 57, no. 2 (1967): 347–59.

Dove, Michael R. "Plants, Politics, and the Imagination over the Past 500 Years in the Indo-Malay Region." *Current Anthropology* 60, no. S20 (2019): S309–20. https://doi.org/10.1086/702877.

Dove, Michael R. "A Revisionist View of Tropical Deforestation and Development." *Environmental Conservation* 20, no. 1 (1993): 17–24, 56. https://doi.org/10.1017/S0376892900037188.

Dove, Michael R. "Theories of Swidden Agriculture, and the Political Economy of Ignorance." *Agroforestry Systems* 1, no. 2 (1983): 85–99. https://doi.org/10.1007/BF00596351.

Dressler, Wolfram, Bram Büscher, Michael Schoon, Dan Brockington, Tanya Hayes, Christian A. Kull, James McCarthy, and Krishna Shrestha. "From Hope to Crisis and Back Again? A Critical History of the Global CBNRM Narrative." *Environmental Conservation* 37, no. 1 (2010): 5–15. https://doi.org/10.1017/S0376892910000044.

Druck, Maria da Graça. "Os sindicatos, os movimentos sociais e o governo Lula: Cooptação e resistência" [Unions, social movements and the Lula government: Cooptation and resistance]. *OSAL (Observatorio social de América Latina)* 6, no. 19 (2006): 329–40.

Duchelle, Amy E., Marina Cromberg, Maria Fernanda Gebara, Raissa Guerra, Tadeu Melo, Anne Larson, Peter Cronkleton, et al. "Linking Forest Tenure Reform, Environmental Compliance, and Incentives: Lessons from REDD+ Initiatives in the Brazilian Amazon." *World Development* 55 (March 2014): 53–67. https://doi.org/10.1016/j.worlddev.2013.01.014.

Duchelle, Amy E., Maron Greenleaf, Denyse Mello, Maria Fernanda Gebara, and Tadeu Melo. "Acre's State System of Incentives for Environmental Services (SISA), Brazil." In *REDD+ on the Ground*, edited by Erin O. Sills, Stibniati Atmadja, Claudio de Sassi, Amy E Duchelle, Demetrius Leo Kweka, Ida Aju Pradnja Resosudarmo, William D. Sunderlin, and Eskil Mattson, 33–50. Bogor, Indonesia: Center for International Forestry Research, 2014. https://www.jstor.org/stable/resrep02148.9.

Dunlap, Alexander, and James Fairhead. "The Militarisation and Marketisation of Nature: An Alternative Lens to 'Climate-Conflict.'" *Geopolitics* 19, no. 4 (2014): 937–61. https://doi.org/10.1080/14650045.2014.964864.

Dunlop, Tessa, and Esteve Corbera. "Incentivizing REDD+: How Developing Countries Are Laying the Groundwork for Benefit-Sharing." *Environmental Science and Policy* 63 (September 2016): 44–54. https://doi.org/10.1016/j.envsci.2016.04.018.

Eiró, Flávio, and Martijn Koster. "Facing Bureaucratic Uncertainty in the Bolsa Família Program: Clientelism beyond Reciprocity and Economic Rationality." *Focaal* 2019, no. 85 (2019): 84–96. https://doi.org/10.3167/fcl.2019.850108.

Ekers, Michael, and Scott Prudham. "The Socioecological Fix: Fixed Capital, Metabolism, and Hegemony." *Annals of the American Association of Geographers* 108, no. 1 (2018): 17–34. https://doi.org/10.1080/24694452.2017.1309963.

Eliasch, Johan. *Climate Change: Financing Global Forests: The Eliasch Review.* London: TSO, 2008.

Elyachar, Julia. *Markets of Dispossession: NGOs, Economic Development, and the State in Cairo.* Durham, NC: Duke University Press, 2005.

Escobar, Arturo. *Encountering Development: The Making and Unmaking of the Third World.* Princeton, NJ: Princeton University Press, 1994.

Escobar, Herton. "Deforestation in the Amazon Is Shooting up, but Brazil's President Calls the Data 'a Lie.'" *Science*, July 28, 2019. http://www.science.org/content/article/deforestation-amazon-shooting-brazil-s-president-calls-data-lie.

Fairhead, James, Melissa Leach, and Ian Scoones. "Green Grabbing: A New Appropriation of Nature?" *Journal of Peasant Studies* 39, no. 2 (2012): 237–61. https://doi.org/10.1080/03066150.2012.671770.

Fearnside, Philip M. "Desmatamento na Amazônia: Dinâmica, impactos e controle" ["Deforestation in the Amazon: Dynamics, impacts and control]. *Acta Amazonica* 36, no. 3 (2006): 395–400. https://doi.org/10.1590/S0044-59672006000300018.

Federici, Silvia. "Social Reproduction Theory." *Radical Philosophy* 2.04 (2019): 55–57.

Ferguson, James. *The Anti-Politics Machine: "Development," Depoliticization, and Bureaucratic Power in Lesotho.* Cambridge: Cambridge University Press, 1990.

Ferguson, James. "The Bovine Mystique: Power, Property and Livestock in Rural Lesotho." *Man* 20, no. 4 (1985): 647–74. https://doi.org/10.2307/2802755.

Ferguson, James. *Give a Man a Fish: Reflections on the New Politics of Distribution.* The Lewis Henry Morgan Lectures. Durham, NC: Duke University Press, 2015.

Ferrante, Lucas, and Philip Martin Fearnside. "The Amazon's Road to Deforestation." Edited by Jennifer Sills. *Science* 369, no. 6504 (2020): 634. https://doi.org/10.1126/science.abd6977.

Ferry, Elizabeth. "Speculative Substance: 'Physical Gold' in Finance." *Economy and Society* 49, no. 1 (2020): 92–115. https://doi.org/10.1080/03085147.2019.1690254.

Fischer, Brodwyn M. *A Poverty of Rights: Citizenship and Inequality in Twentieth-Century Rio de Janeiro.* Stanford, CA: Stanford University Press, 2008.

Fischer, Joern, David J. Abson, Van Butsic, M. Jahi Chappell, Johan Ekroos, Jan Hanspach, Tobias Kuemmerle, Henrik G. Smith, and Henrik von Wehrden. "Land Sparing versus Land Sharing: Moving Forward." *Conservation Letters* 7, no. 3 (2014): 149–57. https://doi.org/10.1111/conl.12084.

Fletcher, Robert, and Jan Breitling. "Market Mechanism or Subsidy in Disguise? Governing Payment for Environmental Services in Costa Rica." *Geoforum*, The

Global Rise and Local Implications of Market-Oriented Conservation Governance, 43, no. 3 (2012): 402–11. https://doi.org/10.1016/j.geoforum.2011.11.008.

Fletcher, Robert, Wolfram Dressler, Bram Büscher, and Zachary R. Anderson. "Questioning REDD+ and the Future of Market-Based Conservation." *Conservation Biology* 30, no. 3 (2016): 673–75. https://doi.org/10.1111/cobi.12680.

Fortun, Mike. *Promising Genomics: Iceland and DeCODE Genetics in a World of Speculation*. Berkeley: University of California Press, 2008.

Foucault, Michel. *"Society Must Be Defended": Lectures at the Collège de France, 1975–1976*. Translated by David Macey. 1st ed. New York: Picador, 2003.

Fraser, Nancy, and Rahel Jaeggi. *Capitalism: A Conversation in Critical Theory*. Medford, MA: Polity, 2018.

French, Jan Hoffman. "Dancing for Land: Law-Making and Cultural Performance in Northeastern Brazil." *PoLAR: Political and Legal Anthropology Review* 25, no. 1 (2002): 19–36. https://doi.org/10.1525/pol.2002.25.1.19.

Freyre, Gilberto. *The Masters and the Slaves*. Translated by Samuel Putnam. Berkeley: University of California Press, [1933] 1987.

Freyre, Gilberto. *Sobrados e mucambos* [The mansions and the shanties]. São Paulo: Global Editora, [1936] 2013.

Friedlingstein, Pierre, Michael O'Sullivan, Matthew W. Jones, Robbie M. Andrew, Luke Gregor, Judith Hauck, Corinne Le Quéré, et al. "Global Carbon Budget 2022." *Earth System Science Data* 14, no. 11 (2022): 4811–4900. https://doi.org/10.5194/essd-14-4811-2022.

Gago, Verónica. *Neoliberalism from Below: Popular Pragmatics and Baroque Economies*. Durham, NC: Duke University Press, 2017.

Gan, Elaine, Nils Bubandt, Anna Lowenhaupt Tsing, and Heather Anne Swanson. "Introduction: Haunted Landscapes of the Anthropocene." In *Arts of Living on a Damaged Planet*, edited by Anna Lowenhaupt Tsing, Nils Bubandt, Elaine Gan, and Heather Anne Swanson, G1–G14. Minneapolis: University of Minnesota Press, 2017.

Garfield, Seth. *Indigenous Struggle at the Heart of Brazil: State Policy, Frontier Expansion, and the Xavante Indians, 1937–1988*. Durham, NC: Duke University Press, 2001.

Garfield, Seth. *In Search of the Amazon: Brazil, the United States, and the Nature of a Region*. Durham, NC: Duke University Press, 2013.

Garrett, Rachael D., Ilyun Koh, Eric F. Lambin, Yann le Polain de Waroux, Jude H. Kastens, and J. Christopher Brown. "Intensification in Agriculture-Forest Frontiers: Land Use Responses to Development and Conservation Policies in Brazil." *Global Environmental Change* 53 (November 2018): 233–43. https://doi.org/10.1016/j.gloenvcha.2018.09.011.

Gay, Robert. "Rethinking Clientelism: Demands, Discourses and Practices in Contemporary Brazil." *Revista Europea de estudios Latinoamericanos y del Caribe / European Review of Latin American and Caribbean Studies*, no. 65 (1998): 7–24.

Geisler, Charles C., and Gail Daneker, eds. *Property and Values: Alternatives to Public and Private Ownership*. Washington, DC: Island Press, 2000.

Geist, Helmut J., and Eric F. Lambin. "Proximate Causes and Underlying Driving Forces of Tropical Deforestation." *BioScience* 52, no. 2 (2002): 143–50. https://doi.org/10.1641/0006-3568(2002)052[0143:PCAUDF]2.0.CO;2.

Ghosh, Amitav. *The Great Derangement: Climate Change and the Unthinkable*. Chicago: University of Chicago Press, 2017.

Gibson-Graham, J. K. *The End of Capitalism (As We Knew It)*. Minneapolis: University of Minnesota Press, 2006.

Gifford, Lauren. "'You Can't Value What You Can't Measure': A Critical Look at Forest Carbon Accounting." *Climatic Change* 161, no. 2 (2020): 291–306. https://doi.org/10.1007/s10584-020-02653-1.

Gilberthorpe, Emma, and Dinah Rajak. "The Anthropology of Extraction: Critical Perspectives on the Resource Curse." *Journal of Development Studies* 53, no. 2 (2017): 186–204. https://doi.org/10.1080/00220388.2016.1160064.

Glaser, Bruno, and William I. Woods, eds. *Amazonian Dark Earths: Explorations in Space and Time*. Berlin: Springer, 2004.

Gluckman, Max. *The Ideas in Barotse Jurisprudence*. Manchester, UK: Manchester University Press, 1972.

Godar, Javier, Toby A. Gardner, E. Jorge Tizado, and Pablo Pacheco. "Actor-Specific Contributions to the Deforestation Slowdown in the Brazilian Amazon." *Proceedings of the National Academy of Sciences* 111, no. 43 (2014): 15591–96. https://doi.org/10.1073/pnas.1322825111.

Goldman, Marcio. *Como funciona a democracia: Uma teoria etnográfica da política* [How democracy works: An ethnographic theory of politics]. Rio de Janeiro: 7 LETRAS, 2006.

Goldstein, Jesse. *Planetary Improvement: Cleantech Entrepreneurship and the Contradictions of Green Capitalism*. Cambridge, MA: MIT Press, 2018.

Gomes, Carlos Valério Aguiar, Ane Alencar, Jacqueline Michelle Vadjunec, and Leonardo Marques Pacheco. "Extractive Reserves in the Brazilian Amazon Thirty Years after Chico Mendes: Social Movement Achievements, Territorial Expansion and Continuing Struggles." *Desenvolvimento e meio ambiente* 48 (November 2018). https://doi.org/10.5380/dma.v48i0.58830.

Gomes, Carlos Valério Aguiar, Stephen G. Perz, and Jacqueline Michelle Vadjunec. "Convergence and Contrasts in the Adoption of Cattle Ranching: Comparisons of Smallholder Agriculturalists and Forest Extractivists in the Amazon." *Journal of Latin American Geography* 11, no. 1 (2012): 99–120.

Gomes, Carlos Valério Aguiar, Jacqueline M. Vadjunec, and Stephen G. Perz. "Rubber Tapper Identities: Political-Economic Dynamics, Livelihood Shifts, and Environmental Implications in a Changing Amazon." *Geoforum* 43, no. 2 (2012): 260–71. https://doi.org/10.1016/j.geoforum.2011.09.005.

Gonçalves, Veronica Korber. "Carbon Offset from the Amazon Forest to Compensate Aviation Emissions: Global Solution, Local Struggles." *Earth System*

Governance 14 (December 2022): 100160. https://doi.org/10.1016/j.esg.2022.100160.

Government of Acre. "Lei N. 2.308, de 22 de Outubro de 2010" [Law number 2.308, of October 22, 2010]. Government of Acre, October 22, 2010. https://www.legisweb.com.br/legislacao/?id=116550.

Graeber, David. *Toward an Anthropological Theory of Value: The False Coin of Our Own Dreams.* New York: Palgrave, 2001.

Grandin, Greg. *Fordlandia: The Rise and Fall of Henry Ford's Forgotten Jungle City.* New York: Picador, 2010.

Greenleaf, Maron. "Beneficiaries of Forest Carbon: Precarious Inclusion in the Brazilian Amazon." *American Anthropologist* 123, no. 2 (2021): 305–17. http://doi.org/10.1111/aman.13543.

Greenleaf, Maron. "Forest Urbanism: Urban Forest Life and Rural Forest Death in the Brazilian Amazon." Society for Cultural Anthropology, July 27, 2021. https://culanth.org/fieldsights/forest-urbanism-urban-forest-life-and-rural-forest-death-in-the-brazilian-amazon.

Greenleaf, Maron. "Rubber and Carbon: Opportunity Costs, Incentives and Ecosystem Services in Acre, Brazil." *Development and Change* 51, no. 1 (2020): 51–72. https://doi.org/10.1111/dech.12543.

Greenleaf, Maron. "Using Carbon Rights to Curb Deforestation and Empower Forest Communities." *New York University Environmental Law Journal* 18, no. 3 (2010): 507–99.

Greenleaf, Maron. "The Value of the Untenured Forest: Land Rights, Green Labor, and Forest Carbon in the Brazilian Amazon." *Journal of Peasant Studies* 47, no. 2 (February 2020): 286–305. https://doi.org/10.1080/03066150.2019.1579197.

Greenleaf, Maron, Jeffrey Hoelle, Magaly Madeiros, and Alberto Tavares. "Forest Policy Innovation at the Subnational Scale: Insights from Acre, Brazil." *Conservation and Society* 21, no. 4 (December 2023): 223–33. https://doi.org/10.4103/cs.cs_3_23.

Gudynas, Eduardo. *Extractivisms: Politics, Economy and Ecology.* Halifax, NS: Fernwood Publishing, 2021.

Guldi, Jo. *The Long Land War: The Global Struggle for Occupancy Rights.* New Haven, CT: Yale University Press, 2022.

Gupta, Aarti, Eva Lövbrand, Esther Turnhout, and Marjanneke J. Vijge. "In Pursuit of Carbon Accountability: The Politics of REDD+ Measuring, Reporting and Verification Systems." *Current Opinion in Environmental Sustainability* 4, no. 6 (December 2012): 726–31. https://doi.org/10.1016/j.cosust.2012.10.004.

Guthman, Julie. *Agrarian Dreams: The Paradox of Organic Farming in California.* Berkeley: University of California Press, 2004.

Guthman, Julie. "Unveiling the Unveiling." In *Frontiers of Commodity Chain Research*, edited by Jennifer Bair, 190–206. Stanford, CA: Stanford University Press, 2009.

Han, Clara. "Precarity, Precariousness, and Vulnerability." *Annual Review of Anthropology*, no. 47 (2018): 331–47.
Hann, Christopher M., ed. *Property Relations: Renewing the Anthropological Tradition*. Cambridge: Cambridge University Press, 1998.
Hansen, Thomas Blom, and Finn Stepputat, eds. *States of Imagination: Ethnographic Explorations of the Postcolonial State*. Durham, NC: Duke University Press, 2001.
Haraway, Donna. "Anthropocene, Capitalocene, Plantationocene, Chthulucene: Making Kin." *Environmental Humanities* 6, no. 1 (2015): 159–65. https://doi.org/10.1215/22011919-3615934.
Haraway, Donna. "Situated Knowledges: The Science Question in Feminism and the Privilege of Partial Perspective." *Feminist Studies* 14, no. 3 (1988): 575–99. https://doi.org/10.2307/3178066.
Haraway, Donna, Noboru Ishikawa, Scott F. Gilbert, Kenneth Olwig, Anna L. Tsing, and Nils Bubandt. "Anthropologists Are Talking—About the Anthropocene." *Ethnos* 81, no. 3 (2016): 535–64. https://doi.org/10.1080/00141844.2015.1105838.
Hardin, Garrett. "The Tragedy of the Commons." *Science* 162, no. 3859 (1968): 1243–48. https://doi.org/10.1126/science.162.3859.1243.
Harris, Nancy L., David A. Gibbs, Alessandro Baccini, Richard A. Birdsey, Sytze de Bruin, Mary Farina, Lola Fatoyinbo, et al. "Global Maps of Twenty-First Century Forest Carbon Fluxes." *Nature Climate Change* 11, no. 3 (2021): 234–40. https://doi.org/10.1038/s41558-020-00976-6.
Harrison, Robert Pogue. *Forests: The Shadow of Civilization*. Chicago: University of Chicago Press, 1993.
Hartwick, Elaine. "Geographies of Consumption: A Commodity-Chain Approach." *Environment and Planning D: Society and Space* 16, no. 4 (1998): 423–37. https://doi.org/10.1068/d160423.
Harvey, David. *A Brief History of Neoliberalism*. Oxford: Oxford University Press, 2005.
Harvey, David. *The New Imperialism*. Oxford: Oxford University Press, 2003.
Harvey, Penny, and Hannah Knox. "The Enchantments of Infrastructure." *Mobilities* 7, no. 4 (2012): 521–36. https://doi.org/10.1080/17450101.2012.718935.
Harvey, Penny, and Hannah Knox. *Roads: An Anthropology of Infrastructure and Expertise*. Expertise. Ithaca, NY: Cornell University Press, 2015.
Hébert, Karen. "In Pursuit of Singular Salmon: Paradoxes of Sustainability and the Quality Commodity." *Science as Culture* 19, no. 4 (2010): 553–81. https://doi.org/10.1080/09505431.2010.519620.
Hecht, Susanna B. "Factories, Forests, Fields and Family: Gender and Neoliberalism in Extractive Reserves." *Journal of Agrarian Change* 7, no. 3 (2007): 316–47. https://doi.org/10.1111/j.1471-0366.2007.00148.x.
Hecht, Susanna B. "From Eco-Catastrophe to Zero Deforestation? Interdisciplinarities, Politics, Environmentalisms and Reduced Clearing in Amazonia."

Hecht, Susanna B. "The Logic of Livestock and Deforestation in Amazonia." *BioScience* 43, no. 10 (1993): 687–95. https://doi.org/10.2307/1312340.

Hecht, Susanna B. *The Scramble for the Amazon and the "Lost Paradise" of Euclides da Cunha.* Chicago: University of Chicago Press, 2013.

Hecht, Susanna B. "Soybeans, Development and Conservation on the Amazon Frontier." *Development and Change* 36, no. 2 (2005): 375–404. https://doi.org/10.1111/j.0012-155X.2005.00415.x.

Hecht, Susanna B., and Alexander Cockburn. *The Fate of the Forest: Developers, Destroyers, and Defenders of the Amazon.* Chicago: University of Chicago Press, [1989] 2010.

Hecht, Susanna, and Raoni Rajão. "From 'Green Hell' to 'Amazonia Legal': Land Use Models and the Re-imagination of the Rainforest as a New Development Frontier." *Land Use Policy* 96 (July 2020): 103871. https://doi.org/10.1016/j.landusepol.2019.02.030.

Hetherington, Kregg. *The Government of Beans: Regulating Life in the Age of Monocrops.* Durham, NC: Duke University Press, 2020.

Hetherington, Kregg. *Guerrilla Auditors: The Politics of Transparency in Neoliberal Paraguay.* Durham, NC: Duke University Press, 2011.

Hetherington, Kregg, ed. *Infrastructure, Environment, and Life in the Anthropocene.* Durham, NC: Duke University Press, 2019.

Hetherington, Kregg. "Surveying the Future Perfect: Anthropology, Development and the Promise of Infrastructure." In *Infrastructures and Social Complexity: A Companion*, edited by Penelope Harvey, Casper Jensen, and Atsuro Morita, 40–50. London: Routledge, 2016.

Hetherington, Kregg. "Waiting for the Surveyor: Development Promises and the Temporality of Infrastructure." *Journal of Latin American and Caribbean Anthropology* 19, no. 2 (2014): 195–211. https://doi.org/10.1111/jlca.12100.

Hetherington, Kregg, and Jeremy M. Campbell. "Nature, Infrastructure, and the State: Rethinking Development in Latin America." *Journal of Latin American and Caribbean Anthropology* 19, no. 2 (2014): 191–94. https://doi.org/10.1111/jlca.12095.

Ho, Karen. *Liquidated: An Ethnography of Wall Street.* Durham, NC: Duke University Press, 2009.

Hochstetler, Kathryn, and Margaret E. Keck. *Greening Brazil: Environmental Activism in State and Society.* Durham, NC: Duke University Press, 2007.

Hoelle, Jeffrey. *Rainforest Cowboys: The Rise of Ranching and Cattle Culture in Western Amazonia.* Austin: University of Texas Press, 2015.

Holston, James. *Insurgent Citizenship: Disjunctions of Democracy and Modernity in Brazil.* Princeton, NJ: Princeton University Press, 2008.

Holston, James. *The Modernist City: An Anthropological Critique of Brasília.* Chicago: University of Chicago Press, 1989.

Horta Duarte, Regina. "Estado, sociedade e meio ambiente no Brasil em 200 anos de independência" [State, society, and environment in Brazil in 200 years of independence]. *Araucaria*, no. 51 (2022): 347–71. https://doi.org/10.12795/araucaria.2022.i51.15.

Huber, Matt. "Resource Geographies I: Valuing Nature (or Not)." *Progress in Human Geography* 42, no. 1 (2018): 148–59. https://doi.org/10.1177/0309132516670773.

Hudson, Ian, and Mark Hudson. "Removing the Veil?: Commodity Fetishism, Fair Trade, and the Environment." *Organization and Environment* 16, no. 4 (2003): 413–30. https://doi.org/10.1177/1086026603258926.

Hughes, David McDermott. *From Enslavement to Environmentalism: Politics on a Southern African Frontier*. Seattle: University of Washington Press, 2011.

Hughes, David McDermott. "Third Nature: Making Space and Time in the Great Limpopo Conservation Area." *Cultural Anthropology* 20, no. 2 (2005): 157–84. https://doi.org/10.1525/can.2005.20.2.157.

Hull, Matthew S. "Documents and Bureaucracy." *Annual Review of Anthropology* 41, no. 1 (2012): 251–67. https://doi.org/10.1146/annurev.anthro.012809.104953.

Hull, Matthew S. *Government of Paper: The Materiality of Bureaucracy in Urban Pakistan*. Berkeley: University of California Press, 2012.

Humphrey, Caroline, and Katherine Verdery. *Property in Question: Value Transformation in the Global Economy*. New York: Routledge, 2004.

Hunter, Wendy. *The Transformation of the Workers' Party in Brazil, 1989–2009*. Cambridge: Cambridge University Press, 2010.

Iglesias, Marcelo Piedrafita. *Os Kaxinawá de Felizardo: Correrias, trabalho e civilização no Alto Juruá* [The Kaxinawá of Felizardo: Busyness, work, and civilization in Alto Juruá]. Brasília, DF: Paralelo 15, 2010.

Igoe, Jim. *The Nature of Spectacle: On Images, Money, and Conserving Capitalism*. Tucson: University of Arizona Press, 2017.

Igoe, Jim, Katja Neves, and Dan Brockington. "A Spectacular Eco-Tour around the Historic Bloc: Theorising the Convergence of Biodiversity Conservation and Capitalist Expansion." *Antipode* 42, no. 3 (2010): 486–512. https://doi.org/10.1111/j.1467-8330.2010.00761.x.

Iriarte, Jose, Sarah Elliott, S. Yoshi Maezumi, Daiana Alves, Regina Gonda, Mark Robinson, Jonas Gregorio de Souza, Jennifer Watling, and Josephine Handley. "The Origins of Amazonian Landscapes: Plant Cultivation, Domestication and the Spread of Food Production in Tropical South America." *Quaternary Science Reviews* 248 (2020): 106582. https://doi.org/10.1016/j.quascirev.2020.106582.

Jaffee, Daniel. *Brewing Justice: Fair Trade Coffee, Sustainability, and Survival*. Berkeley: University of California Press, 2014.

Jaramillo-Giraldo, Carolina, Britaldo Soares Filho, Sónia M. Carvalho Ribeiro, and Rivadalve Coelho Gonçalves. "Is It Possible to Make Rubber Extraction

Ecologically and Economically Viable in the Amazon? The Southern Acre and Chico Mendes Reserve Case Study." *Ecological Economics* 134 (2017): 186–97. https://doi.org/10.1016/j.ecolecon.2016.12.035.

Jasanoff, Sheila. "A New Climate for Society." *Theory, Culture and Society* 27, no. 2–3 (2010): 233–53. https://doi.org/10.1177/0263276409361497.

Jegathesan, Mythri. "Black Feminist Plots before the Plantationocene and Anthropology's 'Regional Closets.'" *Feminist Anthropology* 2, no. 1 (2021): 78–93. https://doi.org/10.1002/fea2.12037.

Jobson, Ryan Cecil. "Road Work: Highways and Hegemony in Trinidad and Tobago." *Journal of Latin American and Caribbean Anthropology* 23, no. 3 (2018): 457–77. https://doi.org/10.1111/jlca.12345.

Jodoin, Sébastien. *Forest Preservation in a Changing Climate: REDD+ and Indigenous and Community Rights in Indonesia and Tanzania*. Cambridge: Cambridge University Press, 2017.

Junge, Benjamin. "NGOs as Shadow Pseudopublics: Grassroots Community Leaders' Perceptions of Change and Continuity in Porto Alegre, Brazil." *American Ethnologist* 39, no. 2 (2012): 407–24. https://doi.org/10.1111/j.1548-1425.2012.01372.x.

Kaimowitz, David, and Arild Angelsen. "Will Livestock Intensification Help Save Latin America's Tropical Forests?" *Journal of Sustainable Forestry* 27, no. 1–2 (2008): 6–24. https://doi.org/10.1080/10549810802225168.

Kainer, Karen A., Marianne Schmink, Arthur Cezar Pinheiro Leite, and Mário Jorge da Silva Fadell. "Experiments in Forest-Based Development in Western Amazonia." *Society and Natural Resources* 16, no. 10 (2003): 869–86. https://doi.org/10.1080/716100619.

Karl, Terry Lynn. *The Paradox of Plenty: Oil Booms and Petro-States*. Berkeley: University of California Press, 1997.

Kawa, Nicholas C. *Amazonia in the Anthropocene: People, Soils, Plants, Forests*. Austin: University of Texas Press, 2016.

Keck, Margaret E. "Social Equity and Environmental Politics in Brazil: Lessons from the Rubber Tappers of Acre." *Comparative Politics* 27, no. 4 (1995): 409–24. https://doi.org/10.2307/422227.

Kingsbury, Donald V. "Latin American Extractivism and (or after) the Left." *Latin American Research Review* 56, no. 4 (2021): 977–87. https://doi.org/10.25222/larr.1668.

Kirksey, Eben, and Stefan Helmreich. "The Emergence of Multispecies Ethnography." *Cultural Anthropology* 25, no. 4 (2010): 545–76. https://doi.org/10.1111/j.1548-1360.2010.01069.x.

Knox, Hannah. "Affective Infrastructures and the Political Imagination." *Public Culture* 29, no. 2 (May 2017): 363–84. https://doi.org/10.1215/08992363-3749105.

Knox, Hannah. *Thinking like a Climate: Governing a City in Times of Environmental Change*. Durham, NC: Duke University Press, 2020.

Knox-Hayes, Janelle. "The Spatial and Temporal Dynamics of Value in Financialization: Analysis of the Infrastructure of Carbon Markets." *Geoforum* 50 (December 2013): 117–28. https://doi.org/10.1016/j.geoforum.2013.08.012.

Kohn, Eduardo. *How Forests Think: Toward an Anthropology beyond the Human.* Berkeley: University of California Press, 2013.

Kongsager, Rico, and Esteve Corbera. "Linking Mitigation and Adaptation in Carbon Forestry Projects: Evidence from Belize." *World Development* 76 (December 2015): 132–46. https://doi.org/10.1016/j.worlddev.2015.07.003.

Kopytoff, Igor. "The Cultural Biography of Things: Commoditization as Process." In *The Social Life of Things: Commodities in Cultural Perspective*, edited by Arjun Appadurai, 64–92. Cambridge: Cambridge University Press, 1986.

Kosoy, Nicolás, and Esteve Corbera. "Payments for Ecosystem Services as Commodity Fetishism." *Ecological Economics* 69, no. 6 (April 2010): 1228–36. https://doi.org/10.1016/j.ecolecon.2009.11.002.

Koster, Martijn. "Mediating and Getting 'Burnt' in the Gap: Politics and Brokerage in a Recife Slum, Brazil." *Critique of Anthropology* 32, no. 4 (2012): 479–97. https://doi.org/10.1177/0308275X12456643.

Kreibich, Nicolas, and Lukas Hermwille. "Caught in Between: Credibility and Feasibility of the Voluntary Carbon Market Post-2020." *Climate Policy* 21, no. 7 (2021): 939–57. https://doi.org/10.1080/14693062.2021.1948384.

Krenak, Ailton. *Ideas to Postpone the End of the World.* Translated by Anthony Doyle. Toronto: Anansi International, 2020.

Kröger, Markus. "Deforestation, Cattle Capitalism and Neodevelopmentalism in the Chico Mendes Extractive Reserve, Brazil." *Journal of Peasant Studies* 47, no. 3 (2020): 464–82. https://doi.org/10.1080/03066150.2019.1604510.

Lansing, David M. "Performing Carbon's Materiality: The Production of Carbon Offsets and the Framing of Exchange." *Environment and Planning A: Economy and Space* 44, no. 1 (2012): 204–20. https://doi.org/10.1068/a44112.

Lansing, David M. "Realizing Carbon's Value: Discourse and Calculation in the Production of Carbon Forestry Offsets in Costa Rica." *Antipode* 43, no. 3 (2011): 731–53. https://doi.org/10.1111/j.1467-8330.2011.00886.x.

Lansing, David M. "Understanding Linkages between Ecosystem Service Payments, Forest Plantations, and Export Agriculture." *Geoforum* 47 (June 2013): 103–12. https://doi.org/10.1016/j.geoforum.2013.03.009.

Larkin, Brian. *Signal and Noise: Media, Infrastructure, and Urban Culture in Nigeria.* Durham, NC: Duke University Press, 2008.

Larson, Anne M., Maria Brockhaus, William D. Sunderlin, Amy Duchelle, Andrea Babon, Therese Dokken, Thu Thuy Pham, et al. "Land Tenure and REDD+: The Good, the Bad and the Ugly." *Global Environmental Change* 23, no. 3 (June 2013): 678–89. https://doi.org/10.1016/j.gloenvcha.2013.02.014.

Latour, Bruno, Isabelle Stengers, Anna Tsing, and Nils Bubandt. "Anthropologists Are Talking—About Capitalism, Ecology, and Apocalypse." *Ethnos* 83, no. 3 (2018): 587–606. https://doi.org/10.1080/00141844.2018.1457703.

Laveaga, Gabriela Soto. *Jungle Laboratories: Mexican Peasants, National Projects, and the Making of the Pill*. Durham, NC: Duke University Press, 2009.

Law, Elizabeth A., and Kerrie A. Wilson. "Providing Context for the Land-Sharing and Land-Sparing Debate." *Conservation Letters* 8, no. 6 (2015): 404–13. https://doi.org/10.1111/conl.12168.

Lazar, Sian. *The Anthropology of Citizenship: A Reader*. Hoboken, NJ: Wiley-Blackwell, 2013.

Leach, Melissa, and Ian Scoones. "Carbon Forestry in West Africa: The Politics of Models, Measures and Verification Processes." *Global Environmental Change* 23, no. 5 (2013): 957–67. https://doi.org/10.1016/j.gloenvcha.2013.07.008.

Levi, Primo. *The Complete Works of Primo Levi*. New York: Liveright Publishing Corporation, 2015.

Li, Tania Murray. "Dynamic Farmers, Dead Plantations, and the Myth of the Lazy Native." *Journal of Peasant Studies* 50, no. 2 (2023): 519–38. https://doi.org/10.1080/03066150.2022.2163629.

Li, Tania Murray. "Fixing Non-market Subjects: Governing Land and Population in the Global South." *Foucault Studies*, no. 18 (2014): 34–48. https://doi.org/10.22439/fs.v0i18.4650.

Li, Tania Murray. "Practices of Assemblage and Community Forest Management." *Economy and Society* 36, no. 2 (2007): 263–93. https://doi.org/10.1080/03085140701254308.

Li, Tania Murray. *The Will to Improve: Governmentality, Development, and the Practice of Politics*. Durham, NC: Duke University Press, 2007.

Li, Tania Murray, and Pujo Semedi. *Plantation Life: Corporate Occupation in Indonesia's Oil Palm Zone*. Durham, NC: Duke University Press, 2021.

Little, Paul. "Territórios sociais e povos tradicionais no Brasil: Por uma antropologia da territorialidade" [Social territories and traditional peoples in Brazil: Toward an anthropology of territoriality]. *Anuário antropológico* 28, no. 1 (2003): 251–90.

Locke, John. *Second Treatise of Civil Government*. Indianapolis, IN: Hackett, [1689] 1980.

Lohmann, Larry. "Marketing and Making Carbon Dumps: Commodification, Calculation and Counterfactuals in Climate Change Mitigation." *Science as Culture* 14, no. 3 (2005): 203–35. https://doi.org/10.1080/09505430500216783.

Lohmann, Larry. "Uncertainty Markets and Carbon Markets: Variations on Polanyian Themes." *New Political Economy* 15, no. 2 (2010): 225–54. https://doi.org/10.1080/13563460903290946.

Lopes, Eva, Britaldo Soares-Filho, Franco Souza, Raoni Rajão, Frank Merry, and Sónia Carvalho Ribeiro. "Mapping the Socio-Ecology of Non Timber Forest Products (NTFP) Extraction in the Brazilian Amazon: The Case of Açaí (Euterpe Precatoria Mart) in Acre." *Landscape and Urban Planning* 188 (August 2019): 110–17. https://doi.org/10.1016/j.landurbplan.2018.08.025.

Lorde, Audre. *The Master's Tools Will Never Dismantle the Master's House: Audre Lorde*. London: Penguin Classics, 2018.

Lovejoy, Thomas E., and Carlos Nobre. "Amazon Tipping Point: Last Chance for Action." *Science Advances* 5, no. 12 (2019): eaba2949. https://doi.org/10.1126/sciadv.aba2949.

Lovell, Heather. "Climate Change, Markets and Standards: The Case of Financial Accounting." *Economy and Society* 43, no. 2 (2014): 260–84. https://doi.org/10.1080/03085147.2013.812830.

Lovell, Heather, Harriet Bulkeley, and Diana Liverman. "Carbon Offsetting: Sustaining Consumption?" *Environment and Planning A: Economy and Space* 41, no. 10 (2009): 2357–79. https://doi.org/10.1068/a40345.

Lovell, Heather, and Diana Liverman. "Understanding Carbon Offset Technologies." *New Political Economy* 15, no. 2 (2010): 255–73. https://doi.org/10.1080/13563460903548699.

Lu, Flora, Gabriela Valdivia, and Néstor L. Silva, eds. *Oil, Revolution, and Indigenous Citizenship in Ecuadorian Amazonia*. New York: Palgrave Macmillan, 2017.

Lund, Jens Friis, Eliezeri Sungusia, Mathew Bukhi Mabele, and Andreas Scheba. "Promising Change, Delivering Continuity: REDD+ as Conservation Fad." *World Development* 89 (January 2017): 124–39. https://doi.org/10.1016/j.worlddev.2016.08.005.

Luttrell, Cecilia, Lasse Loft, Maria Fernanda Gebara, Demetrius Kweka, Maria Brockhaus, Arild Angelsen, and William D. Sunderlin. "Who Should Benefit from REDD+? Rationales and Realities." *Ecology and Society* 18, no. 4 (2013). http://www.jstor.org/stable/26269421.

Luxemburg, Rosa. *The Accumulation of Capital*. Translated by Agnes Schwarzschild. Mansfield Center, CT: Martino Fine Books, [1913] 2015.

Lyall, Angus, and Gabriela Valdivia. "The Speculative Petro-State: Volatile Oil Prices and Resource Populism in Ecuador." *Annals of the American Association of Geographers* 109, no. 2 (2019): 349–60. https://doi.org/10.1080/24694452.2018.1531690.

Lyons, Kristen, and Peter Westoby. "Carbon Colonialism and the New Land Grab: Plantation Forestry in Uganda and Its Livelihood Impacts." *Journal of Rural Studies* 36 (October 2014): 13–21. https://doi.org/10.1016/j.jrurstud.2014.06.002.

Macedo, Marcia N., Ruth S. DeFries, Douglas C. Morton, Claudia M. Stickler, Gillian L. Galford, and Yosio E. Shimabukuro. "Decoupling of Deforestation and Soy Production in the Southern Amazon during the Late 2000s." *Proceedings of the National Academy of Sciences* 109, no. 4 (2012): 1341–46. https://doi.org/10.1073/pnas.1111374109.

MacKenzie, Donald. *An Engine, Not a Camera: How Financial Models Shape Markets*. Cambridge, MA: MIT Press, 2008.

MacKenzie, Donald. "Making Things the Same: Gases, Emission Rights and the Politics of Carbon Markets." *Accounting, Organizations and Society* 34, no. 3–4 (2009): 440–55. https://doi.org/10.1016/j.aos.2008.02.004.

Mahanty, Sango, Wolfram Dressler, Sarah Milne, and Colin Filer. "Unravelling Property Relations around Forest Carbon." *Singapore Journal of Tropical Geography* 34, no. 2 (2013): 188–205.

Malhi, Yadvinder, Daniel Wood, Timothy R. Baker, James Wright, Oliver L. Phillips, Thomas Cochrane, Patrick Meir, et al. "The Regional Variation of Aboveground Live Biomass in Old-Growth Amazonian Forests." *Global Change Biology* 12, no. 7 (2006): 1107–38. https://doi.org/10.1111/j.1365-2486.2006.01120.x.

Malinowski, Bronislaw. *Argonauts of the Western Pacific: An Account of Native Enterprise and Adventure in the Archipelagoes of Melanesian New Guinea*. London: Routledge, [1922] 2014.

Marcus, George E. "Ethnography in/of the World System: The Emergence of Multi-Sited Ethnography." *Annual Review of Anthropology* 24 (1995): 95–117.

Marigo, Paola, Jeffrey D. Cummings, Jason Lee Curtis, Jason Kalirai, Yang Chen, Pier-Emmanuel Tremblay, Enrico Ramirez-Ruiz, et al. "Carbon Star Formation as Seen through the Non-monotonic Initial–Final Mass Relation." *Nature Astronomy* 4, no. 11 (November 2020): 1102–10. https://doi.org/10.1038/s41550-020-1132-1.

Marras, Stelio, and Renzo Taddei, eds. *O Antropoceno: Sobre modos de compor mundos*. Belo Horizonte: Fino Traço Editora, 2022.

Martins, José de Souza. "Representing the Peasantry? Struggles for/about Land in Brazil." *Journal of Peasant Studies* 29, no. 3–4 (2002): 300–335. https://doi.org/10.1080/03066150412331311099.

Marx, Karl. *Capital*. Vol. 1, *A Critique of Political Economy*. New York: Penguin Books, [1867] 1992.

Masco, Joseph. "The Crisis in Crisis." *Current Anthropology* 58, no. S15 (2017): S65–76. https://doi.org/10.1086/688695.

Mathews, Andrew S. "Anthropology and the Anthropocene: Criticisms, Experiments, and Collaborations." *Annual Review of Anthropology* 49, no. 1 (2020): 67–82. https://doi.org/10.1146/annurev-anthro-102218-011317.

Mathews, Andrew S. "Imagining Forest Futures and Climate Change: The Mexican State as Insurance Broker and Storyteller." In *Climate Cultures: Anthropological Perspectives on Climate Change*, edited by Jessica Barnes and Michael R. Dove, 199–220. New Haven, CT: Yale University Press, 2015.

Mathews, Andrew S. *Instituting Nature: Authority, Expertise, and Power in Mexican Forests*. Cambridge, MA: MIT Press, 2011.

Mathews, Andrew S. "Scandals, Audits, and Fictions: Linking Climate Change to Mexican Forests." *Social Studies of Science* 44, no. 1 (2014): 82–108.

Matos, Marcos de Almeida. "A Comissão Pró-Índio do Acre e as línguas indígenas acreanas" [The Acre Pro-Indian Commission and Acrean Indigenous languages]. MOARA—*Revista eletrônica do programa de pós-graduação em letras* 2, no. 32 (2016): 87–108. https://doi.org/10.18542/moara.v2i32.3645.

Mauss, Marcel. *The Gift: The Form and Reason for Exchange in Archaic Societies.* London: Routledge, [1925] 2001.

McAfee, Kathleen. "The Contradictory Logic of Global Ecosystem Services Markets." *Development and Change* 43, no. 1 (2012): 105–31. https://doi.org/10.1111/j.1467-7660.2011.01745.x.

McAfee, Kathleen. "Selling Nature to Save It? Biodiversity and Green Developmentalism." *Environment and Planning D: Society and Space* 17, no. 2 (1999): 133–54.

McAfee, Kathleen, and Elizabeth N. Shapiro. "Payments for Ecosystem Services in Mexico: Nature, Neoliberalism, Social Movements, and the State." *Annals of the Association of American Geographers* 100, no. 3 (2010): 579–99. https://doi.org/10.1080/00045601003794833.

McCarthy, James. "A Socioecological Fix to Capitalist Crisis and Climate Change? The Possibilities and Limits of Renewable Energy." *Environment and Planning A: Economy and Space* 47, no. 12 (2015): 2485–2502. https://doi.org/10.1177/0308518X15602491.

McCarthy, James, and Scott Prudham. "Neoliberal Nature and the Nature of Neoliberalism." *Geoforum* 35, no. 3 (2004): 275–83. https://doi.org/10.1016/j.geoforum.2003.07.003.

McElwee, Pamela D. *Forests Are Gold: Trees, People, and Environmental Rule in Vietnam.* Seattle: University of Washington Press, 2016.

McElwee, Pamela D. "Payments for Environmental Services as Neoliberal Market-Based Forest Conservation in Vietnam: Panacea or Problem?" *Geoforum,* The Global Rise and Local Implications of Market-Oriented Conservation Governance, 43, no. 3 (2012): 412–26. https://doi.org/10.1016/j.geoforum.2011.04.010.

McElwee, Pamela D., Bernhard Huber, and Thị Hải Vân Nguyễn. "Hybrid Outcomes of Payments for Ecosystem Services Policies in Vietnam: Between Theory and Practice." *Development and Change* 51, no. 1 (2020): 253–80. https://doi.org/10.1111/dech.12548.

McElwee, Pamela D., Tuyen Nghiem, Hue Le, Huong Vu, and Nghi Tran. "Payments for Environmental Services and Contested Neoliberalisation in Developing Countries: A Case Study from Vietnam." *Journal of Rural Studies* 36 (2014): 423–40. https://doi.org/10.1016/j.jrurstud.2014.08.003.

McKinsey & Company. *Pathways to a Low-Carbon Economy: Version 2 of the Global Greenhouse Gas Abatement Cost Curve.* McKinsey & Company, 2009. https://www.mckinsey.com/~/media/mckinsey/dotcom/client_service/sustainability/cost%20curve%20pdfs/pathways_lowcarbon_economy_version2.ashx.

McKittrick, Katherine. "Plantation Futures." *Small Axe: A Caribbean Journal of Criticism* 17, no. 3 (2013): 1–15. https://doi.org/10.1215/07990537-2378892.

Medeiros, Leonilde Servolo de. "Social Movements and the Experience of Market-Led Agrarian Reform in Brazil." *Third World Quarterly* 28, no. 8 (2007): 1501–18. https://doi.org/10.1080/01436590701637359.

Medina, Eden, Ivan da Costa Marques, and Christina Holmes. "Introduction: Beyond Imported Magic." In *Beyond Imported Magic*, edited by Eden Medina, Ivan da Costa Marques, and Christina Holmes, 1–24. Cambridge, MA: MIT Press, 2014.

Méndez, V. Ernesto, Christopher M. Bacon, Meryl Olson, Katlyn S. Morris, and Annie Shattuck. "Agrobiodiversity and Shade Coffee Smallholder Livelihoods: A Review and Synthesis of Ten Years of Research in Central America." *Professional Geographer* 62, no. 3 (2010): 357–76. https://doi.org/10.1080/00330124.2010.483638.

Mendoza, Marcos. *The Patagonia Sublime: The Green Economy and Postneoliberal Politics*. New Brunswick, NJ: Rutgers University Press, 2018.

Mendoza, Marcos, Maron Greenleaf, and Eric H. Thomas. "Green Distributive Politics: Legitimizing Green Capitalism and Environmental Protection in Latin America." *Geoforum* 126 (November 2021): 1–12. https://doi.org/10.1016/j.geoforum.2021.07.012.

Mertz, Ole, Christine Padoch, Jefferson Fox, Rob A. Cramb, Stephen J. Leisz, Nguyen Thanh Lam, and Tran Duc Vien. "Swidden Change in Southeast Asia: Understanding Causes and Consequences." *Human Ecology* 37, no. 3 (2009): 259–64. https://doi.org/10.1007/s10745-009-9245-2.

Meyer, Dagmar Estermann, Maria Cláudia Dal'Igna, Carin Klein, and Catharina da Cunha Silveira. "Políticas públicas: Imperativos e promessas de inclusão social" [Public policies: Imperatives and promises of social inclusion]. *Ensaio: Avaliação e políticas públicas em educação* 22 (December 2014): 1001–26. https://doi.org/10.1590/S0104-40362014000400007.

Mignolo, Walter D. "Introduction: Coloniality of Power and De-colonial Thinking." *Cultural Studies* 21, no. 2–3 (2007): 155–67. https://doi.org/10.1080/09502380601162498.

Miki, Yuko. *Frontiers of Citizenship: A Black and Indigenous History of Postcolonial Brazil*. Cambridge: Cambridge University Press, 2018.

Millar, Kathleen M. *Reclaiming the Discarded: Life and Labor on Rio's Garbage Dump*. Durham, NC: Duke University Press, 2018.

Millar, Kathleen M. "Toward a Critical Politics of Precarity." *Sociology Compass* 11, no. 6 (2017): e12483. https://doi.org/10.1111/soc4.12483.

Millard, Candice. *The River of Doubt: Theodore Roosevelt's Darkest Journey*. New York: Knopf Doubleday Publishing Group, 2009.

Miller, Theresa L. *Plant Kin: A Multispecies Ethnography in Indigenous Brazil*. Austin: University of Texas Press, 2019.

Milne, Sarah. "Grounding Forest Carbon: Property Relations and Avoided Deforestation in Cambodia." *Human Ecology* 40, no. 5 (2012): 693–706. https://doi.org/10.1007/s10745-012-9526-z.

Milne, Sarah, and Bill Adams. "Market Masquerades: Uncovering the Politics of Community-Level Payments for Environmental Services in Cambodia." *Development and Change* 43, no. 1 (January 2012): 133–58. https://doi.org/10.1111/j.1467-7660.2011.01748.x.

Milne, Sarah, Sango Mahanty, Phuc To, Wolfram Dressler, Peter Kanowski, and Maylee Thavat. "Learning from 'Actually Existing' REDD+: A Synthesis of Ethnographic Findings." *Conservation and Society* 17, no. 1 (2019): 84–95.

Mitchell, Sean T. *Constellations of Inequality: Space, Race, and Utopia in Brazil.* Chicago: University of Chicago Press, 2017.

Mitchell, Timothy. *Carbon Democracy: Political Power in the Age of Oil.* New York: Verso, 2011.

Mitchell, Timothy. "Society, Economy, and the State Effect." In *The Anthropology of the State: A Reader*, edited by Aradhana Sharma and Akhil Gupta, 169–86. Malden, MA: Blackwell Publishing, 2006.

Miyazaki, Hirokazu. *Arbitraging Japan: Dreams of Capitalism at the End of Finance.* Berkeley: University of California Press, 2013.

Moberg, Mark. *Fair Trade and Social Justice: Global Ethnographies.* New York: NYU Press, 2010.

Molé, Noelle J. "Precarious Subjects: Anticipating Neoliberalism in Northern Italy's Workplace." *American Anthropologist* 112, no. 1 (2010): 38–53.

Moore, Frances C., Justin S. Mankin, and Austin Becker. "Challenges in Integrating the Climate and Social Sciences for Studies of Climate Change Impacts and Adaptation." In *Climate Cultures: Anthropological Perspectives on Climate Change*, edited by Jessica Barnes and Michael R. Dove, 169–95. New Haven, CT: Yale University Press, 2015.

Moore, Jason W. *Capitalism in the Web of Life: Ecology and the Accumulation of Capital.* New York: Verso Books, 2015.

Moore, Jason W. "The Rise of Cheap Nature." In *Anthropocene or Capitalocene?: Nature, History, and the Crisis of Capitalism*, edited by Jason W. Moore, 78–115. Oakland, CA: PM Press, 2016.

Morais, Lecio, and Alfredo Saad-Filho. "Da economia política à política econômica: O Novo-desenvolvimentismo e o governo Lula" [From Political economy to economic policy: New developmentalism and the Lula government]. *Revista de economia política* 31, no. 4 (2011): 507–27. https://doi.org/10.1590/S0101-31572011000400001.

Morais, Maria de Jesus. *Acreanidade: Invenção e reinvenção da identidade acreana* [Acreanidade: invention and reinvention of the Acrean identity]. Rio Branco, AC: EDUFAC, 2016.

Moran, Emilio F. *Developing the Amazon.* Bloomington: Indiana University Press, 1981.

Morris, Meghan L. "Property's Relations: Tracing Anthropology in Property Theory." *Alabama Law Review* 73, no. 4 (2021): 767–82.

Morris, Meghan L. "Speculative Fields: Property in the Shadow of Post-conflict Colombia." *Cultural Anthropology* 34, no. 4 (2019): 580–606. http://dx.doi.org/10.14506/ca34.4.05.

Morton, Gregory Duff. "Managing Transience: Bolsa Família and Its Subjects in an MST Landless Settlement." *Journal of Peasant Studies* 42, no. 6 (2015): 1283–1305. https://doi.org/10.1080/03066150.2014.978298.

Morton, Timothy. "The Dark Ecology of Elegy." In *The Oxford Handbook of the Elegy*, edited by Karen Weisman, 251–71. Oxford: Oxford University Press, 2010.

Moutinho, Sofia, and Herton Escobar. "Falling Star." *Science* 376, no. 6596 (2022): 910–13. https://doi.org/10.1126/science.add1422.

Muehlebach, Andrea. *The Moral Neoliberal: Welfare and Citizenship in Italy*. Chicago: University of Chicago Press, 2012.

Muehlebach, Andrea. "On Precariousness and the Ethical Imagination: The Year 2012 in Sociocultural Anthropology." *American Anthropologist* 115, no. 2 (2013): 297–311. https://doi.org/10.1111/aman.12011.

Münster, Ursula. "Working for the Forest: The Ambivalent Intimacies of Human–Elephant Collaboration in South Indian Wildlife Conservation." *Ethnos* 81, no. 3 (2016): 425–47. https://doi.org/10.1080/00141844.2014.969292.

Myers, Rodd, Anne M. Larson, Ashwin Ravikumar, Laura F. Kowler, Anastasia Yang, and Tim Trench. "Messiness of Forest Governance: How Technical Approaches Suppress Politics in REDD+ and Conservation Projects." *Global Environmental Change* 50 (May 2018): 314–24. https://doi.org/10.1016/j.gloenvcha.2018.02.015.

Nascimento, Abdias do. "The Myth of Racial Democracy." In *The Myth of Racial Democracy*, 445–48. Durham, NC: Duke University Press, 2018.

Nel, Adrian. "Contested Carbon: Carbon Forestry as a Speculatively Virtual, Falteringly Material and Disputed Territorial Assemblage." *Geoforum* 81 (May 2017): 144–52. https://doi.org/10.1016/j.geoforum.2017.03.007.

Nelson, Gerald C., and Daniel Hellerstein. "Do Roads Cause Deforestation? Using Satellite Images in Econometric Analysis of Land Use." *American Journal of Agricultural Economics* 79, no. 1 (1997): 80–88. https://doi.org/10.2307/1243944.

Nepstad, Daniel, David McGrath, Claudia Stickler, Ane Alencar, Andrea Azevedo, Briana Swette, Tathiana Bezerra, et al. "Slowing Amazon Deforestation through Public Policy and Interventions in Beef and Soy Supply Chains." *Science* 344, no. 6188 (2014): 1118–23. https://doi.org/10.1126/science.1248525.

Nepstad, Daniel, Steve Schwartzman, Barbara Bamberger, Márcio Santilli, David Ray, Peter Schlesinger, Paul Lefebvre, et al. "Inhibition of Amazon Deforestation and Fire by Parks and Indigenous Lands." *Conservation Biology* 20, no. 1 (2006): 65–73. https://doi.org/10.1111/j.1523-1739.2006.00351.x.

Nepstad, Daniel, Britaldo S. Soares-Filho, Frank Merry, André Lima, Paulo Moutinho, John Carter, Maria Bowman, et al. "The End of Deforestation in the Brazilian Amazon." *Science* 326, no. 5958 (2009): 1350–51. https://doi.org/10.1126/science.1182108.

Neumann, Roderick P. *Imposing Wilderness: Struggles over Livelihood and Nature Preservation in Africa*. Berkeley: University of California Press, 2002.

Neumann, Roderick P., and Eric Hirsch. *Commercialisation of Non-timber Forest Products: Review and Analysis of Research*. Bogor, Indonesia: Center for International Forestry Research (CIFOR), 2000. https://doi.org/10.17528/cifor/000723.

Neves, Eduardo G., and Michael J. Heckenberger. "The Call of the Wild: Rethinking Food Production in Ancient Amazonia." *Annual Review of Anthropology* 48, no. 1 (2019): 371–88. https://doi.org/10.1146/annurev-anthro-102218-011057.

Neves, José Anael, Francisco de Assis Guedes de Vasconcelos, Mick Lennon Machado, Elisabetta Recine, Giselle Silva Garcia, and Maria Angélica Tavares de Medeiros. "The Brazilian Cash Transfer Program (Bolsa Família): A Tool for Reducing Inequalities and Achieving Social Rights in Brazil." *Global Public Health* 17, no. 1 (2022): 26–42. https://doi.org/10.1080/17441692.2020.1850828.

Newell, Peter, and Matthew Paterson. *Climate Capitalism: Global Warming and the Transformation of the Global Economy*. Cambridge: Cambridge University Press, 2010.

Nichols, Robert. *Theft Is Property! Dispossession and Critical Theory*. Durham, NC: Duke University Press, 2020.

Nixon, Rob. *Slow Violence and the Environmentalism of the Poor*. Cambridge, MA: Harvard University Press, 2011.

Nogueira, Aico Sipiriano. "Institutionalization of Rural Social Movements in the Lula Government and the Decline of Land Reform in Brazil: Co-option, Political Identity, and Agency." *Análise Social* 53, no. 227 (2018): 362–87.

Nogueira, Ana Karlla Magalhães, and Antônio Cordeiro de Santana. "Benefícios socioeconômicos da adoção de novas tecnologias no cultivo do açaí no estado do Pará" [Socioeconomic benefits of adopting new açaí cultivation technologies in the state of Pará]. *Revista Ceres* 63 (February 2016): 1–7. https://doi.org/10.1590/0034-737X201663010001.

Nugent, Stephen. *Amazonian Caboclo Society: An Essay on Invisibility and Peasant Economy*. London: Routledge, 1993.

Nugent, Stephen. *The Rise and Fall of the Amazon Rubber Industry: An Historical Anthropology*. London: Routledge, 2017.

Nugent, Stephen, and Mark Harris, eds. *Some Other Amazonians: Perspectives on Modern Amazonia*. London: University of London Press, 2004.

O'Connor, James R. *Natural Causes: Essays in Ecological Marxism*. New York: Guilford Press, 1998.

Ogden, Laura A. *Loss and Wonder at the World's End*. Durham, NC: Duke University Press, 2021.

Ogden, Laura A., Billy Hall, and Kimiko Tanita. "Animals, Plants, People, and Things: A Review of Multispecies Ethnography." *Environment and Society* 4, no. 1 (2013): 5–24. https://doi.org/10.3167/ares.2013.040102.

Oliveira, Gustavo. "Land Regularization in Brazil and the Global Land Grab." *Development and Change* 44, no. 2 (2013): 261–83. https://doi.org/10.1111/dech.12009.

Oliveira, Gustavo, and Susanna Hecht. "Sacred Groves, Sacrifice Zones and Soy Production: Globalization, Intensification and Neo-Nature in South America." *Journal of Peasant Studies* 43, no. 2 (2016): 251–85. https://doi.org/10.1080/03066150.2016.1146705.

Ondetti, Gabriel. *Land, Protest, and Politics: The Landless Movement and the Struggle for Agrarian Reform in Brazil*. University Park, PA: Penn State Press, 2010.

Osborne, Tracey. "Tradeoffs in Carbon Commodification: A Political Ecology of Common Property Forest Governance." *Geoforum* 67 (December 2015): 64–77. https://doi.org/10.1016/j.geoforum.2015.10.007.

Osborne, Tracey, and Elizabeth Shapiro-Garza. "Embedding Carbon Markets: Complicating Commodification of Ecosystem Services in Mexico's Forests." *Annals of the American Association of Geographers* 108, no. 1 (January 2018): 88–105. https://doi.org/10.1080/24694452.2017.1343657.

Ostos, Natascha Stefania Carvalho de. "O Brasil e suas naturezas possíveis (1930–1945)" [Brazil and its possible natures (1930–1945)]. *Revista de Indias* 72, no. 255 (2012): 581–614. https://doi.org/10.3989/revindias.2012.019.

Ostrom, Elinor. *Governing the Commons: The Evolution of Institutions for Collective Action*. Cambridge: Cambridge University Press, 1990.

Ostrom, Elinor, Joanna Burger, Christopher B. Field, Richard B. Norgaard, and David Policansky. "Revisiting the Commons: Local Lessons, Global Challenges." *Science* 284, no. 5412 (1999): 278–82.

Pacheco, Pablo. "Smallholder Livelihoods, Wealth and Deforestation in the Eastern Amazon." *Human Ecology* 37, no. 1 (2009): 27–41. https://doi.org/10.1007/s10745-009-9220-y.

Padoch, Christine, Eduardo Brondizio, Sandra Costa, Miguel Pinedo-Vasquez, Robin Sears, and Andrea Siqueira. "Urban Forest and Rural Cities: Multi-Sited Households, Consumption Patterns, and Forest Resources in Amazonia." *Ecology and Society* 13, no. 2 (July 2008). https://doi.org/10.5751/ES-02526-130202.

Pádua, José Augusto. "Natureza e território na construção do Brasil" [Nature and territory in the construction of Brazil]. *RCC Perspectives*, no. 7 (2013): 33–40.

Pádua, José Augusto. *Um sopro de destruição: Pensamento político e crítica ambiental no Brasil escravista, 1786–1888* [A breath of destruction: Political thought and environmental criticism in slave-owning Brazil, 1786-1888]. São Paulo: Zahar, 2002.

Palmeira, Moacir. "Voto: Racionalidade ou significado?" [Vote: Rationality or meaning?]. *Revista Brasileira de ciências sociais* 7, no. 20 (1992): 26–30.

Palmer, Charles, Luca Taschini, and Timothy Laing. "Getting More 'Carbon Bang' for Your 'Buck' in Acre State, Brazil." *Ecological Economics* 142 (December 2017): 214–27. https://doi.org/10.1016/j.ecolecon.2017.06.024.

Pandian, Anand. *A Possible Anthropology: Methods for Uneasy Times.* Durham, NC: Duke University Press, 2019.

Parker, Eugene. "Forest Islands and Kayapó Resource Management in Amazonia: A Reappraisal of the Apêtê." *American Anthropologist* 94, no. 2 (1992): 406–28.

Patel, Raj. "The Long Green Revolution." *Journal of Peasant Studies* 40, no. 1 (2013): 1–63. https://doi.org/10.1080/03066150.2012.719224.

Paula, Elder Andrade de. *(Des)envolvimento insustentável na Amazônia Ocidental: Dos missionários do progresso aos mercadores da natureza* [Unsustainable development in the Western Amazon: From missionaries of progress to nature merchants]. Rio Branco, AC: EDUFAC, 2005.

Paxson, Heather. *The Life of Cheese: Crafting Food and Value in America.* Berkeley: University of California Press, 2012.

Pelletier, Johanne, Hambulo Ngoma, Nicole M. Mason, and Christopher B. Barrett. "Does Smallholder Maize Intensification Reduce Deforestation? Evidence from Zambia." *Global Environmental Change* 63 (July 2020): 102127. https://doi.org/10.1016/j.gloenvcha.2020.102127.

Peluso, Nancy Lee, and Peter Vandergeest. "Genealogies of the Political Forest and Customary Rights in Indonesia, Malaysia, and Thailand." *Journal of Asian Studies* 60, no. 3 (2001): 761–812. https://doi.org/10.2307/2700109.

Perry, Keisha-Khan Y. "The Resurgent Far Right and the Black Feminist Struggle for Social Democracy in Brazil." *American Anthropologist* 122, no. 1 (2020): 157–62. https://doi.org/10.1111/aman.13358.

Phalan, Ben, Malvika Onial, Andrew Balmford, and Rhys E. Green. "Reconciling Food Production and Biodiversity Conservation: Land Sharing and Land Sparing Compared." *Science* 333, no. 6047 (2011): 1289–91. https://doi.org/10.1126/science.1208742.

Pham, Thu Thuy, Maria Brockhaus, Grace Wong, Le Ngoc Dung, Januarti Sinarra Tjajadi, Lasse Loft, Cecilia Luttrell, and Samuel Assembe Mvondo. "Approaches to Benefit Sharing: A Preliminary Comparative Analysis of 13 REDD+ Countries." Bogor, Indonesia: Center for International Forestry Research (CIFOR), 2013. https://doi.org/10.17528/cifor/004102.

Phelps, Jacob, Edward L. Webb, and Arun Agrawal. "Does REDD+ Threaten to Recentralize Forest Governance?" *Science* 328, no. 5976 (2010): 312–13. https://doi.org/10.1126/science.1187774.

Pierson, Paul. *Dismantling the Welfare State?: Reagan, Thatcher and the Politics of Retrenchment.* Cambridge: Cambridge University Press, 1994.

Pistorius, Till. "From RED to REDD+: The Evolution of a Forest-Based Mitigation Approach for Developing Countries." *Current Opinion in Environmental Sustainability* 4, no. 6 (2012): 638–45. https://doi.org/10.1016/j.cosust.2012.07.002.

Polanyi, Karl. *The Great Transformation: The Political and Economic Origins of Our Time.* New York: Beacon Press, 2001.

Pollock, Anne, and Banu Subramaniam. "Resisting Power, Retooling Justice: Promises of Feminist Postcolonial Technosciences." *Science, Technology, and Human Values* 41, no. 6 (2016): 951–66. https://doi.org/10.1177/0162243916657879.

Porter, Doug, and David Craig. "The Third Way and the Third World: Poverty Reduction and Social Inclusion in the Rise of 'Inclusive' Liberalism." *Review of International Political Economy* 11, no. 2 (2004): 387–423.

Posey, Darrell A. "Indigenous Knowledge, Biodiversity, and International Rights: Learning about Forests from the Kayapó Indians of the Brazilian Amazon." *Commonwealth Forestry Review* 76, no. 1 (1997): 53–60.

Prevost, Gary, Carlos Oliva Campos, and Harry E. Vanden, eds. *Social Movements and Leftist Governments in Latin America: Confrontation or Cooptation?* New York: Zed Books, 2012.

Prudham, Scott. "Pimping Climate Change: Richard Branson, Global Warming, and the Performance of Green Capitalism." *Environment and Planning A: Economy and Space* 41, no. 7 (July 1, 2009): 1594–1613. https://doi.org/10.1068/a4071.

Raffles, Hugh. *In Amazonia: A Natural History*. Princeton, NJ: Princeton University Press, 2003

Rangel, Alberto. *Inferno verde: Scenas e scenarios do Amazonas* [Green hell: Scenes and scenarios of Amazonas]. Genoa: S.A.I. Cliches Celluloide Bacigalupi, 1908.

Rapozo, Pedro. "Necropolitics, State of Exception, and Violence against Indigenous People in the Amazon Region during the Bolsonaro Administration." *Brazilian Political Science Review* 15 (April 2021): e0002. https://doi.org/10.1590/1981-3821202100020003.

Rasmussen, Mattias Borg, and Christian Lund. "Reconfiguring Frontier Spaces: The Territorialization of Resource Control." *World Development* 101 (January 2018): 388–99. https://doi.org/10.1016/j.worlddev.2017.01.018.

Redford, Kent H., Christine Padoch, and Terry Sunderland. "Fads, Funding, and Forgetting in Three Decades of Conservation." *Conservation Biology* 27, no. 3 (2013): 437–38. https://doi.org/10.1111/cobi.12071.

Reich, Charles. "The New Property." *Yale Law Journal* 73, no. 5 (1964): 733–87.

Revkin, Andrew. *The Burning Season: The Murder of Chico Mendes and the Fight for the Amazon Rain Forest*. Washington, DC: Island Press, 2004.

Ribeiro, Darcy. *Os índios e a civilização: A integração das populações indígenas no Brasil moderno* [Indians and civilization: The integration of indigenous populations in modern Brazil]. São Paulo: Global Editora, [1970] 2023.

Rich, Bruce. *Mortgaging the Earth: The World Bank, Environmental Impoverishment, and the Crisis of Development*. Boston: Beacon Press, 1994.

Ricketts, Taylor H., Britaldo Soares-Filho, Gustavo A. B. da Fonseca, Daniel Nepstad, Alexander Pfaff, Annie Petsonk, Anthony Anderson, et al. "Indigenous

Lands, Protected Areas, and Slowing Climate Change." *PLOS Biology* 8, no. 3 (2010): e1000331. https://doi.org/10.1371/journal.pbio.1000331.

Riles, Annelise, ed. *Documents: Artifacts of Modern Knowledge*. Ann Arbor: University of Michigan Press, 2006.

Riofrancos, Thea N. *Resource Radicals: From Petro-Nationalism to Postextractivism in Ecuador*. Durham, NC: Duke University Press, 2020.

Robbins, Bruce. *The Beneficiary*. Durham, NC: Duke University Press, 2017.

Robbins, Richard H. "Coffee, Fair Trade, and the Commodification of Morality." *Reviews in Anthropology* 42, no. 4 (2013): 243–63. https://doi.org/10.1080/00938157.2013.844035.

Robertson, Morgan M. "Measurement and Alienation: Making a World of Ecosystem Services." *Transactions of the Institute of British Geographers* 37, no. 3 (2012): 386–401. https://doi.org/10.1111/j.1475-5661.2011.00476.x.

Robertson, Morgan M. "The Nature That Capital Can See: Science, State, and Market in the Commodification of Ecosystem Services." *Environment and Planning D: Society and Space* 24, no. 3 (2006): 367–87. https://doi.org/10.1068/d3304.

Robles, Wilder. "Revisiting Agrarian Reform in Brazil, 1985–2016." *Journal of Developing Societies* 34, no. 1 (2018): 1–34. https://doi.org/10.1177/0169796X17749658.

Rodrigues Machaqueiro, Raquel. *The Carbon Calculation: Global Climate Policy, Forests, and Transnational Governance in Brazil and Mozambique*. Tucson: University of Arizona Press, 2023.

Rofel, Lisa, and Sylvia Junko Yanagisako. *Fabricating Transnational Capitalism: A Collaborative Ethnography of Italian-Chinese Global Fashion*. Durham, NC: Duke University Press, 2019.

Rojas, David, Alexandre de Azevedo Olival, and Andrezza Alves Spexoto Olival. "Cultivating Alternatives to Authoritarian Populism in Amazonia." *Journal of Latin American and Caribbean Anthropology* 24, no. 4 (2019): 958–81. https://doi.org/10.1111/jlca.12454.

Roniger, Luis. *Hierarchy and Trust in Modern Mexico and Brazil*. New York: Praeger, 1990.

Rose, Carol M. *Property and Persuasion: Essays on the History, Theory, and Rhetoric of Ownership*. Boulder, CO: Westview Press, 1994.

Rose, Deborah Bird. *Wild Dog Dreaming: Love and Extinction*. Under the Sign of Nature. Charlottesville: University of Virginia Press, 2013.

Ros-Tonen, Mirjam A. F., James Reed, and Terry Sunderland. "From Synergy to Complexity: The Trend toward Integrated Value Chain and Landscape Governance." *Environmental Management* 62, no. 1 (2018): 1–14. https://doi.org/10.1007/s00267-018-1055-0.

Rudel, Thomas K., Laura Schneider, Maria Uriarte, B. L. Turner, Ruth DeFries, Deborah Lawrence, Jacqueline Geoghegan, et al. "Agricultural Intensification and Changes in Cultivated Areas, 1970–2005." *Proceedings of the National*

Academy of Sciences 106, no. 49 (2009): 20675–80. https://doi.org/10.1073/pnas.0812540106.

Sahlins, Marshall. "The Return of the Event, Again." In *Culture in Practice: Selected Essays*, 293–352. New York: Zone Books, 2005.

Salimon, Cleber I., Francis E. Putz, Luis Menezes-Filho, Anthony Anderson, Marcos Silveira, I. Foster Brown, and Luís Cláudio de Oliveira. "Estimating State-Wide Biomass Carbon Stocks for a REDD Plan in Acre, Brazil." *Forest Ecology and Management* 262, no. 3 (2011): 555–60. https://doi.org/10.1016/j.foreco.2011.04.025.

Salisbury, David S., and Marianne Schmink. "Cows versus Rubber: Changing Livelihoods among Amazonian Extractivists." *Geoforum* 38, no. 6 (2007): 1233–49. https://doi.org/10.1016/j.geoforum.2007.03.005.

Santilli, Márcio, Paulo Moutinho, Stephan Schwartzman, Daniel Nepstad, Lisa Curran, and Carlos Nobre. "Tropical Deforestation and the Kyoto Protocol." *Climatic Change* 71, no. 3 (2005): 267–76. https://doi.org/10.1007/s10584-005-8074-6.

Santos Rocha da Silva, Marcelo, and Joel Edward Correia. "A Political Ecology of Jurisdictional REDD+: Investigating Social-Environmentalism, Climate Change Mitigation, and Environmental (in)Justice in the Brazilian Amazon." *Journal of Political Ecology* 29, no. 1 (2022). https://doi.org/10.2458/jpe.4713.

Sarmiento Barletti, Juan Pablo, and Anne M. Larson. "Environmental Justice in the REDD+ Frontier: Experiences from the Amazon and Beyond." In *Landscapes of Inequity: Environmental Justice in the Andes/Amazon Region*, edited by Nicholas Robbins and Barbara Fraser, 167–98. Lincoln: University of Nebraska Press, 2020.

Sarmiento Barletti, Juan Pablo, and Anne M. Larson. *Rights Abuse Allegations in the Context of REDD+ Readiness and Implementation: A Preliminary Review and Proposal for Moving Forward*. Bogor, Indonesia: Center for International Forestry Research (CIFOR), 2017. https://doi.org/10.17528/cifor/006630.

Sax, Marieka, and Daniel Tubb. "The Buzz Phase of Resource Extraction: Liquefied Natural Gas in Kitimat, British Columbia." *Extractive Industries and Society* 8, no. 3 (September 2021): 100938. https://doi.org/10.1016/j.exis.2021.100938.

Scheper-Hughes, Nancy. *Death without Weeping: The Violence of Everyday Life in Brazil*. Berkeley: University of California Press, 1993.

Schmink, Marianne. "Forest Citizens: Changing Life Conditions and Social Identities in the Land of the Rubber Tappers." *Latin American Research Review* 46 (2011): 141–58.

Schmink, Marianne, and Mâncio Lima Cordeiro. *Rio Branco: A cidade da Florestania*. Belém, Pará: Editora Universitária UFPA, 2008.

Schmink, Marianne, Amy E. Duchelle, Jeffrey Hoelle, Flávia Leite, Marcu Vanicio N. D'Oliveira, Jacqueline Vadjunec, Judson F. Valentim, and Richard

Wallace. "Forest Citizenship in Acre, Brazil." In *Forests Under Pressure—Local Responses to Global Issues*, edited by Pia Katila, Glenn Galloway, Wil de Jong, Pablo Pacheco, and Gerardo Mery, 31–47. IUFRO World Series, 2014.

Schmink, Marianne, and Charles Wood. *Contested Frontiers in Amazonia*. New York: Columbia University Press, 1992.

Schroeder, Heike, and Constance McDermott. "Beyond Carbon: Enabling Justice and Equity in REDD+ across Levels of Governance." *Ecology and Society* 19, no. 1 (2014): 31.

Scott, James C. *The Art of Not Being Governed: An Anarchist History of Upland Southeast Asia*. New Haven, CT: Yale University Press, 2009.

Scott, James C. *The Moral Economy of the Peasant: Rebellion and Subsistence in Southeast Asia*. New Haven, CT: Yale University Press, 1976.

Seymour, Frances, and Jonah Busch. *Why Forests? Why Now?: The Science, Economics, and Politics of Tropical Forests and Climate Change*. Washington, DC: Brookings Institution Press, 2016.

Shapiro-Garza, Elizabeth. "Contesting the Market-Based Nature of Mexico's National Payments for Ecosystem Services Programs: Four Sites of Articulation and Hybridization." *Geoforum* 46 (May 2013): 5–15. https://doi.org/10.1016/j.geoforum.2012.11.018.

Sills, Erin, and Shubhayu Saha. "Subsidies for Rubber: Conserving Rainforests While Sustaining Livelihoods in the Amazon?" *Journal of Sustainable Forestry* 29, no. 2–4 (2010): 152–73. https://doi.org/10.1080/10549810903543907.

Sivaramakrishnan, K. *Modern Forests: Statemaking and Environmental Change in Colonial Eastern India*. Stanford, CA: Stanford University Press, 1999

Skidmore, Marin Elisabeth, Fanny Moffette, Lisa Rausch, Matthew Christie, Jacob Munger, and Holly K. Gibbs. "Cattle Ranchers and Deforestation in the Brazilian Amazon: Production, Location, and Policies." *Global Environmental Change* 68 (2021): 102280. https://doi.org/10.1016/j.gloenvcha.2021.102280.

Slater, Candace. *Entangled Edens: Visions of the Amazon*. Berkeley: University of California Press, 2001.

Slater, Candace. "Visions of the Amazon: What Has Shifted, What Persists, and Why This Matters." *Latin American Research Review* 50, no. 3 (2015): 3–23. https://doi.org/10.1353/lar.2015.0039.

Smith, Nigel J. H. *Rainforest Corridors: The Transamazon Colonization Scheme*. Berkeley: University of California Press, 1982.

Soares, Fábio Veras, Rafael Perez Ribas, and Rafael Guerreiro Osório. "Evaluating the Impact of Brazil's Bolsa Família: Cash Transfer Programs in Comparative Perspective." *Latin American Research Review* 45, no. 2 (2010): 173–90.

Soares, Leonardo Barros, Catarina Chaves Costa, Marina de Barros Fonseca, and Victor Amaral Costa. "Fatores explicativos das demarcações de terras indígenas: Uma revisão de literatura" [Explanatory factors of Indigenous land demarcations: A literature review]. *Revista Brasileira de informação bibliográfica em ciências sociais*, no. 96 (September 2021). https://bibanpocs.emnuvens.com.br/revista/article/view/7.

Soares-Filho, Britaldo, Paulo Moutinho, Daniel Nepstad, Anthony Anderson, Hermann Rodrigues, Ricardo Garcia, Laura Dietzsch, et al. "Role of Brazilian Amazon Protected Areas in Climate Change Mitigation." *Proceedings of the National Academy of Sciences* 107, no. 24 (2010): 10821–26. https://doi.org/10.1073/pnas.0913048107.

Sodikoff, Genese Marie. *Forest and Labor in Madagascar: From Colonial Concession to Global Biosphere*. Bloomington: Indiana University Press, 2012.

Soto, Hernando de. *The Mystery of Capital: Why Capitalism Triumphs in the West and Fails Everywhere Else*. New York: Basic Books, 2003.

Souza, Carlos Alberto Alves de. *História do Acre: Novos temas e nova abordagem* [The history of Acre: New themes and a new approach]. Rio Branco, AC: MM Paim, 2002.

Souza, Luís Gustavo de Souza e, and Max Randson de Souza e Souza. "Crescimento da produção de açaí e castanha-do-Brasil no Acre" [Growth in the production of açaí and Brazil nuts in Acre]. *Revista de administração e negócios da Amazônia* 10, no. 3 (2018): 157. https://doi.org/10.18361/2176-8366/rara.v10n3p157-171.

Stainova, Yana. "Enchantment as Method." *Anthropology and Humanism* 44, no. 2 (2019): 214–30. https://doi.org/10.1111/anhu.12251.

Star, Susan Leigh. "The Ethnography of Infrastructure." *American Behavioral Scientist* 43, no. 3 (1999): 377–91. https://doi.org/10.1177/00027649921955326.

Stickler, Claudia, Olivia David, Charlotta Chan, Juan Pablo Ardila, and Tathiana Bezerra. "The Rio Branco Declaration: Assessing Progress toward a Near-Term Voluntary Deforestation Reduction Target in Subnational Jurisdictions across the Tropics." *Frontiers in Forests and Global Change* 3 (June 2020): 50. https://doi.org/10.3389/ffgc.2020.00050.

Strassburg, Bernardo B. N., Thomas Brooks, Rafael Feltran-Barbieri, Alvaro Iribarrem, Renato Crouzeilles, Rafael Loyola, Agnieszka E. Latawiec, et al. "Moment of Truth for the Cerrado Hotspot." *Nature Ecology and Evolution* 1, no. 4 (2017): 0099. https://doi.org/10.1038/s41559-017-0099.

Streck, Charlotte. "Financing REDD+: Matching Needs and Ends." *Current Opinion in Environmental Sustainability* 4, no. 6 (2012): 628–37. https://doi.org/10.1016/j.cosust.2012.10.001.

Sullivan, LaShandra. "Identity, Territory and Land Conflict in Brazil." *Development and Change* 44, no. 2 (2013): 451–71. https://doi.org/10.1111/dech.12010.

Sullivan, LaShandra. "The Overseen and Unseen: Agribusiness Plantations, Indigenous Labor, and Land Struggle in Brazil." *American Anthropologist* 123, no. 1 (2021): 82–95. https://doi.org/10.1111/aman.13519.

Sunderlin, William D., Claudio de Sassi, Erin O. Sills, Amy E. Duchelle, Anne M. Larson, Ida Aju Pradnja Resosudarmo, Abdon Awono, Demetrius Leo Kweka, and Thu Ba Huynh. "Creating an Appropriate Tenure Foundation for REDD+: The Record to Date and Prospects for the Future." *World Development* 106 (2018): 376–92. https://doi.org/10.1016/j.worlddev.2018.01.010.

Svampa, Maristella. *Neo-Extractivism in Latin America: Socio-Environmental Conflicts, the Territorial Turn, and New Political Narratives*. Cambridge: Cambridge University Press, 2019.

Svampa, Maristella. "Resource Extractivism and Alternatives: Latin American Perspectives on Development." *Journal für entwicklungspolitik* 28, no. 3 (2012): 43–73. https://doi.org/10.20446/JEP-2414-3197-28-3-43.

Tarlau, Rebecca. "Coproducing Rural Public Schools in Brazil: Contestation, Clientelism, and the Landless Workers' Movement." *Politics and Society* 41, no. 3 (2013): 395–424. https://doi.org/10.1177/0032329213493753.

Taussig, Michael. "The Beach (A Fantasy)." *Critical Inquiry* 26, no. 2 (2000): 249–78.

Taussig, Michael. *The Magic of the State*. New York: Routledge, 1997.

Thaler, Gregory M. "The Land Sparing Complex: Environmental Governance, Agricultural Intensification, and State Building in the Brazilian Amazon." *Annals of the American Association of Geographers* 107, no. 6 (2017): 1424–43. https://doi.org/10.1080/24694452.2017.1309966.

Thorkelson, Eli. "Precarity Outside: The Political Unconscious of French Academic Labor." *American Ethnologist* 43, no. 3 (2016): 475–87. https://doi.org/10.1111/amet.12340.

Ticktin, Tamara. "The Ecological Implications of Harvesting Non-timber Forest Products." *Journal of Applied Ecology* 41, no. 1 (2004): 11–21. https://doi.org/10.1111/j.1365-2664.2004.00859.x.

Tocantins, Leandro. *Formação histórica do Acre* [Historical formation of Acre]. Rio de Janeiro: Civilização Brasileira, 1979.

Toni, Fabiano, Jair Carvalho dos Santos, Ronei Sant′Ana de Menezes, Charles H. Wood, and Henrique Sant′Anna, eds. *Expansão e trajetórias da pecuária na Amazônia: Acre, Brasil* [Expansion and trajectories of livestock farming in the Amazon: Acre, Brazil]. Brasília, DF: Editora UnB, 2007.

Tsing, Anna Lowenhaupt. *Friction: An Ethnography of Global Connection*. Princeton, NJ: Princeton University Press, 2005.

Tsing, Anna Lowenhaupt. *The Mushroom at the End of the World: On the Possibility of Life in Capitalist Ruins*. Princeton, NJ: Princeton University Press, 2015.

Tsing, Anna Lowenhaupt, Nils Bubandt, Elaine Gan, and Heather Anne Swanson, eds. *Arts of Living on a Damaged Planet*. Minneapolis: University of Minnesota Press, 2017.

Tsing, Anna Lowenhaupt, Andrew S. Mathews, and Nils Bubandt. "Patchy Anthropocene: Landscape Structure, Multispecies History, and the Retooling of Anthropology." *Current Anthropology* 60, no. S20 (2019): S186–97. https://doi.org/10.1086/703391.

Turnhout, Esther, Aarti Gupta, Janice Weatherley-Singh, Marjanneke J. Vijge, Jessica de Koning, Ingrid J. Visseren-Hamakers, Martin Herold, and Markus

Lederer. "Envisioning REDD+ in a Post-Paris Era: Between Evolving Expectations and Current Practice." *WIREs Climate Change* 8, no. 1 (2017): e425. https://doi.org/10.1002/wcc.425.

Twine, France Winddance. *Racism in a Racial Democracy: The Maintenance of White Supremacy in Brazil*. New Brunswick, NJ: Rutgers University Press, 1998.

Vadjunec, Jacqueline M., Marianne Schmink, and Carlos Valério A. Gomes. "Rubber Tapper Citizens: Emerging Places, Policies, and Shifting Rural-Urban Identities in Acre, Brazil." *Journal of Cultural Geography* 28, no. 1 (2011): 73–98. https://doi.org/10.1080/08873631.2011.548481.

Vadjunec, Jacqueline M., Marianne Schmink, and Alyson L. Greiner. "New Amazonian Geographies: Emerging Identities and Landscapes." *Journal of Cultural Geography* 28, no. 1 (2011): 1–20. https://doi.org/10.1080/08873631.2011.548477.

Valentim, Judson Ferreira, and Carlos Mauricio Soares de Andrade. "Tendências e perspectivas da pecuária bovina na Amazônia Brasileira" [Trends and perspectives of cattle ranching in the Brazilian Amazon]. *Amazônia: Ciencia e desenvolvimento* 4, no. 8 (2009): 273–83.

Van Hecken, Gert, Vijay Kolinjivadi, Catherine Windey, Pamela McElwee, Elizabeth Shapiro-Garza, Frédéric Huybrechs, and Johan Bastiaensen. "Silencing Agency in Payments for Ecosystem Services (PES) by Essentializing a Neoliberal 'Monster' into Being: A Response to Fletcher and Büscher's 'PES Conceit.'" *Ecological Economics* 144 (February 2018): 314–18. https://doi.org/10.1016/j.ecolecon.2017.10.023.

Vandergeest, Peter, and Nancy Lee Peluso. "Territorialization and State Power in Thailand." *Theory and Society* 24, no. 3 (1995): 385–426.

Vega, Ailén, James Angus Fraser, Maurício Torres, and Rosamaria Loures. "Those Who Live like Us: Autodemarcations and the Co-becoming of Indigenous and Beiradeiros on the Upper Tapajós River, Brazilian Amazonia." *Geoforum* 129 (February 2022): 39–48. https://doi.org/10.1016/j.geoforum.2022.01.003.

Verdery, Katherine. *The Vanishing Hectare: Property and Value in Postsocialist Transylvania*. Ithaca, NY: Cornell University Press, 2003.

Vijge, Marjanneke J., and Aarti Gupta. "Framing REDD+ in India: Carbonizing and Centralizing Indian Forest Governance?" *Environmental Science and Policy* 38 (April 2014): 17–27. https://doi.org/10.1016/j.envsci.2013.10.012.

Viveiros de Castro, Eduardo. "Images of Nature and Society in Amazonian Ethnology." *Annual Review of Anthropology* 25, no. 1 (1996): 179–200. https://doi.org/10.1146/annurev.anthro.25.1.179.

Walker, Robert, Emilio Moran, and Luc Anselin. "Deforestation and Cattle Ranching in the Brazilian Amazon: External Capital and Household Processes." *World Development* 28, no. 4 (2000): 683–99. https://doi.org/10.1016/S0305-750X(99)00149-7.

Walker, Robert T. "Collision Course: Development Pushes Amazonia toward Its Tipping Point." *Environment: Science and Policy for Sustainable Development* 63, no. 1 (2021): 15–25. https://doi.org/10.1080/00139157.2021.1842711.

Walker, Wayne S., Seth R. Gorelik, Alessandro Baccini, Jose Luis Aragon-Osejo, Carmen Josse, Chris Meyer, Marcia N. Macedo, et al. "The Role of Forest Conversion, Degradation, and Disturbance in the Carbon Dynamics of Amazon Indigenous Territories and Protected Areas." *Proceedings of the National Academy of Sciences* 117, no. 6 (February 11, 2020): 3015–25. https://doi.org/10.1073/pnas.1913321117.

Wallace, Richard H., Carlos Valério A. Gomes, and Natalie A. Cooper. "The Chico Mendes Extractive Reserve: Trajectories of Agro-Extractive Development in Amazonia." *Desenvolvimento e meio ambiente* 48 (2018): 184–213. https://doi.org/10.5380/dma.v48i0.58836.

Wang, Yiting, and Catherine Corson. "The Making of a 'Charismatic' Carbon Credit: Clean Cookstoves and 'Uncooperative' Women in Western Kenya." *Environment and Planning A: Economy and Space* 47, no. 10 (2015): 2064–79. https://doi.org/10.1068/a130233p.

Watling, Jennifer, José Iriarte, Francis E. Mayle, Denise Schaan, Luiz C. R. Pessenda, Neil J. Loader, F. Alayne Street-Perrott, Ruth E. Dickau, Antonia Damasceno, and Alceu Ranzi. "Impact of Pre-Columbian 'Geoglyph' Builders on Amazonian Forests." *Proceedings of the National Academy of Sciences* 114, no. 8 (February 21, 2017): 1868–73. https://doi.org/10.1073/pnas.1614359114.

Watts, Michael. "Resource Curse? Governmentality, Oil and Power in the Niger Delta, Nigeria." *Geopolitics* 9, no. 1 (2004): 50–80. https://doi.org/10.1080/14650040412331307832.

Weatherley-Singh, Janice, and Aarti Gupta. "Drivers of Deforestation and REDD+ Benefit-Sharing: A Meta-Analysis of the (Missing) Link." *Environmental Science and Policy* 54 (December 2015): 97–105. https://doi.org/10.1016/j.envsci.2015.06.017.

Weber, Max. *Economy and Society; An Outline of Interpretive Sociology*. New York: Bedminster Press, [1921] 1968.

Weeks, Kathi. *The Problem with Work: Feminism, Marxism, Antiwork Politics, and Postwork Imaginaries*. Durham, NC: Duke University Press, 2011.

Weinstein, Barbara. *The Amazon Rubber Boom, 1850–1920*. Stanford, CA: Stanford University Press, 1983.

Weinstein, Stephanie, and Susan Moegenburg. "Açaí Palm Management in the Amazon Estuary: Course for Conservation or Passage to Plantations?" *Conservation and Society* 2, no. 2 (2004): 315–46.

Welch, Clifford Andrew, and Sérgio Sauer. "Rural Unions and the Struggle for Land in Brazil." *Journal of Peasant Studies* 42, no. 6 (2015): 1109–35. https://doi.org/10.1080/03066150.2014.994511.

West, Paige. *Conservation Is Our Government Now: The Politics of Ecology in Papua New Guinea*. Durham, NC: Duke University Press, 2006.

West, Paige. *Dispossession and the Environment: Rhetoric and Inequality in Papua, New Guinea*. New York: Columbia University Press, 2016.

West, Paige. *From Modern Production to Imagined Primitive: The Social World of Coffee from Papua New Guinea*. Durham, NC: Duke University Press, 2012.

West, Paige, James Igoe, and Dan Brockington. "Parks and Peoples: The Social Impact of Protected Areas." *Annual Review of Anthropology* 35 (2006): 251–77.

West, Thales A. P., Sven Wunder, Erin O. Sills, Jan Börner, Sami W. Rifai, Alexandra N. Neidermeier, Gabriel P. Frey, and Andreas Kontoleon. "Action Needed to Make Carbon Offsets from Forest Conservation Work for Climate Change Mitigation." *Science* 381, no. 6660 (August 25, 2023): 873–77. https://doi.org/10.1126/science.ade3535.

Weszkalnys, Gisa. "Cursed Resources, or Articulations of Economic Theory in the Gulf of Guinea." *Economy and Society* 40, no. 3 (2011): 345–72. https://doi.org/10.1080/03085147.2011.580177.

Weszkalnys, Gisa. "Geology, Potentiality, Speculation: On the Indeterminacy of First Oil." *Cultural Anthropology* 30, no. 4 (2015): 611–39. https://doi.org/10.14506/ca30.4.08.

Whitington, Jerome. "Carbon as a Metric of the Human." *PoLAR: Political and Legal Anthropology Review* 39, no. 1 (2016): 46–63. https://doi.org/10.1111/plar.12130.

Whyte, Kyle. "Settler Colonialism, Ecology, and Environmental Injustice." *Environment and Society* 9, no. 1 (2018): 125–44. https://doi.org/10.3167/ares.2018.090109.

Williams, Michael. *Deforesting the Earth: From Prehistory to Global Crisis: An Abridgment*. Chicago: University of Chicago Press, 2006.

Williams, Raymond. *The Country and the City*. Oxford: Oxford University Press, 1975.

Wittman, Hannah, Lisa Jordan Powell, and Esteve Corbera. "Financing the Agrarian Transition? The Clean Development Mechanism and Agricultural Change in Latin America." *Environment and Planning A: Economy and Space* 47, no. 10 (2015): 2031–46. https://doi.org/10.1068/a130218p.

Wolff, Cristina Scheibe. *Mulheres da floresta: Uma história. Alto Juruá, Acre, 1890–1945* [Women of the forest: A history. Alto Juruá, Acre, 1890-1945]. São Paulo: Editora Hucitec, 1999.

Wolford, Wendy. "Land Reform in the Time of Neoliberalism: A Many-Splendored Thing." *Antipode* 39, no. 3 (2007): 550–70. https://doi.org/10.1111/j.1467-8330.2007.00539.x.

Wolford, Wendy. "Participatory Democracy by Default: Land Reform, Social Movements and the State in Brazil." *Journal of Peasant Studies* 37, no. 1 (2010): 91–109. https://doi.org/10.1080/03066150903498770.

Wolford, Wendy. "The Plantationocene: A Lusotropical Contribution to the Theory." *Annals of the American Association of Geographers* 111, no. 6 (2021): 1622–39. https://doi.org/10.1080/24694452.2020.1850231.

Wolford, Wendy. *This Land Is Ours Now: Social Mobilization and the Meanings of Land in Brazil*. Durham, NC: Duke University Press, 2010.

Wong, Grace Yee, Cecilia Luttrell, Lasse Loft, Anastasia Yang, Thuy Thu Pham, Daisuke Naito, Samuel Assembe-Mvondo, and Maria Brockhaus. "Narratives in REDD+ Benefit Sharing: Examining Evidence within and beyond the Forest Sector." *Climate Policy* 19, no. 8 (2019): 1038–51. https://doi.org/10.1080/14693062.2019.1618786.

Wynter, Sylvia. "Novel and History, Plot and Plantation." *Savacou* 5, no. 1 (1971): 95–102.

Yanagisako, Sylvia. *Producing Culture and Capital: Family Firms in Italy*. Princeton, NJ: Princeton University Press, 2002.

Zaloom, Caitlin. "How to Read the Future: The Yield Curve, Affect, and Financial Prediction." *Public Culture* 21, no. 2 (2009): 245–68. https://doi.org/10.1215/08992363-2008-028.

INDEX

Note: Page numbers followed by f refer to figures.

açaí, 29, 43–44, 60, 64–66, 71–82, 89, 100, 108, 154, 157, 199n1, 203n28, 203n30, 207n57; palms, 20, 65–66, 69, 72, 74–76, 78–82, 203–4nn33–34, 206n53; prices, 206n56; production, 31, 81, 200n2, 203n31, 204n34, 204n38, 206n53, 207n56; research, 205n40

accumulation, 4, 17; by dispossession, 9; land, 81; primitive, 229n21

Acre, 2–4, 7–20, 24–26, 28, 34–43, 45–47, 49f, 50, 56, 89, 104, 108–10, 114–15, 120, 131–38, 143–45, 150–53, 156, 159, 163, 172–73n44, 176n65, 179n86, 189n21, 191n33, 192n43, 207–8n1, 208n1, 211n26, 217n79, 224n23; açaí and, 65–67, 71, 73–74, 77, 79, 206n53; aquaculture in, 101–2; BR-364 and, 29–32, 82–85, 113f, 130; cattle in, 198n4; citizenship in, 116; crime in, 223n16; da Cunha's visit to, 180n93, 190n29; deforestation in, 2, 8, 17–18, 23, 26, 45, 62, 93, 102, 105, 107, 154–55, 162, 179n81, 197n2; Embrapa and, 76, 205n44; forest carbon's valorization in, 44; forest offsets in, 186n4; GCF Task Force and, 193n49; gender relations in, 189n20; geoglyphs in, 184n126; history of, 189n22; Indigenous people in, 178n80, 213n41, 222n1, 223n20; land occupations in, 214n45; land ownership in, 92–94; migration to, 28, 61, 87, 106, 135, 188n20, 189n25; mucuna in, 220n17; politics in, 190n37; private forest carbon projects in, 47, 94–98, 107; PT's loss in, 160–61, 228n15; REDD+ and, 168n12, 214n51; REDD Offset Working Group and, 195n58; regulation in, 52; roads in, 130; rubber economy in, 221n24; rubber tapper movement in, 177n71; rubber tappers in, 189n24; SISA and, 98; state forest and, 179n82. *See also* Feijó; Jurupari River; Manaus; Rio Branco; urban forest; Xapuri

Acrean government, 70f, 98, 149, 170n30, 181n104; cattle ranchers and, 42; deforestation and, 19, 30, 89, 154, 161; forest-based governance strategies of, 172n44; forest carbon credits and, 3, 162; Indigenous peoples and, 181n106; KfW and, 7, 34, 160; rubber and, 38. *See also* SISA

Acreanidade (concept), 43, 134–36, 139, 157, 172n44, 223n12

Acreanidade (publication), 134–35

agribusiness, 43, 78, 105, 161–62, 205n44, 214n48; economic growth and, 2, 161; monocrop, 18; monoculture, 80; politicians and, 8

agricultural extension technicians, 21, 30–32, 96, 118–22, 125–28, 219n12, 220nn15–16

agriculture, 71, 76–78, 80, 205n46; Cerrado, 205n44; Green Revolution, 206n54; industrial, 77, 150; low-yield, 61; mechanized, 220n17; monocrop, 66–67; plantation, 205n47; slash-and-burn, 68, 201n15; small-scale, 29; subsistence, 58, 79, 82; sustainable, 202n22; swidden, 23, 26, 59, 68, 82, 207n60; traditional, 121

agroforestry, 44, 80, 181n106, 206n53, 224n20

Amazonia, 26, 61; açaí in, 64, 199n1; bioeconomy in, 157; monoculture in, 58; property in, 110

Anand, Nikhil, 117, 178n77, 221n30

Anthropocene, 4, 26, 127, 158, 168n15; environmental harms of, 11; governance, 115; landscapes of, 71, 76; mutual dependence and, 32; plantation and, 203n33

anthropology, 4, 90, 165n2; economic, 219n4; property and, 209n9; of roads, 185n137

Appel, Hannah, 4, 169n20, 174n50, 178n77, 183n116, 183n121

Apurinã, 37, 178n80

aquaculture, 100–102, 150, 157, 161

Araújo, J. S., 41–42, 44

Arena of the Forest, 138, 151–52

associations. *See* rural producers: associations

Bakx, Keith, 93, 189n24, 208n1

Bear, Laura, 14, 50, 187n10, 187n13, 195n65, 228n14

Becker, Bertha, 132, 222n3, 225n38

beef, 18, 62, 85, 145, 163, 198n4

beneficiaries, 101, 116–19, 122, 126–27, 163; capitalism and, 218–19n3; cash payments and, 24; forest, 4, 116–17, 127, 132, 148, 203n27, 220n15; forest protection and, 127; official workers and, 100; rural forest, 140, 142, 158; state and, 109; technicians and, 119, 122, 127, 220n16; urban forest, 133, 139

benefits, 4, 20, 22, 24, 28, 62, 69, 98–99, 101, 105, 110–11, 115–28, 141, 148–49, 151, 154, 160, 181n106, 212n37, 220n13, 220n15, 221n28; environmental, 186n2, 200n8; as exchange, 221n24; formal, 215n63; forest, 22, 24, 31–32, 75, 97, 102, 104–5, 116–18, 120–21, 125–27, 158; smallholder, 70, 75, 88–89, 108–9

Besky, Sarah, 13, 52, 177n75, 188n19, 196n71

biodiversity, 35, 81, 150, 207n57; Amazonian, 40; loss, 18; of Serra do Divisor National Park, 185n139; socio-biodiversity, 167n11

BlackRock, 8, 162

Blanchette, Alex, 13, 177n75

Boa Vista association, 120–22

Bolivia, 2, 15, 17, 37, 93, 189n21

Bolsa Família, 88, 149, 208n3

Bolsonaro, Jair, 8, 18, 32, 154, 160–61, 170n29, 185n139

BR-364, 29–32, 42, 56–60, 62–63, 67–68, 83, 85–87, 94, 110, 112–13, 116, 118–19, 143, 156–67, 185n142, 212n39; açaí and, 64–65, 71–76, 78–80, 82; benefits and, 212n37; Embrapa and, 205n44; extending of, 185n139; fishponds and,

101–2; flooding of, 136; Indigenous communities and, 222n1; paving of, 129, 138, 141–42, 149, 152, 161, 194n55; repairs on, 130; smallholders and, 96, 111, 128, 197n77; urban forest and, 132, 151
Brasília, 15, 96, 119, 148, 179n89
Brazilian dictatorship, 17, 42–43, 63, 77, 135, 191n32, 191n37, 208n1
Brazilian Federal Forest Code, 13, 106, 197n2, 213n44
Brazil nuts, 29, 43–44
Brum, Eliane, 18, 180n95
business as usual (BAU), 6, 10, 19, 50–51, 169n23, 195n60

Cabrujas, José Ignacio, 7, 170n27
California, 31, 33, 35–36, 44–46, 56, 157, 169n22, 186n1; cap-and-trade program, 35, 160, 172n38, 186n2, 187n4, 187n12; carbon market of, 3, 34, 46, 52–54, 159–60, 186n4; GCS Taskforce and, 193n49. See also CARB (California Air Resources Board)
California Public Utilities Commission, 160, 162
Callon, Michel, 9, 36, 118n19
Cameli, Gladson, 160–62, 229n27
Campbell, Jeremy, 93, 110, 176n63, 211n30, 213n41, 214n48, 221n25
cap-and-trade programs, 5, 33, 35, 160, 169n22, 172n38, 186n2, 187n4, 187n12
capital, 82, 174n47, 187n13; cultural, 35; investment, 166n6; natural, 7; owners of, 90
capitalism, 4–5, 9, 14, 66, 90, 131, 155, 164, 168n18, 174n50; Anthropocene and, 11, 158; beneficiaries and, 218n3; commodification and, 177n75; deforestation and, 179n88; extractive, 17, 20, 36, 216n68; global, 16–17, 187n8, 219n3; neoliberal, 25, 32, 109, 127
CARB (California Air Resources Board), 33–34, 36, 44, 46, 50, 54, 169n22, 172n38, 186n1, 186n4

carbon credits, 5, 6, 8–9, 45, 169n21, 174n47. See also forest carbon: credits
carbon dioxide, 6, 56; emissions, 201n15; offsets, 5
carbon markets, 4, 18, 36, 53, 92, 109, 156, 169n23; carbon credits and, 6, 169n21; climate legislation and, 5; compliance, 5–6, 8 (see also cap-and-trade programs); offsets and, 2, 159, 169n22; REDD+ emissions reductions and, 7, 172n38; role of states in, 168n18; UNFCC, 188n17; voluntary, 160, 170–71n37, 212n37. See also California
carbon neutrality, 5, 8, 171n37
carbon offsets, 3, 5–6, 8–9, 11, 19, 33–36, 48, 50, 52–54, 69, 79, 90–91, 95, 97, 99, 106, 158–60, 169n21, 169n22, 186n2, 186n4, 209n5; commodification and, 90, 105; forest protection and, 171n37; private projects for, 212n37; property size and, 214n49; rubber's socioenvironmental relations to, 47. See also forest carbon: offsets
carbon sequestration, 2, 4, 9, 13–14, 23, 71, 100, 157, 176n65, 183n120; climate frontiers and, 18; forest, 35, 44, 56, 66, 78, 88–89, 98, 104, 115, 118, 127, 158–59, 162, 181n106; future, 196n66; inclusion and, 20, 24; labor and, 175n57; SISA and, 167n11; socioenvironmental relations and, 7, 52
Castree, Noel, 92, 177n75
cattle, 41, 62–63, 87, 94, 102–3, 120, 145–47, 149–50, 155–56, 198n4; culture, 43; deforestation and, 3, 11, 26, 68, 101, 162, 199n12; forest carbon and, 160; pastures, 26, 114, 154; rubber tappers and, 55. See also beef; ranchers; ranching
caucho, 40, 190n29
Ceará, 87, 188–89n20
Cerrado, 132, 167n6, 205n44
Chico Mendes Extractive Reserve, 55, 147, 191n33

INDEX 273

Chico Mendes Law, 44, 47
CIFOR (Center for International Forestry Research), 28, 30, 45, 47, 60–61, 71, 104, 140, 148; Global Comparative Study on REDD+, 184n129
citizenship, 24, 43, 122, 127, 151, 164, 219n9; cultural, 222n8; deforestation and, 11; democratic, 117, 126; environmental, 116; environmentally negotiated, 4, 32; politics of, 155; poor rural people and, 14, 23; private property ownership and, 210n16; rural campesino, 182n108; as socioenvironmental relationship, 89, 116, 118; water and, 221n30. See also *florestania* (forest citizenship)
Climate, Community, and Biodiversity Alliance (CCBA), 54, 91, 181n104, 209n5
climate change, 2, 125, 155, 162–63, 165nn1–2, 166n4, 186n4, 193n53; California and, 186n1; early, 10, 155–57; market failure of, 4
climate mitigation, 1–2, 9, 33–34, 68, 166n2, 173n46
coffee, 17, 61, 188n19, 196n69
colonialism, 179n85, 202n21; green, 9
colonists (*colonos*), 29, 93
colonization, 11, 17, 93, 156, 168n15, 178n80; of the Amazon, 18; Brazilian, 87, 145; European, 91; rubber-based, 45
commodification, 9, 48, 53, 90, 92, 177n75; of carbon, 6, 36, 176n69, 217n80; of forest carbon, 50, 176n65
community-based natural resource management (CBNRM), 66–67
Company for the Development of Environmental Services, 107, 175n54
conservation, 66, 69, 99–100, 196n73, 221n27; Amazonian, 161, 167n6; areas, 211n26, 217n74; community-focused, 201n10; forest, 2–3, 6, 18, 34, 46, 67, 133, 137, 200n7, 213n44;

funding, 7; governance, 42; policies, 13; of production potential, 207n60; REDD+ and, 1, 8; soil, 167n11; spectacle and, 193n50; trends, 6; wildlife, 196n66
corporations, 4, 14, 82
credit, 18, 110; Acre's, 137, 223n17; carbon, 36, 169n21
cultivation, 59, 61, 65, 76, 179n88, 211n29; açaí, 79, 81, 204n34; forest, 26; forest-based, 75; Indigenous Amazonian, 184n126; mucuna, 121; swidden agriculture and, 68; traditional, 154

da Cunha, Euclides, 40, 94, 180n93, 189n24, 190n29
Dagnino, Evelina, 133, 222n8, 225n32
Dalsgaard, Steffen, 12, 176n67, 176n69, 188n18
da Silva, Lula. *See* Lula
debt: peonage, 37, 189n24; state, 223n17
de Deus, Carlos Edegard, 23, 25, 153
deforestation, 1–4, 8, 11–14, 17–20, 23, 26, 36, 45, 48–51, 66–71, 93, 99, 115–19, 146, 152, 155, 157–59, 161–62, 167n6, 173n45, 179n81, 179n88, 185n42, 193n54, 201n12, 213n44; açaí and, 78; Bolsa Família and, 149; Bolsonaro and, 170n29; BR-364 and, 29–32, 129, 194n55; cattle ranching and, 34, 42, 62, 94, 102, 104, 199n12; causes of, 198n4; colonization via, 87; cultivation and, 211n29; development and, 179n90; emissions from, 175n59; *fazandeiros* and, 143; fines, 176n62; forgoing, 2–3, 71, 115, 124–27, 157, 203n25; Indigenous people and, 177n74, 181n106; labor, 121, 125; policies, 199n11; property rights and, 89–90; reducing, 49f, 50, 66, 69, 98–101, 105–8, 119, 124, 154, 156, 159, 166–67n2, 200–201n8, 215n54, 216n69; restrictions on, 150; rubber extraction and, 190n28; rubber trees and, 41; sat-

274 INDEX

ellite imagery and, 194n56; SISA and, 59; tCO₂e and, 176n65; tropical, 46, 198n4; Vaccaro and, 76
degradation: air quality, 154; environmental, 4, 91, 210n17; forest, 1, 26, 98
democracy, 127, 191n32, 208n1; carbon, 24, 127, 183n120; participatory, 225n32; racial, 184n131
dependence, 118, 127; mutual, 32, 118, 127
dispossession, 25, 92, 110; accumulation by, 9, 229n21; BR-364 and, 129; cattle ranching and, 34; of Indigenous, 17, 67; neoliberal, 4; rubber tapper, 17, 42
do Rêgo, José Fernandes, 43–44
Dove, Michael, 67, 82, 207n60

economic development, 19, 48, 182n110
economics, 5, 91, 168n19; narrative, 187n10
ecotourism, 8, 69, 147, 195n62
El Dorado, 7, 18
Embrapa, 76–77, 204n36, 205n44, 206n53, 220n17
Emergent, 8, 162, 172n42
emissions reductions, 6–11, 20, 34, 91, 98, 100, 107–8, 173n45, 174n47, 186n1, 215n54; payments for, 3, 7, 11, 13, 36, 44, 48–50, 52, 54, 70, 79, 92, 99, 106, 150, 161, 169n21, 175n54 (*see also* KfW); REDD+, 172n38. *See also* forest emissions; greenhouse gases (GHGs): emissions
Environmental Defense Fund, 99, 192n44, 215n58
environmentalism, 42, 166n6, 191n34; neoliberal, 4–5, 9, 106, 110
environmentalists, 6, 27, 42, 67, 191n34
environmentality, 23, 219n7
environmental markets, 5, 109
environmental organizing, 18, 29
environmental protection, 25, 44, 69, 98, 106, 191n33, 192n39, 217n81

environmental services, 8, 47
estradas, 40, 94
extraction, 18, 36, 68, 74, 216n68; açaí, 75; of carbon, 24, 127; forest, 75, 77; of forest carbon, 10; fossil fuel, 173n46; future, 35, 52; resource, 14, 37, 52, 145; rubber, 40–41, 57, 77, 155, 188n20, 189n24, 190nn28–29, 195n62; timber, 205n44; of trees, 17, 52
extractive reserves, 42, 54, 96–97, 147, 191n33, 221n21. *See also* Chico Mendes Extractive Reserve
extractivism, 36, 43, 82, 97, 192n43, 197n79; forest-based, 76, 80; neoextractivism, 105, 216n68; rubber, 55

Facebook, 55, 84–85, 130, 141
fazendeiros, 28–29, 63, 129, 143, 149, 199nn10–11
Feijó, 30–32, 57–61, 111–12, 119, 128–29, 142–43, 185n42, 219n13; açaí and, 64–65, 72–76, 78–79, 203n30; adult education in, 116; land occupations and, 214n45; land tenure in, 96; rural producers associations and, 118, 148; smallholders in, 28, 31, 58, 86, 96, 102, 128, 146, 185n142, 197n77, 212n39. *See also* BR-364
Feliz Rural Producers Association, 114, 117, 120–21, 124
Ferguson, James, 24, 89, 100, 117, 127, 198n3, 202n25, 217n82
fishponds, 20, 69, 75, 88, 101–2, 104, 123, 150. *See also* aquaculture
florestania (forest citizenship), 32, 43, 116–18, 126, 134–35, 139, 144, 151–52, 157, 164; beneficiaries and, 120, 122; socioenvironmental relations of, 127
forest benefits, 22, 24, 75, 89, 102, 116–18, 120, 125–26; Acre's, 127; BR-364 and, 31–32; forest carbon sequestration and, 158; land rights and, 125; mucuna seeds as, 121

INDEX   275

forest carbon, 2–4, 6–10, 14, 22–25, 27–36, 38, 40–42, 49f, 53–56, 163–64, 177n76, 194n57, 215n66; commodification of, 10, 50, 53, 91, 176n65, 177n73, 217n80; credits, 3, 8, 162, 172n42; distribution funded by, 89; emissions, 3, 107; monetization of, 50, 90; money, 108–9; offsets, 8, 10, 13, 34–36, 44–46, 50, 54, 90–91, 150, 156–57, 159, 163, 172n38, 186n4; privatization of, 4; programs, 3, 19, 25, 33, 40, 92, 105, 115, 217n80; projects, 13, 31, 47, 49f, 89, 94–98, 107, 212nn36–37, 217n75; project developers, 30; as public wealth, 109; sequestration of, 44, 88–89, 104, 118, 127, 133, 157–58, 162, 183n120; SISA program, 115; standardization of, 4; territorialization and, 217n74; tropical, 97, 172n38; valorization of, 4, 12–13, 19–20, 32, 42, 44, 47–48, 66, 79, 92, 110, 155, 175n54, 203n25; valuation of, 218n84; value of, 7, 10, 20, 22, 24, 32, 34–35, 44, 46, 51–53, 55–56, 70, 89–90, 97–100, 104–10, 116–17, 133, 136–37, 158–60, 162, 168n14, 196n73, 217n80
forest carbon offsets, 13, 44–46, 90–91, 150, 156–57, 159, 163; Acrean, 8, 10, 35–36, 45, 54, 157, 186n4; tropical, 34, 46, 50, 172n38
forest emissions, 7, 9–10, 34, 49f, 68, 161
forest protection, 3, 13, 19, 23–25, 31, 46, 52, 67–68, 150–51, 160; benefits and, 88, 122–23, 126–27; citizenship and, 116, 133; climate change and, 125; fish farming-based, 104; forest carbon sequestration and, 44, 99, 109, 171n37; forest communities as vanguard of, 53; funders, 34; intensification and, 200n8; requirements, 42; rubber and, 54; smallholders and, 118–19, 126; women and, 219n12

fossil fuels, 9, 12, 155, 176n69, 183n120; economies of, 34; emissions from, 9, 18; energy production based on, 24; extraction of, 173n46

Gago, Verónica, 109, 193n51
GCF (Governors' Climate and Forest) Task Force, 46–47, 56, 97, 136, 192n49, 229n27
geoglyphs, 26, 184n126
Global North, 9, 14, 18, 20, 26, 35, 37–38, 216n68; neoliberalism in, 217n81
Global South, 9, 14, 18, 24, 89, 109, 178n78
governance, 47, 115, 196n73; BR-364 and, 31; Brazilian, 37, 42; carbon democracy and, 24; environmental, 18; forest, 14, 19, 45, 48, 53, 150, 172n44, 217n80; inclusion as strategy of, 23; participatory, 19; REDD+ and, 168n12; SISA, 161–62
Government of the Forest, 19, 26, 35, 43–45, 47, 51, 106–7, 134, 136–37, 156, 158–59, 161, 164, 177n71, 181n101, 191n38; government employment and, 224n30; state debt and, 223n17. *See also* SISA
greenhouse gases (GHGs), 6, 53, 68, 165n1, 176n65, 176n69; emissions, 1–3, 5–6, 11–12, 14, 33, 97–98, 155, 169nn21–22, 186n1
Green Revolution, 82, 200n8, 206n54

Hardin, Garret, 91, 210n17
Harvey, David, 92, 174n48
Hecht, Susanna, 40, 67, 180n93, 190n29, 201n12, 205n46, 221n21
Hetherington, Kregg, 19, 116, 161, 182n108, 206n54, 219n7, 220n14, 224n30, 227n11
Hoelle, Jeff, 54, 62, 145, 199n10
Holston, James, 93, 211n30
Hughes, David McDermott, 54, 69, 195n62, 196n66
Huni kuĩ, 37, 178n80
hunting, 29, 42, 87, 124

IMC (Instituto de Mudanças Climáticas e Serviços Ambientais), 27, 107, 161–63, 181n103, 204n36, 217n75
inclusion, 20, 23, 25–26, 66, 81–82, 89, 117, 151, 159; BR-364 and, 31, 113, 129; in Cap-and-Trade Program, 187n4; deforestation and, 154; digital, 138; forest-protective, 53, 122; green capitalist, 31, 67; material, 120, 123, 126; productivism and, 29; smallholders and, 24, 110, 123; social, 24–25, 44, 46, 109, 115, 133, 151–52, 182n114
Indigenous communities, 29, 96–97, 100, 178n80, 202n22; BR-364 and, 185n139, 222n1; extraction and, 37; land rights and, 218n84; organizing, 25; REDD+ program and, 174n48
Indigenous peoples, 69, 94, 134, 137, 163, 177n74, 178n80, 212–13nn40–41; Brazilian racial democracy and, 184n131; carbon sequestration and, 181n106; dispossession of, 17; enslavement of, 40; exploitation of, 190n28; murder of, 37; violence against, 161, 191n36. *See also* Apurinã; Huni kuĩ; Yawanawa
inequality, 31–32, 97, 105, 128
infrastructure, 83, 136, 139, 178n77; deforestation and, 11, 18, 67, 179n90; emissions reduction and, 202n15; green capitalism and, 157; urban forest and, 151; water, 104
INPE (Instituto Nacional de Pesquisas Espaciais), 98, 170n36, 179n81; PRODES system, 50, 194n56
International Forest Day, 2, 19
IPAM (Amazonian Environmental Research Institute), 81, 206n56
Iteracre, 95–96

Jurupari River, 72, 74, 82

KfW, 19–20, 99, 192n44, 215n55; payments to Acrean government, 7, 34, 44–45, 70, 159–60, 181–82n106
Kyoto Protocol, 5, 166n2

labor, 29, 35, 37, 69, 94, 99–100, 121; benefits linked to, 101; coffee and, 188n19; commodification of, 177n73; conditions, 189n24; day, 60; deforestation, 121, 125; of ecosystem services, 109; exploitation, 67; forest carbon's value to, 110; forest protective, 108, 201n9; gendered dynamics of, 221n21; hired, 68; lack of, 175n57; local, 23; Lockean views of, 215n61; plantations and, 204n38; of rubber tappers, 190n29; wage, 124
landholders, 99, 214n48, 216n73; Amazonian, 13, 93, 106; large, 22–23, 95, 124, 212n37; ranching, 102
Landless Workers Movement (MST), 213–14n45, 218n83
land rights, 32, 42, 91–94, 96–97, 110, 125, 191n33, 213n41; carbon rights and, 210n20; forest carbon's valuation and, 218n84; inaccessibility of secure, 155; Landless Workers Movement (MST) and, 213n45, 218n83; private, 109; Quilombola, 213n41; REDD+ and, 105; rubber tappers and, 18, 29
land tenure, 91, 95–97, 99, 207n56, 209n13, 214n49
latex, 38, 40, 58
law, 5, 90; environmental, 5, 146, 168n19, 194n56; federal, 208n1. *See also* Chico Mendes Law; SISA
Li, Tania, 23, 68, 82, 202n17, 202n21
Liverman, Diana, 36, 174n47, 210n22
Locke, John, 100, 210n16, 211n29
loggers, 29, 43, 96
Lula, 43, 135, 161, 208n3

Machado, Altino, 16f
Manaus, 84, 132, 138
manioc, 22, 23f, 29, 59, 87, 95, 103; fields, 114; flour, 122; products, 85
marginalization, 25, 31–32, 149
Markit: Environmental Registry, 45; Global Carbon Index, 170n30, 175n54
Marx, Karl, 100, 229n21

Mathews, Andrew, 51, 99, 108, 182n11, 188n16, 203n33, 215n54, 227n11
McKinsey & Company, 68, 97, 175n56, 201–2n15
Méadel, Cécile, 36, 188n19
Mendes, Chico, 34, 43, 45–47, 53, 135, 137–38, 143–44, 192n43; BR-364 and, 29. *See also* Chico Mendes Extractive Reserve; Chico Mendes Law; rubber tapper movement
Mexico, 56, 227n11; Chiapas, 186n4, 217n80; forest carbon projects in, 13; MOU and, 186n4; REDD+ program in, 99
middlemen, 72, 75, 122, 124
migration, 17, 28, 189n25
miners, 29, 43
Mitchell, Timothy, 24, 183n120
monocrop, 4, 11, 26, 57, 66–67, 71, 81; açaí as, 80, 206n53; agribusiness and, 18; carbon as, 79; plantations, 69, 77. *See also* soy (soybeans)
Morais, Maria de Jesus, 133, 137, 144, 223n19
mucuna, 20–23, 25, 27, 31, 123–24, 126, 220n17, 221n20; as problem, 220–21n19; seeds, 20–22, 69, 75, 121, 157, 182n107

Natex, 38, 40, 42, 44, 46, 55
nature-based solutions, 8, 157
neoliberalism, 89–91, 109, 120, 200n6, 217n82; in Argentina, 193n51; in the Global North, 217n81; inclusion and, 24
NGOS, 14, 19, 47, 70, 82, 106, 144; açaí and, 66, 74; deforestation and, 157; environmental, 140–41; forest benefits and, 32; forest protection and, 219n12; forest valorization and, 3; neoliberalism and, 217n81; private REDD+ projects and, 209n5; SISA and, 181n104
NTFPS (non-timber forest products), 66–67, 200n7

Osborne, Tracey, 13, 177n73, 217n76, 217n80

Pará, 11, 74, 76, 81–82, 206n56; açaí in, 199n1, 204n34, 205n40; floodplains (*várzea*) of, 76, 82, 203n31, 204n34, 206n53, 207n57; southern, 45
Paraguay, 116, 182n108; soy production in, 206n54, 227n11
Paxson, Heather, 48, 193n52
pay for ecosystem services (PES), 4, 91, 215n66
Peixes da Amazônia, 101–2, 161
Peru, 2, 56, 185n139, 229n27
petro-states, 24, 183n116
plantations, 66, 69, 71, 157; açaí, 76–78, 82, 203n33, 204n38, 205n40; palm, 202n17, 202n20, 220n19; rubber, 37, 40, 55, 77
polluters, 2, 6, 9, 34, 95; regulated, 5, 33
posseiros, 93–94, 97, 100; rights, 105, 107, 212n37
poverty, 120, 144, 150, 196n69; Amazonian, 143; reduction of, 2, 46, 105, 156, 208n3; rubber tapping and, 54, 145; rural, 32, 151, 160
PPCD (Plan for the Prevention and Control of Deforestation), 48, 193–94nn54–55
private property, 14, 90–92, 97–100, 155, 210n16; in Acre, 211n26; forest carbon as, 4, 99; green capitalism and, 24; state and, 32, 89–90
privatization, 92, 157, 211n23; of forest carbon, 4
productivism, 25, 29; green, 122; inclusive, 22, 31, 66, 71, 79–80, 82, 100, 117; rural, 67, 71
property relations, 14, 92, 177n76, 217n80
property rights, 17–18, 89–92, 99–100, 109; Coase theorem and, 168n19; commons and, 210n17; environmental harms and, 5; regulation of, 97; rural people's private, 99; rural smallholders and, 110, 115
public property, 4, 89

quilombolas, 96, 213nn40–41

Rabeharisoa, Vololoa, 36, 188n19
Raffles, Hugh, 27, 201n14
ranchers, 17–18, 23, 42–43, 45, 87, 94, 96, 105, 135; deforestation and, 51, 62, 68, 101, 125, 134, 136, 144; *fazandeiros*, 28, 63; fields of, 27
ranching, 63, 79, 145; deforestation and, 20, 34, 62, 67–68, 162, 198n4, 199n11; economy, 23; industrial, 150; landholders, 102; rise of, 151; rubber tapping and, 67; rural producers and, 29, 62
REDD+ (reduction in emissions from deforestation and forest degradation), 1, 3, 6–9, 11, 19, 26, 28, 36, 45, 46, 48, 68–69, 101, 167nn3–4, 168n12, 188n16, 202n15, 210n15, 215n54, 217n78; advocacy for, 175n56; CIFOR's Global Comparative Study on, 184n129; critics of, 201n15; critique of top-down characterizations of, 187n7; Early Movers Programme (REM), 7, 19–20, 161–62; emissions and, 173n46; emissions reductions, 172n38; Environmental Excellence Standard, 171n37; Feijó and, 185n142; funding, 71; global efforts, 159; Indigenous communities and, 174n48, 183n122, 218n84; jurisdictional, 98, 106–7, 214n51; land rights and, 105; land tenure and, 97; in Mexico, 99; Offset Working Group, 50, 195n58, 195n60; private projects, 108, 209n5, 212n36; property-based, 100; property relations and, 177n76; property rights and, 91; quantitative model projections and, 51; tropical forest regions and, 178n78
redistribution, 14, 24, 89–90, 104, 108–10, 115, 133; green, 155; PT and, 151
regulations, 5, 13
Rio+20, 2, 7, 14, 19

Rio Branco, 15, 27, 32, 45, 61, 82, 95, 106, 115, 125, 131, 143, 145–48, 150, 153, 163, 176n65, 218n1; açaí and, 69, 74–76, 108; Arena of the Forest, 151; BR-364 and, 30, 42, 57, 65, 112, 113f, 119, 129, 142; California Public Utilities Commission and, 160; Digital Forest Wi-Fi network, 138, 139f, 151–52; Embrapa and, 205n44; fishpond construction in, 103; flooding in, 84–85; Forest People's House, 132, 134, 136–37, 139, 152, 157; Forest People's Plaza, 137, 151; Governors' Climate and Forest (GCF) Task Force meeting in, 46, 56, 97; mayor of, 135; Peixes da Amazonia and, 101; rubber trees in, 38, 39f, 54–55; state-building in, 216n71. *See also* urban forest; Viana, Jorge
Rio de Janeiro, 2, 15–16, 179n89
Rio Madeira, 83, 85, 207n1
roads, 27, 60, 130, 151, 178n77; anthropology of, 185n137; deforestation and, 29; lack of reliable, 62; side, 31, 111, 118, 128, 141, 212n39; socioenvironmental relations and, 14, 139
Rofel, Lisa, 35, 168n18, 187nn8–9, 209n8, 211n23, 228n14
Rondônia, 29–31, 40–42, 45, 62, 83–84, 129, 161
rubber, 20, 31, 34–38, 43–45, 47–48, 53, 60, 77–79, 109, 116, 134, 136–38, 143, 155, 160, 179n86, 188n20, 197n77; açaí and, 65, 74, 78; barons, 37, 41, 55, 58–59, 87, 94, 119; boom, 17, 34, 40, 42, 54, 93, 137–38, 188–89n20, 190n28; decline of, 54–56, 58–59, 65, 77, 109, 116, 151, 208n1; economy, 17–18, 37–38, 54, 59, 87, 94, 104, 189n25, 208n1, 221n24; estates, 37, 44, 58, 86–87, 93–94; extraction, 40–41, 57, 77, 155, 188n20, 189n24, 190nn28–29, 195n62; as NTFP, 200n7; plantations, 37, 40, 55, 77; political ecology of, 93; production, 38, 40,

rubber (continued)
47, 77, 188n20, 189n24, 195n62, 205n44, 221n22; subsidies, 20, 38, 41, 44, 47, 54, 69, 192n41, 202n22; tapping, 44, 54–55, 67, 125, 145, 213n43; trees, 17, 34, 37–41, 55, 58, 68, 77–78, 86–87, 94, 157, 189n24. See also *caucho*; latex; Natex

rubber narrative, 31, 35–37, 44–48, 51–57, 135, 155–56, 192n43; investment and, 195n64; performances of, 142; rubber tapper movement and, 45, 53, 193n51, 196n73

rubber tapper movement, 26, 34, 42–43, 45, 47, 53–55, 94, 118, 135, 138, 144, 156, 190–91n32; Acrean identity and, 213n43; environmentalism and, 191n34; as environmental justice movement, 187n12; history of, 137, 193n51; land rights and, 96; Workers' Party (PT) and, 61, 118; rural workers and, 28, 100, 185n133; socioenvironmental and, 177n71

rubber tappers, 18, 28, 37–38, 40–46, 52, 54–55, 62, 68, 77–78, 134, 189n24, 190n29, 221n22; descendants of, 61, 141; dispossession of, 17; ecotourism and, 147; environmental conservation and, 196n73; former, 63, 65–66, 87, 136, 149, 184n32; inclusion of, 23; Indigenous women and, 189n20; land rights and, 29, 96; migrant, 37, 58, 180n93, 192n41; PT and, 144; rubber barons and, 59, 94, 221n24; statues celebrating, 132, 137; as traditional people, 213n40

rural development, 3, 35, 43, 46, 61, 67, 168n14, 202n24; low-carbon, 8, 44, 158; rubber and, 54

rural producers, 28–29, 58, 61–62, 96–97, 100, 115, 120–21, 125, 195n62; associations, 75, 96, 114–15, 117–21, 148, 213n42, 220n15; de Deus and, 153; environmental agencies and, 161; mucuna and, 124, 221n20

rural workers, 28, 93, 100, 185n132; unions, 28, 42, 55, 61, 75, 185n133

Schmink, Marianne, 19, 139–40, 224n22, 224n30
SEAPROF (Secretaria de Extensão Agroflorestal e Produção Familiar), 27–28, 70f, 74, 77–78, 118, 120–23, 221n22
Semedi, Pujo, 82, 202n17, 202n20
settlers, 23, 93
Shapiro-Garza, Elizabeth, 13, 177n73
SISA (System of Incentives for Environmental Services/Sistema de Incentivos a Serviços Ambientais), 3, 7–8, 10, 19, 45–48, 69–70, 74, 97–100, 106–9, 136, 149, 159–63, 167n11, 181nn103–4, 192n43, 214n51, 217n75; açaí and, 78; benefits, 101, 105, 154, 212n37; deforestation and, 59, 133, 161, 185n142, 194n56, 215n54; Embrapa and, 204n36; forest carbon program, 33, 115, 175n54, 186n3; Indigenous peoples and, 181–82n106; Institute for Climate Change and Environmental Services, 27; rubber subsidies and, 54; Scientific Committee, 50; state-building efforts under, 217n80; State Policy for the Valorization of Forest Environmental Assets and, 44, 198n2. See also IMC; REDD+
slavery, 183n124; abolition of, 93
smallholders, 4, 17, 20, 27–28, 61–62, 70–71, 82, 93, 134, 146, 157–58, 163; açaí and, 74–76, 78–81, 203n31; aquaculture and, 101–2, 104; benefits and, 69, 88–89, 108–9, 115–16, 118–20, 123, 126–27, 202n22; BR-364 and, 30–31, 58, 72, 86, 96, 111, 128, 142, 197n77; deforestation and, 12, 23, 66, 68, 114, 125, 156, 197–98n2, 199n11; *florestania* and, 32; Forest School and, 44; land rights and, 96–97, 105, 110; mucuna and, 21–22, 121, 124, 220n17;

rural producers and, 29, 184n132; SISA and, 3; socioenvironmental relations and, 14; soy production in Paraguay and, 206n54

social movements, 54, 183n120, 183n122, 213–14n45, 215n61; PT and, 118. *See also* rubber tapper movement

socioenvironmental relations, 13–14, 26, 30, 50, 80, 89–90, 92, 98, 110, 131, 133, 152, 157, 187n9; Acrean identity and, 139; Acrean rubber and, 35, 40–42, 44, 47, 53, 56; beneficiaries and, 115–17, 126–27; benefits and, 32, 117; carbon emissions reduction and, 34, 51; carbon sequestration and, 7, 52, 66, 158; citizenship and, 116–18; deforestation and, 50–51, 162; forest carbon valorization and, 155; offsets and, 36, 53, 91; in rural areas, 151; socially inclusive, 54; urban forest and, 150

sovereignty, 17, 122, 218n84; Brazilian, 116, 180n90; food, 173n44; resource, 216n68; state, 23, 105

soy (soybeans), 18, 26, 58, 77, 160, 162–63, 201n8, 205n46; in Paraguay, 206n54, 227n11

Spain, 56, 93

Stoltenberg, Jens, 1–2, 8, 10–11, 14, 157, 166n2, 167n4

subsidies, 88, 179n90, 215n66; for aquaculture, 157; for emissions reductions, 108; for ponds, 101–2; rubber, 20, 38, 41, 44, 47, 54, 69, 192n41, 202n22

sustainability, 8, 23, 25, 70

sustainable production (*roçada sustentável*), 121, 195n62

System of Incentives for Environmental Services. *See* SISA

taxes, 5, 18

tCO$_2$e (ton of carbon dioxide equivalent), 6, 176n65

technicians. *See* agricultural extension technicians

tropical forests, 5–8, 14, 48, 52, 54, 178n78; carbon emissions from, 167n4; carbon sequestration and, 18, 23; deforestation and, 2, 11; land ownership in, 92; land use practices in, 68, 201n15; offsets, 2, 33–34, 36, 45–46, 50, 97, 172n38, 186n4; rubber and, 36

Tropical Forest Standard (TFS), 54, 196n74

Tsing, Anna Lowenhaupt, 14, 35, 52, 92, 195n64, 201n10, 203n33, 227n3

unions, 28, 42, 185n133

United Nations, 43; climate conference, 1, 161n1

urban forest, 32, 132–33, 140, 145, 150–51, 155–57, 225n38

Vaccaro, Eloy Luiz, 76–77, 205n40

valorization, 41–42; of carbon, 108; cultural, 29, 32, 125, 131, 133, 144, 151–52; forest, 3–4, 19, 25, 32, 90, 131, 139; forest carbon, 4, 12–13, 20, 32, 44, 48, 79, 92, 110, 155, 203n25; of intensive agriculture, 67

Viana, Jorge, 19, 43–44, 46–47, 55–56, 134, 140, 161, 191nn37–38, 223n20, 228n15; BR-364 and, 130. *See also* Government of the Forest

Viana, Tião, 2–3, 19, 55, 76, 191nn37–38, 228n15. *See also* Government of the Forest

violence, 29, 42, 166n3; in Acre, 67, 94, 161, 173n44, 228n15; against Indigenous peoples, 29, 161, 177n74, 181n106, 191n36; of large-scale forest clearing, 26; racial, 203n33; of rubber extraction, 155; rural, 67, 191n36; socioenvironmental, 229n21

welfare state, 24, 106; environmental, 121, 157; environmentally premised, 4, 9, 32, 90, 108, 120, 155; policies, 116, 121, 185n132

West, Paige, 92, 188n19, 196n69

Workers Party (PT), 19, 23–24, 61, 108, 123, 142–43, 161, 191n37; backlash against, 228n15; benefits and, 104–5, 115; clientelist politics and, 219n11; *florestania* and, 116; forest protection and, 160; politicians, 133; politics of, 109; rubber tappers and, 42, 144. *See also* Government of the Forest; inclusion: social; Lula; Viana, Jorge; Viana, Tião

World Bank, 29, 42, 96; Forest Carbon Partnership Facility, 7, 91, 209n13
World War II, 17, 37, 77, 94, 189n25

Xapuri, 38, 42, 44, 145–46; Rural Workers Union, 55, 185n133

Yanagisako, Sylvia, 10, 14, 35, 168n18, 187n9, 209n8, 211n23, 222n1
Yawanawa, 37, 178n80